Constructivist Instructional Design (C-ID)

Foundations, Models, and Examples

Constructivist Instructional Design (C-ID)

Foundations, Models, and Examples

edited by

Jerry W. Willis
Manhattanville College

Information Age Publishing, Inc.
Charlotte, North Carolina • www.infoagepub.com

Library of Congress Cataloging-in-Publication Data

Constructivist instructional design (C-ID) : foundations, models, and
examples / edited by Jerry W. Willis.
    p. cm. -- (Qualitative research methods in education and educational
technology)
 Includes bibliographical references.
 ISBN 978-1-930608-60-3 (pbk.) -- ISBN 978-1-930608-61-0 (hardcover)
 1. Instructional systems--Design. 2. Constructivism (Education) I.
Willis, Jerry, 1942-
  LB1028.38.C664 2008
  371.3--dc22

                                                  2009002234

CONTENTS

PREFACE

I have had this book in the back of my mind for over 10 years. And, on a regular basis, it was in the front of my mind. Over and over again, I would start working on it, then conclude that I was not yet ready because the development of thinking about instructional design from a constructivist perspective was rapidly evolving and maturing. Anything I said would be outmoded in a year or two because not only was the field in general changing, my own views of constructivist instructional design (C-ID) were in flux.

That excuse supported my inactivity on this book for many years but it was eventually reinforced and supported by another excuse—there were a growing number of papers and reports in the literature that did an outstanding job of addressing various issues related to C-ID. Why write a chapter on a topic when it would not be as good as what someone else had already done?

These two excuses allowed me to write three other books while the idea of "a book on C-ID" remained in my mind as both a concept and a desire. Now, in the summer of 2008, I have overcome both those excuses and have developed the book. I believe we are at a point in the growth and development of C-ID that a book on the subject is appropriate. Several, in fact, would be welcome. In addition, I solved the issue of whether to write new chapters when there are excellent papers on the same subjects already in the literature. This book is of mixed parentage. I have written a number of the chapters and all the introductions to sections. Many of the chapters, however, are revisions or reprints of papers already published in a wide array of journals, or presented at one of the many conferences on educational technology. I have looked at over a 1,000 papers and selected those I feel are the strongest for inclusion in a book that introduces the foundations, concepts, models, and practices of C-ID. There are probably more than 2,000 papers and presentations related to this topic and I was simply not able to find or access all of them. Thus, I apologize in advance to authors I should have included in this book but did not. However, I believe the chapters in this book do a good job of presenting the developing field of C-ID.

The book is divided into four sections, each of which deals with one broad aspect of the topic. *Section I, The Many Foundations and Faces of Instructional Design,* addresses the theoretical and conceptual foundations of C-ID. *Section II, Family Resemblances of C-ID is* more to the point. Papers in this section introduce readers to some of the major organizing ideas and practices that guide how instructional designers do constructivist ID. Reflective practice, recursive or iterative concepts of design, emergent approaches to design, and participatory design are all introduced along with broader ideas that are compatible with C-ID.

Section III, R2D2 and Other C-ID Models, includes chapters on several models or frameworks for doing C-ID. All of these chapters are efforts to develop meaningful models others can consider when designing instructional resources. Two chapters cover my own C-ID model, the *Recursive and Reflective Design and Development* model (R2D2), but constructivist designer theorists do not usually present their own work as definitive or canonical. Instead, they offer it to the community of designers as one possibility they are welcome to adopt, adjust, or borrow from as it appears worthwhile in their own design work. Therefore, several other C-ID models are also presented in Section III with the idea that all of them are well worth your consideration and evaluation. Perhaps you will find my C-ID model, R2D2, worthwhile, in whole or in part. On the other hand, one of the other models featured in this section, such as Katherine Cennamo's Layers of Negotiation Model, may be more suited to the type of design work you do. Or, you may find aspects of several models appealing and create your own C-ID framework, adapting and adopting aspects of several models.

The final section of the book, *Section IV, C-ID in Practices: Examples from the Field,* contains chapters that tell the stories of how groups of designers used C-ID principles and models to develop a particular learning resource. Most of these papers also address topics covered in earlier sections, but the focus is on how those theories, ideas, guidelines, and concepts are put to work in the real world of designing.

I hope that you find this book useful as you explore current thinking on what C-ID is and how it can be used in professional practice. I would welcome your comments, suggestions for future editions (including papers that would warrant inclusion), and information about how you have incorporated C-ID into your own work. My most stable e-mail address is one I have had for over 15 years, jwillis@aol.com.

Jerry Willis
Professor and Coordinator of the Doctoral Program
Manhattanville College, Purchase, New York

SECTION I

The Many Foundations and Faces of Instructional Design

Jerry Willis

There are millions of people with the word "design" in their title or job description. Perhaps the most famous members of the design guild are those that create what we will be wearing next year (or at least design what we will be watching other people wear). Over the last 50 years names like Christian Dior, Donatella Versache, Giorgio Armani, Donna Karan, Tommy Hilfiger, Yves St. Laurent, and Vera Wang have become familiar to billions of people. Less well know and less often in the fashion news are designers of jewelry and housewares, but names like Tiffany, Faberge, Paloma Picasso, Marcel Breuer, Charles Eames, and Pablo Pardo have achieved considerable worldwide recognition.

There are, of course, other genres of design with well known practitioners. In architecture, for example, there is Richard Buckminster Fuller, I. M. Pei, Frank Lloyd Wright, Adolf Loos, Zaha Hadid, and Arata Isozaki. There are, in fact, so many different fields of design that it would probably be foolish to attempt a definition of design that includes all the different fields. Some would, almost inevitably be left out. In the introduction to the *International Journal of Design*, the editors neatly sidestepped that problem this way:

1

The *International Journal of Design* is a peer-reviewed, open-access journal devoted to publishing research papers in all fields of design, including industrial design, visual communication design, interface design, animation and game design, architectural design, urban design, and other design related fields.

The term "open access" means the journal charges no subscription fee and is freely available online (http://www.ijdesign.org/ojs/index/index.php/ IJDesign) without charge. The editors included the catchall phrase "other design related fields" to cover everything they did not mention explicitly. Omitted, for example, were the hundreds of thousands of industrial designers who design everything from toasters to staircases as well as designers of dinnerware, silverware, automobiles, and educational materials. This last category, which hardly ever gets mentioned in the wider world of design, is the focus of this book.

Today the process of designing educational materials is called many things: curriculum development, instructional design, instructional systems design, and even teaching. I will call it instructional design or ID in this book, probably because I work in the field called educational technology (or instructional technology) and ID is the term used in my field. Although ID has led a relatively solitary life, it has many close as well as distant relatives. Some of the close relatives are illustrated in Figure I-1.

Because ID is associated strongly with the integration of information and educational technologies into the classroom, it has much in common

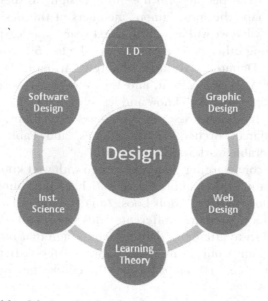

Figure I.1.   Fields of design closely related to instructional design.

with software design and computer interface design as well as Web design. ID professionals often borrow techniques and ideas from professionals in these fields. It is also closely allied with fields like instructional science and learning theory because those fields are concerned with how humans learn and how they can be taught.

There are many other, mostly distant, relatives of ID. They include industrial design, architecture, graphic design, computer science, most fields of engineering, some areas in business such as marketing, some areas of communication and journalism, and commercial art.

Design as a task that needs to be done has probably been around almost as long as humans, and as a formal specialty it has existed for thousands of years. Archaeologists, for example, can often establish a relatively accurate cultural context and timeline for a settlement being excavated by studying the pottery and other artifacts being unearthed. With what they already know about pottery designs, and what culture produced them, they can tell us a great deal about the settlement.

With a little effort we can claim a very long history for instructional design as well. For example, Plato's book, *The Republic*, contains much that would be considered instructional design. In Part II 4 Plato presents the results of his efforts to design the educational experiences of the "philosopher kings" who would rule the ideal Greek city state (*polis*). However, despite thousands of years of work on designing both educational systems and educational materials, the formal discipline called instructional design is less than 50 years old.

Formalized ID has been around since the early 1960s, when the disciplines of educational technology and educational psychology began to be interested in how you could design effective instruction as well as the materials to support such instruction. Over the years an amazing number of models for how to do instructional design have been proposed. Most of the models are American, and because behaviorism dominated American psychology in that period, most of the ID models have more or less been based on behaviorist theories of learning (Dick & Carey, 1996). The approach to design reflected in these ID models are expressions of a particular family of theories and paradigms that are anchored in empiricist views of knowledge, and behaviorist models of teaching and learning. Several chapters in this first section will explore the foundations of traditional ID models and offer criticisms of both those foundations and the ID models that are based on them.

More recently, several approaches to ID have been developed that use an interpretive rather than an empiricist paradigm of knowledge and constructivist rather than behavioral theories of learning. One of them is the reflective and recursive design and development (R2D2) model (Willis, 1995; Willis, 2000; Willis & Wright, 2000). You will learn more about Constructivist-ID in this section, with a particular emphasis on conceptual and theoretical foundations.

THE CHAPTERS IN SECTION I

The section begins with an introductory chapter about the recent history and current status of instructional design. "Three Trends in Instructional Design" (Jerry Willis) is an adaptation of a chapter in another book I wrote that deals with qualitative research in educational technology (*Qualitative Research Methods in Education and Edcuational Technology,* 2008). The first section of this chapter discusses the two uses of the term ID. Many of the papers on ID assume that designing instructional resources is primarily a process of selecting the most effective teaching methods and applying them to the current teaching task. This approach, *Pedagogical ID,* is, I argue, a weak way of viewing ID for many reasons. It tends to reduce ID to selecting a particular theory of learning (e.g., behaviorism, cognitive science, cognitive constructivism, social constructivism) and then selecting the "right" pedagogical strategy from the stable of strategies that have been developed on the foundation of that particular theory of learning.

A more comprehensive and more productive way of thinking about ID is to view it as a process. Most Process ID approaches treat the design of educational materials as a social process that calls for the participation of many different people—from instructional designers and teachers to students and parents (or workers, trainers, and managers). Most of the ID models introduced in this book will be Process ID models.

The second section of the chapter proposes that the field of ID today has three somewhat contradictory but influential trends. The most established is the work on traditional instructional design models based on a positivist epistemology and a family of learning theories that includes behaviorism. In this chapter I argue that this approach to ID has a less promising future than two others. One of those newer approaches is the Design Based Research (DBR) movement and the other is the Constructivist-ID movement. While I believe DBR is still much too influenced by the same positivist epistemology and behavioral theories of learning as traditional ID theory, I believe it offers some appealing perspectives for contemporary instructional designers. I am even more optimistic about C-ID models because they are based on different and more worthwhile philosophical and theoretical grounds that break away from the limitations of traditional empiricism and behaviorism. In the last section of the chapter I propose that some aspects of DBR be integrated into C-ID models and that the result would be both better models of design and better instructional resources.

The second chapter in this section was written by Karin Wiburg at New Mexico State University. In "Instructional Design: Is It Time to Exchange Skinner's Teaching Machine for Dewey's Toolbox?," Wiburg focuses specifically on whether instructional designers should abandon the behavioral theories of learning developed by B. F. Skinner that dominated ID for most of the last half of the twentieth century. She discusses many of the products of behavioral learning theory—from programmed learning and Madeline Hunter lesson plans to behavioral objectives and Robert Gagnè's conditions of learning. Dr. Wiburg makes, I believe, a persuasive argument against using behavioral learning theories as a foundation for ID and proposes instead that we consider other options such as contemporary expressions of John Dewey's perspectives on teaching and learning in a democratic society.

The next chapter, "Constructivism, Instructional Design, and Technology: Implications for Transforming Distance Education," was written by Maureen Tam at the University of Hong Kong. Like many of the papers in the literature, Dr. Tam does not attempt to provide a universal answer to the question of what type of ID model should be used to develop instructional materials. Instead, she focuses on one context, distance education, and she presents her views on appropriate instructional design for distance education in an ordered and systematic way. As you read this chapter, note that Dr. Tam is writing about ID from a pedagogical perspective. She says very little about the *process* of ID. Instead, she presents the basics of constructivist learning theory, explores the basic guidelines for teaching and learning from a constructivist perspective, and compares constructivist approaches to pedagogies based on behavioral theories of learning. While I am not enthusiastic about limiting our concepts of ID solely to pedagogical issues, it is nevertheless true that pedagogy is a component of the ID process. This chapter provides a succinct overview of both constructivist learning theory and constructivist pedagogy.

In the fourth chapter in this section, "Foundations of Instructional Design: What's Worth Talking About and What Isn't," I explore the issue of what characterizes strong contributions to the scholarship and professional practice literature of ID. For example, until recently, the literature on C-ID had many more general, conceptual, and theoretical papers than it did ID models, examples of application, and discussions of problems and issues. The literature on C-ID is becoming both broader and deeper today, and there has also been a change in the type of papers that criticize C-ID. Braden's strident (1996) paper on why traditional, linear ISD is the right model for all instructional designers is not typical of the papers being published today. Contemporary papers are more likely to suggest more flexible, more open, and more ecumenical approaches to the debate about ID models. In this chapter, I suggest that we need to distinguish between issues related to particular design or teaching strategies, which are less important, and debates over principles that guide our professional practice as designers. Debates over whether direct instruction methods are "better" than constructivist or discovery or student-centered methods sometimes ignore the fact that many contemporary theorists are in the process of broadening their repertoire of instructional strategies to include those from many different theoretical frameworks. On the other hand, debates over the principles derived from a broad paradigm or theoretical framework, are more likely to be enlightening and to further our efforts to enhance both the quality of ID and the quality of student learning.

In this chapter I recommend William Winn's (1997) idea of *first principles* as a way of linking theory/paradigm and practice. Winn's first principles are not the same as Aristotle's first principles. Aristotle's first principles were a few laws about the world that were so basic and so foundational that they could not be questioned or demonstrated to be true or false. Winn's first principles are really more like "Guiding, Tentative Principles" that help us make decisions in a particular context but which are not to be elevated to the level of invioable laws that cannot be questioned or changed. This chapter makes several suggestions about how to approach the scholarly and professional discussion of contemporary issues about ID, particularly C-ID.

The emphasis on foundational issues continues in chapter 5, "Constructivism in instructional design theory," when Frank Dinter, a professor at the Dresden University of Technology, lays out his view of how to approach foundational philosophical issues that are important to ID. Dinter organizes his discussion around the questions of one important branch of philosophy—epistemology. Epistemology attempts to specify what methods are appropriate to use in our efforts to better understand both ourselves and the world we live in. Modernist perspectives have often been based on an empiricist epistemology that asserts the best, if not the only, reliable way to know how the deterministic and material world works is to use the scientific method.

Dinter criticizes instructional design as a field because it is reluctant to explore the foundational assumptions of different approaches to ID, and he suggests that without such careful and thoughtful exploration of our foundations in philosophy we will not make much progress toward developing more sophisticated conceptions of ID. He furthers that process by linking some important epistemological issues to our decisions about ID. He notes, for example, that there are many forms of realism and that different ones are more or less compatible with constructivist theory. Strong empiricists believe humans can discover and learn truths about the world, and that is a major foundation for ID with a goal of creating instructional resources for teaching students the way the world "really is." On the other hand, constructivists assume we have no mind-independent way of getting at the world. The "facts" we gather about the world will always come from interactions that are influenced by and interpreted through our mental activities. Designers who make this assumption will always be uncomfortable with ID models that seem organized to bring students to the "right answer" about something. Instead, concepts like "multiple perspectives," interpretation, standpoints, and paradigms will creep into design conversations.

Discussions of the underlying epistemologies of design practice may seem odd and out of place to many instructional designers who were taught modernist models that did not question the philosophical assumptions those models are based on. However, as Dinter points out, such discussions are part of the necessary work that all disciplines (and professions) must do if they are to move beyond simply doing what has been done in the past with little more reason than "that is the way we have always done it." Dinter also points out that some of the epistemological positions of radical constructivism will, if taken seriously, lead us into *cul-de-sacs* that are difficult, if not impossible, to escape without giving up the whole constructivist theory. He makes it clear, however, that such positions are not required of constructivist theory and that there are a wide range of epistemological positions in constructivism just as there are in other paradigms that can be used to guide ID work. Dinter's examination of the potential pitfalls of some constructivist epistemological positions is another strength of his work.

Near the end of his paper, Dinter takes a position that I disagree with. He argues that "there are no direct implications derivable from epistemological considerations for instructional design theories and models. The impact of epistemological considerations is on the construction of psychological theories of learning and instruction which serve as meta-theories for instructional design." Dinter then notes that other constructivist instructional designers disagree with his position. I agree that the philosophical foundations he discusses do influence ID indirectly, through the impact they have on psychological theories that are closely tied to different ID models. However, I do not believe this indirect path is the sole means of influencing ID. Dinter's position is easier to accept if we restrict our concept of ID to pedagogy. However, if you think about ID as a *process* of designing instruction, there are many ways constructivist epistemological theory may influence ID directly. For example, if humans do not have direct, "objective," access to the world, then their accepted "truths" will never be completely free of subjective, contextual, and experiential influences. Therefore, the idea of an instructional designer who can play the role of an "expert" who can communicate that expert knowledge to others is questionable. Given the lack of access to objective knowledge, an instructional designer might better play the role of organizer and supporter of a participating design team, with each member having something worthwhile to contribute.

Dinter explored the philosophical foundations of C-ID, but his paper was not a full exploration of philosophical positions and their implications for the professional practice of ID. Dinter did not because he does not believe it is appropriate. The influences of philosophical assumptions are, according to Dinter, filtered through psychological theories of learning and instruction. In the next chapter I will argue that the implications of philosophical positions for ID are both direct and indirect, and that they form a rich and promising foundation for new ways of both thinking about and doing ID. Chapter 6, "Considering the Philosophies of Wittgenstein, Dewey, and Rorty as Potential Foundations for C-ID," is the last in this section. It offers my view of how ID might be done if the models and the practice of ID were thoroughly infused with the theories of three twentieth century philosophers—the Austrian Ludwig Wittgenstein who taught and worked for much of his life in Vienna and then Cambridge University in England, the American John Dewey who was both a philosopher and educator in the first half of the twentieth century, and Richard Rorty who was also American and who had a profound influence on both philosophy and the social sciences in the last half of the twentieth century. Ironically, more than one philosopher has declared Wittgenstein to be the most influential philosopher of the twentieth century. However, Dewey is often described as the most influential and popular American philosopher of the twentieth century. The laboratory school he established at the University of Chicago is still in operation and his influence while at both the University of Chicago and Columbia University in New York are unsurpassed. Richard Rorty, who died in 2007, spent most of his academic career at Princeton, the University of Virginia and then at Stanford, but he did his undergraduate work at the University of Chicago and grew up in New York where his parents were activists in the labor movement and also communists. They moved to the relatively isolated area around Delaware Gap, New Jersey after his parents broke with American communism because of the leadership's support of Stalin. Rorty held professorships in philosophy (Princeton), humanities (Virginia), and comparative literature (Stanford) and many would argue that he has been the most influential American philosopher in the last 60 years, although much of his influence has been in the social sciences and the humanities rather than in philosophy.

REFERENCES

Duffy, T., & Cunningham D. (1996). Constructivism: Implications for the design and delivery of instruction. In D. H. Jonassen (Ed.), *Handbook of Research for Educational Communications and Technology* (pp. 170–198). New York: Simon & Schuster.

Smith, P. L., & Ragan, T. J. (2005). *Instructional design* (3rd ed.). Hoboken, NJ: Wiley.

Braden, R. (1996, March/April). The case for linear instructional design and development: A commentary on models, challenges, and myths. *Educational Technology, 35*(2), 5–24.

Willis, J. (2008). *Qualitative research methods in education and educational technology.* Charlotte, NC: Information Age Publishing.

Willis, J. (1995). A recursive, reflective instructional design model based on con-struc-tivist-interpretivist theory. *Educational Technology, 35*(6), 5–23.

Willis, J. (2000). The maturing of constructivist instructional design: Some basic principles that can guide practice. *Educational Technology, 40*(1), 5–16.

Willis, J., & Wright, K. (2000, March/April). A general set of procedures for con-structivist instructional design: The New R2D2. *Educational Technology, 40*(2), 5–20.

Winn, W. (1997, January/February). Advantages of a theory-based curriculum in instructional technology. *Educational Technology, 37*(1), 34–11.

CHAPTER 1

THREE TRENDS IN INSTRUCTIONAL DESIGN

Jerry Willis

Instructional Design, or ID, has a long history in educational technology. At least one course on ID is part of virtually every graduate degree program in educational or instructional technology offered in North America and in most of the rest of the world. ID, however, has an interesting and somewhat puzzling history.

THE TWO MEANINGS OF ID: *PEDAGOGICAL* ID AND *PROCESS* ID

As you read the scholarly literature on ID, you will find two related but very different uses of the term. Both have the core idea that ID involves the creation of instructional materials and educational resources. However, one use of the term *ID* is to indicate the theories of learning, principles of teaching, and the pedagogical strategies that *should* be used in creating those educational materials and resources. In this tradition, the core responsibility of the instructional designer is to make the "right" choices about:

- guiding theories of learning (e.g., behaviorism, cognitive science, information processing theory, constructivism, or ?),
- general strategies for teaching and learning (e.g., direct instruction, student-centered instruction), and
- pedagogies (e.g., problem-based learning, anchored instruction, tutorials, simulations, and so on).

This use of the term ID I will call *Pedagogical ID* because it emphasizes the responsibility of the instructional designer to choose, or help choose, the right way to teach the content that has been selected. In the Pedagogical ID literature, the instructional designer is someone:

- with knowledge of scientifically proven theories that tell us what teaching strategies should be used to teach each type of content,
- who uses a technical-rational approach to designing instructional materials, and
- who serves as the leader and expert on the design team because the designer has special knowledge that others do not.

The work of the designer is in the *technical-rational tradition* because it involves using scientifically established rules about which learning theories, and what types of teaching strategies, are best for teaching various types of content to various types of students. In the Pedagogical ID tradition this is at the heart of what it means to design instructional materials.

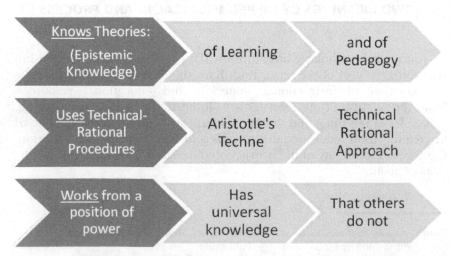

Figure 1.1.   Characteristics of a designer from a pedagogical ID perspective.

The underpinnings of the Pedagogical ID perspective are concisely expressed by M. David Merrill and the ID2 Research Group (1996) at Utah State University:

- The discipline of ID.
- There is a scientific discipline of instructional design.
- The discipline of instructional design is based on a set of specific assumptions.
- The discipline of instructional design is founded on scientific principles verified by empirical data.

Those persons who claim that knowledge is founded on collaboration rather than empirical science, or who claim that all truth is relative, are not instructional designers. They have disassociated themselves from the discipline of instructional design.

Instruction and Learning

- Instructional design is a technology for the development of learning experiences and environments which promote the acquisition of specific knowledge and skill by students.
- Instructional design is a technology which incorporates known and verified learning strategies into instructional experiences which make the acquisition of knowledge and skill more efficient, effective, and appealing.
- While instruction takes place in a larger organizational context, the discipline of instructional design is concerned only with the development of learning experiences and environments, not with the broader concerns of systemic change, organizational behavior, performance support, and other human resource problems.
- Instruction involves directing students to appropriate learning activities; guiding students to appropriate knowledge; helping students rehearse, encode, and process information; monitoring student performance; and providing feedback as to the appropriateness of the student's learning activities and practice performance. (Merrill and ID2 Research Group, 1996)

This view of ID is widely held and often used as the unstated but implicitly accepted framework for a great many papers, especially case studies, that describe a particular instructional design project. The definition below, which is widely quoted on the Internet, is also a definition of Pedagogical ID:

> Instructional Design is the systematic process of translating general principles of learning and instruction into plans for instructional materials and learning. (elearnspace, 2002)

Pedagogical ID is the foundation for many of the published papers on design in the educational technology literature. For example, in his paper on how to integrate "objectivism" and "constructivism" in instructional design, Johanes Cronjé (2006) focused on the theories of learning and the teaching practices/learning models supported by the two major competing paradigms in instructional design today. After summarizing the theoretical foundations of both objectivism and constructivism Cronjé focused on the general and specific teaching and learning methods that are proposed by objectivists and constructivists. However, he proposes that instructional designers adopt a "middle way" that freely uses professional practices from both ends of the continuum—objectivism and constructivism. Cronjé's paper suggests ways of deciding when instructivist or constructivist teaching and learning strategies should be used in the instructional resources you are designing.

The middle way proposed by Cronjé (2006), who teaches at the University Pretoria in South Africa, is only one of several positions taken in the literature relative to how design decisions should be made within a Pedagogical ID framework. Others, such as Merrill and the ID2 Group (1996) at the University of Utah, have taken a position supporting the instructional methods of one particular paradigm or theory of teaching and learning. Objectivist/behaviorist designers tend to refer to Gagné's (1985) nine events of instruction that are based on behavioral learning theory while designers who advocate constructivist pedagogies tend to base their work on the theories of Jean Piaget, Lev Vygotsky, John Dewey, Jerome Bruner, and progressive as well as constructivist designers and educators (Brooks & Brooks, 2001; Dawson, 2008). One of the most widely cited statements of paradigmatic purity on the constructivist side is the chapter by Bednar, Cunningham, Duffy, and Perry (1992). In that chapter, the authors made a case for the necessity of selecting a particular paradigm such as constructivism and then only using teaching and learning methods that are compatible with that foundation. While Bednar and her colleagues emphasized the conceptual and theoretical issues of instructional design, there are thousands of books and papers in the literature on how to teach from a constructivist perspective. Gagnon and Collay's (2006) book, for example, is a detailed guide to creating a range of constructivist learning environments.

A case study that takes for granted the framework of Pedagogical ID is Gordon Hensley's (2005) article on creating a hybrid college course. Hensley describes the revision of a popular humanities course at Appalachian State University, *Introduction to Theatre*. To deal with the problem of too many students wanting to take the course and not enough classrooms to handle the load, Dr. Hensley decided to create a hybrid version of the course that met face to face only a few times a semester and worked online the rest of the time. In his paper he summarized his prior experience teaching online and reviewed the relevant literature. He found "there are several theories in support of hybrid learning and ample supporting research available." In designing his course at Appalachian State University he used studies of hybrid courses at the University of Central Florida and the University of Wisconsin-Milwaukee as well as the *exemplary* courses on the WebCT site (http://webct.com/exemplary) to guide his thinking. He also attended workshops at Appalachian State on Dreamweaver and WebCT to develop the technical skills he needed to create his online Web course. Essentially, he transferred his face-to-face course to the online site. "I created a simple web site of materials to be covered. I then created study sheets for the quizzes, walked through all of my PowerPoint presentations to make sure the written text was self-explanatory, and wrote a few informal articles on topics—I basically wrote out my lectures in a brief and informal way." He put the course web site on a CD-ROM and distributed the CD to students because many had slow internet access at home. Then he created a set of "discussion posts, live chats, and personal student web pages [as well as] quizzes and assignments for the class in WebCT." Finally, Dr. Hensley created "a small support site to introduce students to hybrid learning."

In the article Dr. Hensley (2005) basically starts with the theoretical foundations he used to create his hybrid course. Then he tells us the end result of his design work. (Hensley does not, however, tell us very much about the design process he used.) The article ends with an overview of what worked, what failed, and some suggestions for other designers who want to create hybrid courses. Figure 1.2 illustrates how Hensley's approach to design fits the pedagogical view of ID.

Hensley's (2005) paper is an illustration of how ID is considered from the perspective of pedagogical ID. It involves the technical-rational application of known truths about teaching and learning to a new but familiar problem that has been addressed before by many others. Those who have come before have developed and validated general theories, specific models, and detailed teaching strategies that can be applied to the *new but familiar* design job. From a pedagogical ID perspective, anyone wanting to criticize the work of Dr. Hensley would do so by questioning the choice of teaching methods, content taught, or

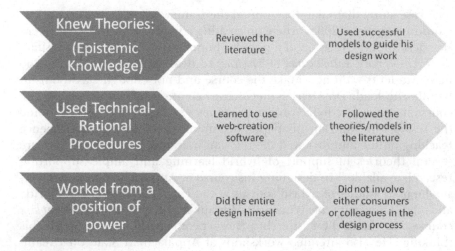

Figure 1.2.   Hensley's case study as an example of pedagogical ID.

technology deployed. There is, however, another way of thinking about ID—the process approach.

ID as a Process

The other use of the term ID is to indicate the *process* by which educational materials and resources are developed. This general definition of ID as a process for accomplishing design work does not refer directly to learning theory or instructional strategies. They are, of course, important to an instructional designer, but they are *not* design. ID is the *process* of designing learning materials, and one aspect of a designer's work is the use of knowledge and theories from many different disciplines. Theories of learning and pedagogy are part of the tool set that instructional designers use, but there is a difference between designing and the tools a designer uses. We do not equate nailing with being a carpenter or knowledge of the load bearing capacities of beams with architecture. The definition below, which is compatible with a view of ID as a process, is from Christa Harrelson at the University of Georgia:

> When people ask me what instructional technology is, instructional design is a large part of the definition. Instructional design is the process by which instruction, computer-based or not, is created. Instructional design provides a framework for the creative process of design, and ensures the learners'

needs are met. (Retrieved 11/18/2009, from http://ttc.coe.uga.edu/christah/ clhport/themes.htm#id)

I am going to use the view that ID is a process and reject the narrower and, I believe, confusing, use of the term ID to emphasize the use of certain psychological and instructional theories to creating instructional material. *ID as Pedagogy* is much too narrow a definition, and it completely ignores the idea that the translation of learning and instructional theory into an instructional resource is not a simple or obvious process. ID is a rather complex process and quite a few models have emerged over the past 60 years to guide that process.

I suggest that doing ID involves (1) developing and using phronetic (local, contextually constrained) knowledge, (2) using a range of technical knowledge and expertise, and (3) working collaboratively with a team of stakeholders (see Figure 1.3). However, I have actually committed an error that I sometimes criticize in others. Figure 1.3 does not actually represent the *general* idea of ID as a process. Instead, it presents ID as a *constructive-interpretive* process. Many of the definitions of ID in the literature are like that. They purport to provide a general definition of ID but are actually advocating a definition that is heavily laden with the preferred theoretical and epistemological positions of the authors. In my case the constructivist-interpretive paradigms are the foundations for my view of ID as a process. I have also framed the view of ID with and emphasis on two of Aristotle's types of knowing: phronetic and technical. Most of the traditional definitions of ID put an emphasis on Aristotle's third type of knowing—epistemic knowledge that is universal and eternal.

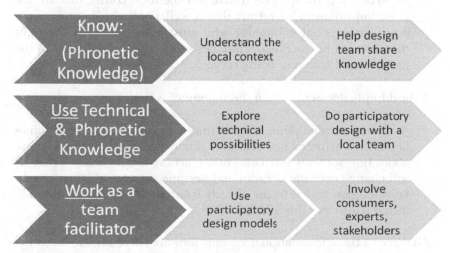

Figure 1.3. ID as a participatory process.

The quote from Merrill and the ID2 (1996) group given earlier, for example, defines ID within a framework that emphasizes both the existence and use of epistemic knowledge. It would take a very long article to unpack all the implicit but foundational assumptions and beliefs that are inherent in the definition presented by Merrill and his colleagues. To illustrate, consider the statement:

The discipline of instructional design is founded on scientific principles verified by empirical data.

There are at least three implicit assumptions in this statement and all of them are hotly contested in education and the social sciences today. The first is that:

- Human behavior can be subjected to scientific research that leads us to laws or law-like generalizations – expressed in theories – that are portable. They travel well from one context to another and are the most important type of knowledge a designer must know.

Many of the other design fields have moved beyond this positivist model of how we come to understand human behavior. A second implicit but fundamental assumption of the Merrill definition is that:

- Variations from one setting to another are of little importance compared to the value of the law-like generalizations that scientific research has given us. Thus, when it comes to creating instructional materials, it is the general truths, not the local truths, that are the most important. This means that a well prepared designer is an *expert* with special knowledge who must make sure that the right theories and instructional strategies guide the development of the instructional materials being created.

A third implicit assumption in the statement is that:

- As a scientific discipline, ID may not be perfect now but it becomes better and better as new scientific research tells us more and more about how students learn and how teachers should teach. Thus, ID will change over the decades and become better and better as new knowledge is added to the already substantial foundation of established truths that guide it today.

All three of these foundational assumptions are currently the subject of intense criticism in education, social science, and educational technology.

There is no body of scientific research that proves these assumptions. And, that brings into question the statement that *"The discipline of instructional design is founded on scientific principles verified by empirical data"* because all three of the assumptions noted above *must* be true before we can accept the statement as even possibly true. As Brent Flyvbjerg (2001) has pointed out, the social sciences (including education) have yet to come up with a single law of human behavior that is reliable and generalizable in the way that laws in physics and chemistry are considered reliable and general (and there is some debate today in the sciences about whether the "natural science" model fits even the research and knowledge of the natural sciences!). The assertion that a particular teaching method, or way of designing instructional materials, is "scientifically proven" is simply not true today, and may never be true. Similarly, criticisms of another approach on the grounds that it is "unscientific" because it is not supported by empirical research lacks foundation because little, if any, of what we believe we know about human behavior, human learning, and teaching is clearly supported by empirical research. As social science moves away from the idea that there are solid, empirical foundations for our theories and beliefs about human behavior, it becomes less and less acceptable to take an aggressive and fundamentalist perspective about any position. We simply do not have that type of research. One of the reasons I say that is because the second implicit assumption noted above is also wrong. When it comes to human behavior, the local context is important. It is often more important than the general laws proposed by experts, and many of the failures of reform and transformation efforts in education have foundered on this very point—the failure to understand the local context and how important that context is. And, if the laws that positivists are so confident of, turn out to be less dependable than they believe, then instructional designers do not have such a strong hold on the title of Expert. They might better consider their role as one of facilitator and organizer than of expert. (Note how I have strongly criticized ID positivists for their almost absolute confidence in certain truths and in the scientific method while doing exactly the same thing myself—claiming that what I happen to believe is *really* the truth and not what opponents say. I do not believe any of us can take a fundamentalist view that we are obviously correct and thus an opponent is obviously wrong. There is much more in the literature about this particular issue and it is up to you as a participant in the ongoing and extensive dialog to make up your own mind. Hopefully you will do it from an informed perspective so that you understand the various positions taken by different scholars before making your own decision.)

The idea that ID is a discipline based on results of scientific research has also been questioned by other educational technology scholars, including Thomas Reeves (2000) at the University of Georgia. A respected and widely read scholar, Reeves commented on the position paper by Merrill and his colleagues (1996) this way:

> Not everyone in or out of academe shares Professor Merrill's positive assessment of instructional design as a scientifically valid technology.... There is an enormous gap between Merrill's identification of instructional design as a robust technology derived from the science of instruction and Resnick's conclusion that instructional design is a field that does not seem to contribute to the solution of educational problems. (para. 2)

The Resnick that Professor Reeves (2000) refers to is Dr. Lauren Resnick from the University of Pittsburgh. He quoted from her 1999 AERA discussion after a presentation:

> We don't have a well-developed design field in education, ... I've looked around at the field called instructional design in which people can get degrees, and so far have not been interested in hiring any of the people with those degrees who have crossed my path. Just didn't look like they were going to add much.

Reeves goes on to point out that many other scholars, in and out of our field, have very little confidence in ID and educational technology research in general. If we are to be successful we must take a different approach to ID. One of the foundations I propose for that different approach is to consider the core of ID as the *process* of designing and developing instruction. ID is not the straightforward application of theories of learning to new but familiar learning situations.

At this point in the history of ID, we have a number of paths that can be pursued. They will be discussed in the next section.

Three Contemporary Threads in ID

In the first decade of the twenty-first century, three broad movements are active in the area we call Instructional Design or ID, and they are not all going in the same direction. The three movements are:

- **Traditional ID scholarship** based on positivist epistemologies. The most popular example of this movement is the Dick and Carey (Dick, Carey, & Carey, 2004) model of ID. The Dick and Carey ID model is the most popular example of a type of ID model often referred to as ISD or "instructional systems design." It is also a specific example of a generic ID model called ADDIE that has five sequential phases: Analysis, Design, Development, Implementation, and Evaluation. A great many ID models in use today are variations on the ADDIE model.

- **The Design-Based Research (DBR) Movement.** This is an effort to integrate design and research in ways that advance both our basic or theoretical knowledge and at the same time create high quality educational resources. Special issues of *Educational Researcher* (Vol. 32, No. 1, 2004) and *Educational Technology* (Vol. 45, No. 1, 2005) contain articles that define DBR and what it attempts to do. The book edited by a group of scholars at the Universities of Twente and Utrecht in the Netherlands (Akker, Gravemeijer, McKenney, & Nieveen, (2006) is one of the first book-length treatments of DBR. What sets DBR apart from other approaches to ID is that it tries to solve a local problem by designing effective instructional resources or procedures while making an effort to create knowledge that has a broader application than the local context. DBR is thus positivist in its view, and advocates typically remain optimistic that the positivist agenda of finding law-like truths about human behavior is possible.

- **Constructivist-ID Models (C-ID).** ID models based on interpretive epistemologies and constructivist theories of teaching and learning have begun to appear in the last two decades (Cennamo, Abell, & Chung, 1996; Cennamo, Abell, Chung, Campbell, & Hugg, 1995; Duffy & Cunningham, 1996; Johari, Chen, & Toh, 2005; Lebow, 1993; Willis, 1995, 2000; Willis & Wright, 2000; Wilson, 1997). These models tend to reject the idea that there are universal "laws of teaching and learning" that are applicable across educational context. C-ID models tend to emphasize the importance of understanding the local context and taking that into consideration when creating new educational resources. C-ID models also tend to emphasize collaborative development procedures, and they typically take a nonlinear and iterative or recursive approach to development. That is, you do not necessarily follow a sequence of steps in which Step 1 *must* come before Step 2, which must come before Step 3. C-ID models generally adopt a process that involves jumping back and forth from one task to another throughout the development process.

In an earlier paper (Willis, 1995) I suggested that traditional (e.g., *objective-rational*) ID models like the one developed by Dick and Carey share a number of important "family characteristics" (see Table 1.1) and that C-ID models also share their own "family characteristics (see Table 1.2).

Table 1.1.   Family Characteristics of
Objective-Rational Instructional Design (ID) Models

1. The Process Is Sequential and Linear
The design process is sequential, objective, and focused on experts who have special knowledge.

2. Planning Is Top Down and "Systematic" Begin with a precise plan of action including clear behavioral objectives. Proceed through the instructional design process in a systematic, orderly, planned manner.

3. Objectives Guide Development
Precise behavioral objectives are essential. Considerable effort should be invested in creating instructional objectives and objective assessment instruments

4. Experts, Who Have Special Knowledge, Are Critical to ID Work
Experts, who know a great deal about the general, universally applicable principles of instructional design, are needed to produce good instruction.

5. Careful Sequencing and the Teaching of Subskills Are Important
Break complex tasks down into subcomponents and teach the subcomponents separately. Pay particular attention to the sequence of the subskills taught as well as the events of instruction.

6. The Goal Is Delivery of Preselected Knowledge
Emphasis is on delivery of "facts" and enhancement of skills selected by experts, which favors drill and practice, tutorial, and other direct instruction methods. The computer takes the roles of a traditional teacher-information deliverer, evaluator, recordkeeper.

7. Summative Evaluation Is Critical
Invest the most assessment effort in the summative evaluation because it will prove whether the material works or not.

8. Objective Data Are Critical
The more data the better, and the more objective the data the better. From identifying entry behaviors to task and concept analysis, pretests, en-bedded tests, and posttests, this model emphasizes collection and analysis of objective data.

Tables 1.1 and 1.2 illustrate the many differences between the traditional and C-ID models, but the great majority of them are related to two fundamental choices—the preferred epistemology and the preferred theory of learning and teaching (see Table 1.3).

Positivist/postpositivist epistemologies generally assume that scientific research can discover universals—laws and rules of human behavior—that can then be generalized to new settings. For that reason, objective-rational ID models such as Dick and Carey and ADDIE tend to assume that the ID process is one of applying know laws and rules to new learning contexts. They tend to see ID from a pedagogical perspective and they tend to select the teaching methods incorporated into their designs from the set of methods developed within a few theories of learning—particularly behaviorism and cognitive science.

Table 1.2.   Family Characteristics of Constructivist-Interpretivist Instructional Design (C-ID) Models

1. The ID Process Is Recursive, Nonlinear, and Sometimes Chaotic
Development is recursive or iterative; you will address the same issues many times. Development is also nonlinear. There is no required beginning task that must be completed before all others. Some problems, improvements, or changes will only be discovered in the context of design and use. Plan for recursive evaluations by users and by experts. Plan for false starts and redesigns as-well as revisions.

2. Planning Is Organic, Developmental, Reflective, and Collaborative
Begin with a vague plan and fill in the details as you progress. "Vision and strategic planning come later. Premature visions and planning can blind" (Fullan, 1993). Development should be collaborative. The design group, which includes many who will use the instructional material, must work together to create a shared vision. That may emerge over the process of development. It cannot be "established" at the beginning. "Today, 'vision' is a familiar concept in corporate leadership. But when you look carefully, you find that most 'visions' are one person's (or one group's) vision imposed on an organization. Such visions, at best, command compliance-not commitment.... If people don't have their own vision, all they can do is 'sign up' for someone else's" (Senge, 1990, pp. 206–211).

3. Objectives Emerge from Design and Development Work
Objectives do not guide development. Instead, during the process of collaborative development, objectives emerge and gradually become clearer.

4. General ID Experts Don't Exist
General ID specialists, who can work with subject matter experts from any discipline, are a myth. You must understand the "game" being played before you can help develop instruction. If specialists are used, they should be immersed in the environment of use before assisting with design. However, for much ID development, the "citizen legislator" model, as opposed to the "professional politician" model, is preferred. Citizen legislators are developers who know and understand the content and context of practice and who pick up the ID skills needed. Professional politicians are ID specialists who are not situated in the learning context, or the content knowledge, domain.

5. Instruction Emphasizes Learning in Meaningful Contexts (The Goal Is Personal Understanding Within Meaningful Contexts)
Standard direct-instruction approaches that focus on teaching content outside a meaningful context often result in "inert" knowledge that is not useful. The instructional emphasis should be on developing understanding in context. This approach favors strategies such as anchored instruction, situated cognition, cognitive apprenticeships, and cognitive flexibility hypertext. Also favored are instructional approaches that pose problems and provide students with access to knowledge needed to solve the problems. This favors development of hypermedia/multimedia information resources, electronic encyclopedias, and a wide range of accessible, navigable electronic information resources.

6. "Formative" Evaluation Is Critical
Invest the most assessment effort in formative evaluations because they are the ones that provide feedback you can use to improve the product. Summative evaluation does nothing to help you improve the product.

Table continues on next page.

Table 1.2. Contnued

7. Subjective Data May Be the Most Valuable
Many important goals and objectives cannot be adequately assessed with multiple choice exams, and exclusive reliance on such measures often limits the vision and value of instruction. Some things can be shown or observed but not quantified. Many types of assessment, including authentic assessment, portfolios, ethnographic studies, and professional opinions, should be considered. In addition, during the instructional design process, there are many points where informal or qualitative approaches, such as interviews, observations, user logs, focus groups, expert critiques, and verbal student feedback, can be much more valuable than a data from a 10 item, Likert-scale, questionnaire.

Table 1.3.   The Choices Made by Developers of Objective-Rational and C-ID Models

Family of ID Models	Epistemology	Learning/Instructional Theory
Objective-rational (e.g., Dick and Carey, ISD, ADDIE)	Positivism, postpositivism	Behaviorism, information processing theory, cognitive science, instructionism, direct instruction
Constructivist Instructional Design Models	Interpretivism, hermeneutics	Constructivism, social constructivism, Deweyian progressive education theories

C-ID models, on the other hand, assume that there are no universals, no absolute laws of human behavior and learning, that can be confidently generalized from one situation to another. These models are based on interpretive epistemologies that assume humans do not have access to a "God's eye view" of the world where they can confidently state laws about why humans behave the way they do. The local context is critically important in determining what works and what does not, and the best we can do is develop a very good understanding of that local context. C-ID models also generally accept the interpretivist assumption that on most major issues there will be multiple perspectives and students are best served by helping them learn, not the Right answer, but what a number of different groups or communities think is true. This leads designers who use C-ID models to include instructional strategies from many different theories of learning, including constructivism. These models typically emphasize helping students "construct" their own understanding of a topic through experience in context (e.g., problem-based learning, authentic instruction).

I will return to C-ID models later, but at this point it is important to explore the other relatively new influence on contemporary ID scholarship, Design-Based Research (DBR).

Design-Based Research: A Promising Model or More of the Same?

In her paper inaugurating a series of articles on how to do research in the field of educational technology, M. D. Robyler (2005) began with the statement "We need a more organized and persuasive body of evidence on technology's benefits to classroom practice." She then discussed the problems of the existing body of research such as "fragmented and uncoordinated approaches to studying technology resources and strategies, methods that lack rigor or are ill-matched to the research questions at hand, and poorly written reports that render problematic subsequent attempts at replication and follow-up." Robyler's analysis of the continuing problems of traditional educational technology research leads her to suggest some solutions. One of them is to acknowledge that the instructional design process is one form of research. ID has been a major component of the field of educational technology virtually from the beginning, but it has typically been viewed as a professional practice activity which is separate from, and different from, educational technology research. But Robyler believes "the field is beginning to resound with the call for a new educational technology research agenda—one that focuses on capturing the unique impact of technology-enhanced instructional designs, rather than the digital technologies themselves." She describes an approach that suggests five pillars for educational technology research:

1. **The Significance Criterion.** "Every educational research study should make a clear and compelling case for its existence." She advises that in an era when scientific evidence of impact is emphasized by federal officials and many others, "articles that report research studies with technology-based teaching strategies should begin by making it clear that they address a significant educational problem, as opposed to a proposed technology solution."

2. **The Rationale Criterion.** Failure to "attend to the need for a solid theory base in research" is very damaging to the field. "If we are to make progress in this field or any other, new research must carefully consider previous lines of research, and each study must be built on a foundation of theory about expected effects derived from past work."

3. **The Design Criterion.** "After establishing research questions, researchers must decide on a research approach and methods that are well-suited to capturing and measuring impact on the variables of interest." Robyler rejects the positivist foundations of the No Child Left Behind model of educational research that considers randomized experimental studies as the gold standard. She argues

that many research designs are appropriate. "What is essential is a design that is a logical choice for the questions under study." She supports the thoughtful use of both *objective, scientific* methods as well as *naturalistic inquiry* which includes a great many traditional qualitative research methods plus methods drawn from the humanities and philosophy.

4. **The Comprehensive Reporting Criterion.** This criterion addresses the problem of how research is to progress toward a goal of better knowledge about important questions and problems. Knowledge should be cumulative across a series of studies on the same issue. Robyler believes this is often difficult or impossible because many research reports do not provide "enough sufficiently detailed information to allow others to analyze and build on previous work." Well organized and detailed reports of research are needed but often not provided.

• **The Cumulativity Criterion.** This criterion also addresses the question of how individual research papers can contribute to cumulativity—to "moving the field forward." Robyler advises that "articles reporting research ideally should make it clear that the study is part of a current or proposed line of research, along with proposed next steps in the line."

• Figure 1.4. Two responses to failure.

In the political sphere, when efforts to solve a major problem have not been very successful, the proposals for correction fall into two broad categories (Figure 1.4).

Some will argue that what has been done in the past was not done well enough, or was not done in the quantity needed. As I write this, the most

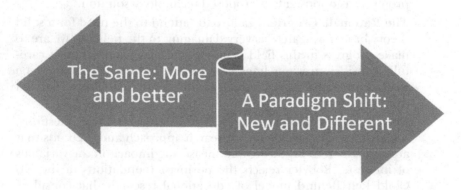

Figure 1.4. Two responses to failure.

obvious example of this tendency is President Bush's "surge" answer to the problems of the war in Iraq. When the military solution to Iraq did not work, he sent more troops. On the other side, virtually all of the Democratic candidates for President took the opposite position—remove the combat troops from Iraq in the near future. Robyler's solution to the problems of educational technology research falls roughly in the "more and better" tradition of dealing with failure. With the exception of her acceptance of natural inquiry models of research, all of her Pillars are compatible with a positivist model of applied research (and many are also compatible with an interpretivist approach as well). The Pillars and the text of the article emphasize the importance of the *Theory-Implications-Practice* link that is at the heart of positivist thinking. She emphasizes over and over again that we should use research methods that "can help researchers understand the effects of technology on student learning." My reading of Robyler is that she believes by doing better research we can discover general rules about technology and learning that are robust enough to be transferred from the context of discovery to other contexts of application. This is a positivist goal that is solidly in the tradition of finding universals that can be generalized to other settings.

In a commentary on the Robyler article, Chris Dede (2005) praised the beginning of the series on research methods and commented on one of the types of research mentioned by Robler: "studies of technology-based instructional designs" which she described as "almost non-existent in the existing literature." Dede argued the reverse is true, "scholars are publishing a growing body of high quality, design-based research studies that address many of the weaknesses of typical scholarship in educational technology." What Dede was referring to is *design-based research* (DBR). He notes that there have been special issues on DBR in the *Journal of Learning Sciences* (2004, Vol. 13, No. 1), *Educational Researcher* (2003, Vol. 32, No 1) and *Educational Technology* (2005, Vol. 45, No. 1). Several authors, including Dede, have used the work of Pasteur as a foundation for DBR. Pasteur's work tended to focus on "difficult, applied, practice-driven questions" that direct the researcher to questions that are, at their base, "fundamental theoretical issues." This idea comes from Stokes' (1997) book about how basic and applied research can be related. Stokes rejected what he considered the simplistic idea that research could be organized by a binary categorization of Basic or Applied. He developed a four quadrant diagram that represented the relative importance of basic knowledge and applied use to the researcher (see Figure 1.5).

The work of Neils Bohr is an example of a research program that has a High interest in Foundational Knowledge and No interest in Applied Use. When he developed a description of the atomic structure of the universe, Bohr was searching for basic, foundational knowledge but he had little or

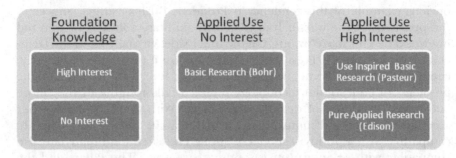

Figure 1.5.   Stokes' four quadrants of basic and applied research.

no interest in how that knowledge might be applied. Thomas Edison's research, on the other hand, is an example of work with a very High interest in Applied Use and No interest in Foundational Knowledge. When he developed the light bulb and the phonograph, his interest was on developing a useful product. And, while he was quite willing to use foundational knowledge developed by others, he had little or no interest in developing basic or foundational knowledge in his own work. He was a *consumer,* not a producer, of foundational knowledge.

Stokes (1997) comments that it would be appropriate to call his bottom right quadrant "Edison's quadrant "in view of how strictly this brilliant inventor kept his coworkers at Menlo Park, in the first industrial research laboratory in America, from pursuing the deeper scientific implications of what they were discovering in their headlong rush toward commercially profitable electric lighting. A great deal of modern research that belongs in this category is extremely sophisticated, although narrowly targeted on immediate applied goals" (p. 74).

Many of the educational technology scholars who appreciate Stokes' (1997) model, believe the work of Louis Pasteur, not Thomas Edison, is the example we should follow. Stokes suggests calling the upper right quadrant Pasteur's quadrant "in view of how clearly Pasteur's drive toward understanding and use illustrates this combination of [basic and applied] goals" (p. 74). Pasteur worked from the context of practice. He looked for problems in practice and then tried to understand the origins and causes of the problem. For example, he became concerned with the practical problem of stopping milk and wine from going sour. His research led to his understanding that active and living microorganisms entered nutrient broths such as milk and wine when they were open to the air. (An earlier explanation was the *spontaneous generation* of living organisms from nonliving matter.) He also showed that these microorganisms were the cause of the problem. That was basic knowledge that was not known before his research. He developed the process of pasteurization to prevent the problem, based on his discovery that even microscopic organisms would die if exposed to enough heat. That was his applied contribution in this case.

Stokes' (1997) model and Pasteur's quadrant appeal to educational technologists who want our discipline to be an applied field that makes a difference in education and training *as well as* a field that contributes to basic knowledge about human learning. Pasteur's quadrant lets the field do just that. Chris Dede (2005) used Stokes model when he proposed that DBR is an example of research in Pasteur's quadrant.

> DBR resembles the scholarly strategy chosen by the scientist Pasteur, in which investigation of difficult, applied, practice-driven questions demands and fosters studies of fundamental theoretical issues. As one illustration, the research my colleagues and I are conducting on multi-user virtual environments (Nelson, Ketelhut, Clarke, Bowman, & Dede, 2005) tests the efficacy of three alternative pedagogical strategies based on different theories about learning: guided social constructivism, expert mentoring and coaching, and legitimate peripheral participation in communities of practice. We are examining which of these pedagogies works best for various types of content and skills, as well as for different kinds of learners.

DBR is the most common name for this type of scholarship but it is not the only one. Other names include *development research, design research, formative research, design experiments* and *educational design research*. All these terms have been used in the literature to refer to design research in Pasteur's quadrant.

Pasteur's quadrant is also the focus of some proponents of change in the general field of educational research. Burkhardt and Schoenfelt (2003), for example, cite the "awful reputation of educational research" and use Pasteur's quadrant to advocate design research based on an engineering model:

> In the educational research community the engineering approach is often undervalued. At major universities only "insight" research in the humanities or science tradition tends to be regarded as true research currency for publication, tenure, and promotion. Yet engineering research has a key role to play in making educational research as a whole more useful. In Pasteur's Quadrant, Stokes (1997) argues that better insights come from situating inquiry in arenas of practice where engineering is a major concern. Stokes's motivating example is Pasteur, whose work on solving real world issues contributed fundamentally to theory while addressing pressing problems such as anthrax, cholera, and food spoilage.... Analogous arguments have been made regarding the potential for such work in education ... and serve as a justification for design experiments. Our point is that the same profitable

dialectic between theory and practice can and should occur (with differing emphases on the R&D components) from the initial stages of design all the way through robust implementation on a large scale. We also argue that success will breed success: Once this approach is shown to produce improved materials that work on a large scale, more funding will become available for it. Such has been the history in other applied fields, such as medicine and consumer electronics. (p. 5)

Design-based research is an example of how Pasteur's quadrant can be used to refocus and reform our approach to research in education and educational technology. As an emerging model for research, many aspects of DBR have not yet been agreed upon, even by those most involved in promoting the approach. However, I believe there is enough agreement to permit a critique of DBR. In the next two sections I will explore desirable and undesirable family characteristics of DBR. I do this from the perspective of an interpretivist and a constructivist and make no pretense at trying to be "objective" about my judgments

Desirable Characteristics of DBR

Two characteristics of DBR seem to add considerably to the possibilities of educational technology research, and particularly to the idea of ID as scholarship.

Emphasize *Use-Inspired Research.* This suggestion, made by Stokes (1997) and many others, makes sense. In education and the social sciences we have failed at the positivist research goal of discovering universals that can guide practice. The result of that failure is the separation of research from practice so that the "success" of a researcher is dependent, not on the contributions made to improving professional practice, but to the judgment of other researchers. When researchers publish mainly for other researchers, and their contributions are evaluated by other researchers, the gap between research and practice becomes wider and wider. One way to narrow that gap is to insist that in an applied field like educational technology, researchers should be encouraged to conduct "use-inspired research." DBR does just that—it focuses research on the design and development of educational resources.

Integrate Research and Development. Reeves (2000) points out that there has been a traditional separation of "research" and "development" (with ID considered development, not research) in our field. It is even evident in the way one of the leading journals is organized. Both the title of the journal, *Educational Technology Research and Development,* and the organization of the journal into two separate sections (one on research, another on development) with different groups making decisions about what is published and what is rejected, highlights the tendency in our field to view research and development as separate and distinct activities.

> Some instructional technologists appear to have great commitment to basic research, regardless of whether it has any practical value, perhaps because basic research seems more scientific or they believe that it is someone else's role to figure out how to apply the findings of basic research. Others seem to believe that the value of basic research in a design field such as IT is limited and that IT research should therefore have direct and clear implications for practice. (Reeves, 2000, para. 7)

Reeves proposes that we place ourselves squarely in Pasteur's quadrant and focus on "use-inspired basic research." He justifies his proposal in several ways, but one of the most important is that "in contemporary science, new technological developments often permit the advancement of new types of research, thus reversing the direction of the basic to applied model." Reeves thus does not give up on discovering basic knowledge; he is simply arguing that we approach the task from a different angle. Instead of doing "basic research" in the Niels Bohr tradition, we should emulate another successful scientist, Louis Pasteur, and work on problems that are important to practitioners. In the process of doing "use-inspired basic research" we may discover more basic knowledge than we have discovered using the Bohr or basic research model.

Summarizing Ann Brown (1992) and Alan Collins (1992), Reeves (2006) specifies three critical characteristic of the type of design experiments he advocates:

- "addressing complex problems in real contexts in collaboration with practitioners,
- integrating known and hypothetical design principles with technological affordances to render plausible solutions to these complex problems, and
- conducting rigorous and reflective inquiry to test and refine innovative learning environments as well as define new design principles"

These three characteristic represent a radical departure from the traditional positivist approach to educational technology research. For example, the first characteristic acknowledges that work in "real contexts" is more important than maintaining tight control of the study by conducting it in tightly controlled "laboratory" conditions. The first characteristic also acknowledges the critical importance of local knowledge by making collaboration with practitioners a fundamental component of design experiments. The second critical characteristic emphasizes the need to work from an informed position, taking into consideration the relevant professional and scholarly literature. The third characteristic emphasizes the need to carefully and thoroughly study the products of design experiments to determine what works and what does not work. Reeves (2006) specifically includes an interpretivist method, reflective inquiry, as an important method of doing that. The only vestige of a positivist approach is his last phrase—"as well as define new design principles." Without this last phase, design experiments as he defines them could just as well fall into Thomas Edison's quadrant of pure basic research as Pasteur's quadrant of use-inspired basic research. However, even here he acknowledges that some people may consider these new design principles not as truths that have been discovered but as ideas readers, rather than the researcher, will decide whether to adopt, adapt, or reject in their own work context. In spite of that concession, Reeves remains enamored with the idea that design experiments will produce generalizable knowledge very similar to the law-like generalizations of positivist research. For example, he describes the work of Jan Herrington and her colleagues at Edith Cowan University as a "long-term effort to develop and apply a model of situated learning theory.... She not only developed a model of the critical factors of situated learning and instantiated these factors in multimedia learning environments, but she tested the model and the technological products in multiple contexts, including preservice teacher education courses and K–12 schools." This sounds very much like research in the positivist tradition and I believe it reflects Reeves' reluctance to give up on the positivist agenda of finding laws and generalizable truths that can be discovered in one context and generalized to other contexts. His diagram of the process of development research (e.g., design experiments or DBR) also reflects this emphasis

The second and fourth boxes reflect a positivist heritage in Reeve's model. Requiring that solutions be developed "with a theoretical framework" harks back to the positivist idea that in *real* research the most important thing is to develop and validate theories that represent our current thinking about what is true and generalizable from one context to another.

In Reeves' (2000) model, there is no opportunity to take a pragmatic approach and ignore theory (or take a theoretically promiscuous approach) while trying to design instructional resources that "work" in the local context. In Reeves model, theory comes in early and stays late. In the fourth step of the process, documentation of what happened when the resource was actually used, and reflection on the process and results, come together to produce "design principles" that can be passed on to other designers. Note that the emphasis is on the creation of design principles,

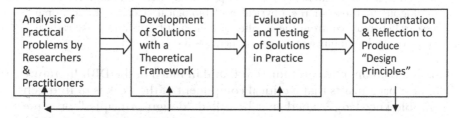

| Analysis of Practical Problems by Researchers & Practitioners | Development of Solutions with a Theoretical Framework | Evaluation and Testing of Solutions in Practice | Documentation & Reflection to Produce "Design Principles" |

Source: Reeves (2000).

Figure 1.6.   Reeves' model of design experiments.

which are based on the theoretical framework established in Step 2. The model thus emphasizes the accomplishment of a theoretical goal—the creation of design principles—rather than the accomplishment of a practical goal—the creation of useful resources for use in the local context. This difference is what separates Edison's quadrant from Pasteur's quadrant.

Another difference reflects a positivist rather than an interpretivist approach to generalization. The researchers doing design experiments who develop the "design principles" and pass them on to practitioners. A more interpretive approach might provide practitioners with enough information about a design experiment to let them make the decision about design principles they want to try out in their own setting rather than accepting those developed by the researcher. Some of Reeve's (2006) commentary suggests just that approach, but other comments are more compatible with a positivist model.

Undesirable Characteristics of DBR

Two characteristics of DBR seem likely to generate very undesirable results. One of them has already been mentioned.

Maintaining the Hope of Finding Laws. As noted earlier Reeves' (2000) model of DBR emphasizes the importance of organizing design work within a theoretical framework and using the applied design work to develop "design principles" that are essentially generalizations in the positivist sense of that term. However, he undercuts the possibility of doing that by distinguishing between research in fields like chemistry and biology, and research in education.

Education is a fundamentally different type of science, if it is a science at all, and educational researchers have never produced discoveries even remotely analogous to those in the physical and biological sciences. Educational researchers must confront the sterility of their past labors and make radical steps to conduct inquiry in more productive ways.

This is a fundamental contradiction found in much of the DBR literature. The approach insists that a critical component of the work is to find generalizable knowledge, whether it be called "design principles" or something else. Yet, at the same time, proponents of DBR criticize the existing research for its failure to find just those sorts of generalizations. Many, like Reeves (2000, 2006) and Flyvbjerg (2001) also doubt that such a goal will ever be accomplished. Seeking law-like generalizations while also developing locally useful educational resources has the potential to reduce the amount and quality of those educational resources while, at the same time, continuing the tradition of positivist research in the social sciences—investing much to find laws and achieving little. The core issue here is the question of whether creating useful educational resources for a specific context, or finding theory-based general principles or laws is the primary focus. Reeves and most proponents of DBR seem to privilege theory-based generalization over the creation of quality education materials. This does two things. It limits the options in creating educational resources because it assumes the "best" solution will always be found in the set of implications based on one of the existing theories. That is not always true, however. The foundatonal question of whether light is a wave or particles illustrates this point. Sometimes light acts like a wave and sometimes like particles. Trying to determine which of the answers offered by existing theories is true may not always be the most fruitful approach to design. Further, focusing on the implications of a theory also turns our attention away from another valuable of potential solutions— the tacit and experience-based knowledge of practitioners. If, as interpretivists argue, the local context is a major factor in "what works" then the knowledge of those who have intimate knowledge of that context is a valid and important source of guidance in design, even if it does not fit neatly into one of the existing theories.

Rejection of Work Based on Interpretivist and Hermeneutic Paradigms. Much of the work on the conceptual and theoretical foundations of design-based research involves a careful balancing act between harsh criticism of the existing body of positivist research in educational technology, and enthusiastic recommendations to pursue a modified version of the positivist agenda. While it is a thoughtful and useful analysis, Reeves' (2000) paper is a good example of this. After criticizing the "poor quality" of the existing body of research on educational technology, he nevertheless rejects the most radical efforts at reform. Interpretivist studies, which are "focused on portraying how education works by describing and interpreting phenomena related to teaching, learning, performance assessment, social interaction, innovation, and so forth" are criticized as producing little results that can be generalized to other settings. He concludes that "there is little evidence that the increasing popularity of qualitative methods will improve the impact of IT research on practice, especially given that the proponents of qualitative approaches make few claims to generalizability." In making this criticism of interpretive/qualitative research Reeves is using a positivist framework that insists the development of generalizable knowledge is the primary goal of a researcher. He ignores the focus in interpretive research on shifting the responsibility for generalization to the consumer of the research. He also seems to assume that generalization can only occur in the form specified by the positivist research model. That form of generalization involves carefully studying a selected sample of subjects drawn from the population the researcher wishes to generalize the results to. If all the requirements of positivist research are met (which almost never happens) the findings of a study should be generalizable from the sample to the population. In addition to relying on a positivist form of generalization that has not worked well in education and the social sciences, Reeves also seems to downplay the importance of success on a local level. If a particular approach to dealing with problems and issues in education actually developed good "local" solutions, that, in itself, would be grounds for celebration. If the state of American education is as bad as critics say it is, a process that helps local participants find good local solutions would be a major contribution to the future of educaton.

Greenwood and Levin (2005) presented another approach to generalization in their discussion of how universities, especially the academic departments of social science, could be reformed and brought into more relevant contact with the societies around them. These authors follow Flyvbjerg's (2001) approach of situating a discussion in Aristotle's framework of three types of knowledge: epistemic, techne, and phronesis:

- *Episteme* is "the conventional and favored form of explicit and theoretical knowledge and the form that currently dominates academic social sciences.... The sources of *episteme* are multiple—speculative, analytical, logical, and experiential—but the focus is always on eternal truths beyond their materialization in concrete situations.... *Episteme* accords rather closely to everyday usage of the term theory" (p. 50).
- *Techne* "is a form of knowledge that is inherently action oriented and inherently productive. *Techne* engages in the analysis of what

should be done in the world in order to increase human happiness.... The courses of techne are multiple, [but all of them require] sufficient experiental engagement in the world to permit the analysis of 'what should be done.' " It is craft and art knowledge, and "as an activity it is concrete, variable, and context-dependent. The objective of techne is application of technical knowledge and skills according to a pragmatic instrumental rationality" (p. 50). While Greenwood and Levin acknowledge that "practitioners of *techne* do engage with local stakeholders, power holders, and other experts, ... they are first and foremost professional experts who do things 'for,' not 'with,' the local stakeholders. They bring general designs and habits of work to the local case and privilege their own knowledge over that of the local stakeholders" (p. 51).

- *Phronesis* "is best understood as the design of action through collaborative knowledge construction with the legitimate stakeholders in a problematic situation. The sources of *phronesis* are collaborative arenas for knowledge development in which the professional researcher's knowledge is combined with the local knowledge of the stakeholders in defining the problem to be addressed. Together, they design and implement the research that needs to be done to understand the problem. They then design the actions to improve the situation together, and they evaluate the adequacy of what has been done. If they are not satisfied, they cycle through the process again until the results are satisfactory to all the parties" (p. 51).

I am not sure that Aristotle would recognize this definition of phronesis if through some magical process this quote could be read to him. It is an expansion of Aristotle's ideas into the modern context of qualitative research in complex twenty-first century democracies, which is no mean feat. However, Greenwood and Levin's (2005) characterization is probably defensible if expansive. They go on to say that phronesis involves creating a "new space for collaborative reflection, the contrast and integrating of many kinds of knowledge systems, the linking of the general and the particular through action and analysis, and the collaborative design of both the goals and the actions aimed at achieving them" (p. 51). This is very close to the core ideas of participatory action research but it can also be applied to an interpretivist form of design experiments. The authors also talk about phronesis developing within egalitarian "communities of practice" that collaboratively develop their own core beliefs that lead to an understanding of "knowing how to act to reach certain goals ... in real-world contexts with real-world materials." This is *contextual knowledge* as Greenwood and Levin point out, but it is also ethical and aesthetic knowledge, which involves making fundamental judgments about what is good for people. This is not emphasized by the authors but it is a part of Aristotle's original conception of phronesis.

Greenwood and Levin (2005) prefer *phronesis* over other forms of knowledge and approach the issue of how qualitative research results can be generalized by establishing their position on the role of epistemic knowledge.

"Knowing how" thus implies knowing how in a given context in which appropriate actions emerge from contextual knowing. The conventional understanding of general knowledge that treats it as supracontextual and thus universally applicable is of very little interest to us because we do not believe that what constitutes knowledge in the social sciences can be addressed usefully from the hothouse of armchair intellectual debate. (pp. 51–52)

The act of generalization in a positivist sense is based on the idea that humans can discover "supracontextual and thus universally applicable" knowledge about humans. Without that foundational assumption, generalization, at least in a positivist sense, is not a concern because it is not possible. Greenwood and Levin argue that social science's inability to discover universal knowledge means the traditional distinction between basic and applied research is untenable.

We believe this division makes social research impossible. Thus, for us, the world divides into action research, which we support and practice, and conventional social research (subdivided into pure and applied social research and organized into professional subgroupings) that we reject on combined epistemological, methodological, and ethical/political grounds. (p. 53)

The authors defend their position against "the dominance of positivistic frameworks and *episteme* in the organization of the conventional social sciences" and against the "hard-line interpretivists" who take a completely relativist position that since nothing is ever known for sure we cannot make sensible decisions about what action to take based on what we have come to know through scholarship.

If universal or epistemic knowledge is not possible in the social sciences, and extreme relativism in which every theory is no better, and no worse, than any other theory, are there any other viable possibilities? There are several, but the one Greenwood and Levin (2005) offer is based on a pragmatic philosophy of social science that was nurtured by three American scholars from the nineteenth and early twentieth centuries: John Dewey, William James, and C. S. Pierce. One of the key points of pragmatism is how it defines truth. In a lecture given in 1906 titled "What Pragmatism Means," William James explained pragmatic truth this way:

> Riding now on the front of this wave of scientific logic Messrs. Schiller and Dewey appear with their pragmatistic account of what truth everywhere signifies…. It means, they say, nothing but this, **that ideas (which themselves are but parts of our experience) become true just in so far as they help us to get into satisfactory relation with other parts of our experience,** to summarise them and get about among them by conceptual short-cuts instead of following the interminable succession of particular phenomena. Any idea upon which we can ride, so to speak; any idea that will carry us prosperously from any one part of our experience to any other part, linking things satisfactorily, working securely, simplifying, saving labor; is true for just so much, true in so far forth, true **instrumentally**. This is the "instrumental" view of truth taught so successfully at Chicago, the view that truth in our ideas means their power to "work," promulgated so brilliantly at Oxford.
>
> Messrs. Dewey, Schiller and their allies, in reaching this general conception of all truth, have only followed the example of geologists, biologists and philologists. In the establishment of these other sciences, the successful stroke was always to take some simple process actually observable in operation—as denudation by weather, say, or variation from parental type, or change of dialect by incorporation of new words and pronunciations—and then to generalise it, making it apply to all times, and produce great results by summarising its effects through the ages.

James (1906) is saying that "what works" is what is true. This is sometimes referred to as the "cash value" of truth. This is not the epistemic truth of positivists; it is the contextual and constrained truth of craft knowledge (techne) and the practical or praxis knowledge of phronesis. James does talk about the ability to generalize this form of truth, but that generalization is not the same as positivist generalization. For the positivists, a discovered law *must* generalize because it is universally true. Thus, if an effort to generalize a universal truth about how children learn mathematics fails, the most likely targets for blame are the teachers; they *must* have implemented the universal truth incorrectly. The generalization of pragmatists is not so assured because their idea of truth is based on the assumption that all distinctions and boundaries are subjective. The idea of *usefulness* is at the heart of pragmatic truth, not a match between what is said and what exists in the physical world. When you combine the assumption that our understanding of the world is always subjective with the concept of truth as *useful knowledge,* the result is an approach to scholarship that is more likely to question the truth than the implementers of truth when something goes wrong. Based on those foundations Greenwood and Levin (2005) propose that we do social science research that:

- Involves the generation of knowledge "through action and experimentation in context" (p. 53).
- Is based on "participative democracy as both a method and a goal" (p. 53).
- Relies on cogenerative inquiry which "aims to solve pertinent problems in a given context through democratic inquiry in which professional researchers collaborate with local stakeholders to seek and enact solutions to problems of major importance to the stakeholders. We refer to this as cogenrative inquiry because it is built on professional researcher-stakeholder collaboration and aims to solve real-life problems in context" (p. 54).
- Combines local knowledge and professional knowledge. That is, cogenerative inquiry involves the collaborative use of both local and professional knowledge and the result is often a unique solution to a problem that relies on that blending of the two forms of knowledge. Proponents of this approach might argue that positivist approaches tend to privilege professional knowledge too much while interpretivist approaches tend to privilege local knowledge too much.

In advocating this form of scholarship the authors reject the dominant, positivist paradigm of social science research. "This positivist credo obviously is wrong, and it leads away from producing reliable information, meaningful interpretations, and social actions in social research" (p. 53).

Blending Constructivist Instructional Design and DBR: An Appealing Alternative

Traditional instructional design models based on positivist assumptions and behavioral theories of learning are increasingly questioned by both practitioners and ID scholars (Cobb, 2002; Crawford, 2004; Häkkinen, 2002; Willis, 1995; Winters & Mor, 2008).

One of the alternatives, *Constructivist Instructional Design*, or C-ID, is emerging as an option for many instructional designers as they shift from positivist and behaviorist paradigms to constructivist and interpretivist or critical paradigms. However, C-ID models have been around for less than 20 years and they are still not widely used, nor have they had time to mature through several generations of use and revision cycles. Further, none of the existing C-ID models have as a purpose the positivist goal of finding universals such as the "right" processes for designing certain types of educational resources, the "best" teaching methods for particular types of content, or the "most effective" pedagogies for gifted, hyperactive, handicapped, immigrant, talented, or learning disabled children. C-ID models may be employed to develop educational resources and curricula for many contexts, but the emphasis is on development work *for the context at hand*. The generalization of design processes, pedagogies, or instructional strategies to other settings is left to those who read about the C-ID projects rather than the scholars and professionals who write papers on those projects. For many instructional technologists this is a weakness. Reeves (2000) is critical of action research because it is "similar to development research (e.g. DBR) except that there is little or no effort to construct theory, models, or principles to guide future design initiatives. The major goal is solving a particular problem in a specific place within a relatively short timeframe. Some theorists maintain that this type of inquiry is not research at all, but merely a form of evaluation." Reeves (2006) does accept action research as a "legitimate form of research provided reports of it are shared with wider audiences who may themselves choose to draw inferences from these reports in a sense similar to reports of interpretivist research." While hardly providing a ringing endorsement of interpretive, critical, and action research, Reeves does at least make some room for these approaches in educational technology. However, his primary focus is on "developmental research" in the form of "use-inspired basic research." As noted earlier, I find the optimism of DBR proponents that law-like generalizations and universal laws can be discovered about education to be too optimistic given the more than 100 years of failure to make even modest strides toward fulfilling this hope (Flyvbjerg, 2001)

Perhaps there is a way of combining the more open and flexible processes of C-ID models (Willis, 2000; Willis & Wright, 2000) with the goal of generalizable conclusions that are often a goal of design-based research (DBR). C-ID proponents would accept that what they learn in the professional practice of design may be of use to other designers, and DBR proponents would accept the idea that the end result will not be positivist laws and universals. Instead the results are kinder, gentler suggestions, ideas, and exemplars that can be considered by other designers, but never unthinkingly adopted without thoughtful and reflective consideration of their relevance for a new and different context. This middle ground asks the constructivist-interpretivists to accept that there can be enough commonality from one context to another to justify thoughtful importation (and adaptation) across the contexts. It asks DBR proponents to check their enthusiasm and their hope that universal laws about human behavior can be discovered and articulated.

A related approach was suggested by a National Science Foundation Senior Program Director, Nora Sabelli, and Chris Dede at the Harvard Graduate School of Education (2001):

The strategy we advocate for increasing the impact of research on education practice goes beyond "transfer" and "action research" towards reconceptualizing the relationship between scholarship and practice.... An analogy for understanding a "scholarship of practice" is to consider the levels of experimentation and applied research that take place after scientific research in the physical sciences is conducted and published, and before the results of this research are applied large-scale in society. The field of education does not provide roles akin to engineering for developing research prototypes into robust practices and products. As discussed later, the outcomes of research include people, not just knowledge, and transfer between research and practice is implemented through both scholarly products and human capacity-building.

Stokes' recent book and related policy papers reinforce this perspective by presenting models for relating fundamental and targeted research, strategies largely absent in educational scholarship. These include the use-driven research model successfully applied by NIH to simultaneously:

- provide resources for fundamental biological research and its medical public health applications,
- resolve persistent problems in medical practice through the adaptation of research findings, and
- develop public and political support for allocating resources to both basic and applied research.

Stokes argues convincingly for the Pasteur (or NIH) model, in which research on use-driven, applied questions demands and fosters studies of fundamental scientific problems associated with that practice. In fact, "use-inspired" basic research has the same quest for fundamental understanding that is present in "pure" basic research, associated in Stokes's analysis with Bohr's work as the defining paradigm. In contrast, in Stokes's analysis Edison's work is shown as driven solely by considerations of use, without a concomitant quest for fundamental understanding. (p. 2)

Sabelli and Dede's proposal is still too positivist for me. For example, it continues to use natural sciences as a model for doing research in the human and social sciences even when a century of research did not produce anything like the expected, and promised, outcome. Not one of the contemporary debates over questions such as charter schools, bilingual education, multicultural education, gender bias, phonics versus whole language, evidence-based education, high stakes testing, accountability and a hundred other major issues in education has been settled by empirical research. Like opposing lawyers in a major court case, each side has its favored experts and cites its favored research, but what is surprising from a positivist view is that there seems to be research that supports all the major sides in any significant debate about education! Consider the question of phonics versus whole language. Research has not settled that long and contentious debate. Over the past 60 or so years, there have been shifts in what approach is emphasized in literacy programs, but those shifts are geneerally linked to shifts in ideology (often liberal versus conservative political ideologies) rather than new empirical research.

Therefore, if we add the goal of generalization to instructional design work based on C-ID models, it is not only reasonable, it is critically important, that the type of generalizations we attempt to make are tentative and contextual rather than assertive and universal. How do we do that? We do it by including in our papers and presentations on ID projects the "thick description" and detail so eloquently called for by Clifford Geertz (1973) so many years ago. Readers cannot make decisions about what to generalize from a study without a rich description of all aspects of an instructional design project. Also, and perhaps even more important, when we write papers about our ID projects we should add to the rich narrative detail about the process, components that take the discussion up a level or two in generality to make suggestions about design procedures that seemed to work well in a particular context (appropriately detailed in the paper), pedagogies and strategies that seemed to work well for students or content or instructional purposes (also all appropriately detailed in the paper). And finally, the third thing we would need to do is make sure we do not fall back into the positivist mode of thinking that allows us to present our generalizations as laws and universals. Presenting them with rich descriptions of the context in which we came to our understanding, and acknowledging the contextual limitations of any generalization about human behavior, models appropriate behavior for readers just as does acknowledging that our own biases, beliefs, and prior experiences (again, detailed in our papers to the best of our abilities) have had a major influence on how we interpret the process and results of our design work.

In Summary

Instructional design has been dominated since its inception by positivist paradigms, empiricist beliefs, and behaviorist theories of learning. However, over the past 30 or so years there have been a number of efforts to move beyond Instructional Systems Design(ISD)/ADDIE models that tend to organize the process of ID into a set of linear steps based on positivist thinking and behavioral learning theory. Many of these models are actually Pedagogical ID models because they emphasize the selection of teaching strategies based on a preferred family of learning theories. Recently, two alternatives, Design-Based Research (DBR) and Constructivist Instructional Design (C-ID), have emerged as appealing options to ISD. Though different, these two approaches to ID both have strengths.

One possible way of harnessing those strengths to (1) develop locally useful educational resources and (2) suggest generalizations that may be applicable in other settings, is to combine the approach to ID as a process that comes from C-ID models with the continuing optimism of DBR that we can yet discover useful generalizations about human behavior. Those generalizations are not, however, the generalizations of positivists. They are not efforts to concoct universal laws; instead, they are the products of reflective thought and analysis of experiences in one local context that may be helpful to other designers as they practice the professional art of instructional design.

The remaining chapters of this book will take you on an exploration of the foundations of C-ID, models for the process of ID based on C-ID concepts, and several examples from the real world of design that tell how designers used C-ID principles and practices to create instructional resources.

REFERENCES

Akker, J., Gravemeijer, K., McKenney, S., & Nieveen, N. (2006). *Educatonal Design Research*. London: Routledge.

Bednar, A., Cunningham, D., Duffy, T., & Perry, J. (1992). Theory into practice: How do we link? In T. Duffy & D. Jonassen (Eds.), *Constructivism and the technology of instruction*. (pp. 17–34). Hillsdale, NJ: Erlbaum.

Brown, A. L. (1992). Design experiments: Theoretical and methodological challenges in creating complex interventions in classroom settings. *The Journal of the Learning Sciences, 2*(2), 141–178.

Brooks, J., & Brooks, M. (2001). *In search of understanding: The case for constructivist classrooms*. Alexandria, VA: ASCD.

Burkhardt, H., & Schoenfeld, A. (2003). Improving educational reserach: Toward a more useful, more influential and better funded enterprise. *Educational Researcher, 32*(9), 3–14.

Cennamo, K. S., Abell, S. K., & Chung, M. L. (1996, July–August). A "Layers of Negotiation" model for designing constructivist learning materials. *Educational Technology, 39*–48.

Cennamo, K., Abell, S., Chung, M., Campbell, L., & Hugg, W. (1995). A "Layers of Negotiation" model for designing constructivist learning materials. *Proceedings of the Annual National Convention of the Association for Educational Communications and Technology (AECT), Anaheim, CA, 95*, 32–42.

Cobb, P. (2002). Theories of knowledge and instructional design: A response to Colliver. *Teaching and Learning in Medicine, 14*(1), 52–55.

Collins, A. (1992). Towards a design science of education. In E. Scanlon & T. O'Shea (Eds.), *New directions in educational technology* (pp. 15–22). Berlin: Springer.

Crawford, C. (2004). Non-linear instructional design model: Eternal, synergistic design and development. *British Journal of Educational Technology, 35*(4), 414–420.

Cronjé, J. (2006). Paradigms regained: Toward integrating objectivism and constructivism in instructional design and the learning sciences. *Educational Technology Research and Design, 54*(4), 387–416.

Dawson, K. (2008). *Technology and constructivist learning environments.* Charlotte, NC: Information Age Publishing.

Dick, W., Carey, L., & Carey, J. (2004). *The systematic design of instruction* (6th ed). Boston: Allyn & Bacon.

Dede, C. (2005). Commentary: The growing utilization of design-based research. *Contemporary Issues in Technology and Teacher Education* [Online serial], *5*(3/4). Retrieved from, http://www.citejournal.org/vol5/iss3/seminal/article1.cfm

Duffy, T., & Cunningham D. (1996). Constructivism: Implications for the design and delivery of instruction. In D. H. Jonassen (Ed.), *Handbook of Research for Educational Communications and Technology* (pp. 170–198). New York: Simon & Schuster.

elearnspace. (2002). *Instructional design in elearning.* Retrieved from http://www.elearnspace.org/Articles/InstructionalDesign.htm

Flyvbjerg, B. (2001). *Making social science matter: Why social inquiry fails and how it can succeed again.* Cambridge, England: Cambridge University Press.

Fullan, M. (1993). *Change forces: Probing the depths of educational reform.* Bristol, PA: The Falmer Press.

Gagne, R. (1985). *The conditions of learning* (4th ed.) New York: Holt, Rinehart and Winston.

Gagnon, G., & Collay, M. (2006). *Constructivist learning design.* Thousand Oaks, CA: Corwin.

Geertz, C. (1973). Thick description: Toward an interpretive theory of culture. In *The interpretation of cultures* (pp. 3–32). New York: Harper.

Greenwood, D., & Levin, M. (2005). Reform of the social sciences, and of universities through action research. In N. Denzin & Y. Lincoln (Eds.), *Sage Handbook of Qualitative Research* (pp. 43–64). Thousand Oaks, CA: SAGE.

Häkkinen, P. (2002). Challenges for design of computer-based learning environments. *British Journal of Educational Technology, 33*(4), 461–469.

Harrelson, C. (2003). *Comment on instructional technology and instructional design.* Retrieved November 18, 2008, from http://ttc.coe.uga.edu/christah/clhport/

Hensley, G. (2005). Creating a hybrid college course: Instructional design notes and recommendations for beginners. *Journal of Online Teaching, 2*(1). Retrieved June 11, 2007, from http://jolt.merlot.org/vol1_no2_hensley.htm

James, W. (1906). *Lecture II: What pragmatism means.* Retrieved June 13, 2007 from http://www.marxists.org/reference/subject/philosophy/works/us/james.htm

Johari, A., Chen, C., & Toh, S. (2005, March). A feasible constructivist instructional development model for virtual reality (VR)-based learning environments: Its efficacy in the novice car driver instruction of Malaysia. *Educational Technology Research and Development, 53*(1), 111–123.

Lebow, D. (1993). Constructivist values for instructional systems design: Five principles toward a new mindset. *Educational Technology Research and Development*, *41*(3), 4–16.

Merrill, M. D. & ID2 Research Group. (1996). Reclaiming the Discipline of Instructional Design. *ITForum*. Retrieved June 12, 2007, from http://itech1.coe.uga.edu/itforum/extra2/extra2.html

Reeves, T. (2000). *Enhancing the worth of instructional technology research through "design experiments" and other development research strategies.* Paper presented at Annual Meeting of the American Educational Research Association, New Orleans, LA. Retrieved June 12, 2007 from http://it.coe.uga.edu/~treeves/AERA2000Reeves.pdf

Reeves, T. (2006). Design research from a technology perspective. In J Akker, K. Gravemeijer, S. McKenney, & N. Nieveen (Eds.), *Educational design research.* (pp. 52–66). London: Routledge.

Roblyer, M. D. (2005). Educational technology research that makes a difference: Series introduction. *Contemporary Issues in Technology and Teacher Education* [Online serial], *5*(2). Retrieved from http://www.citejournal.org/vol5/iss1/seminal/article1.cfm and http://www.citejournal.org/vol5/iss2/seminal/article1.cfm

Sabelli, N., & Dede, C. (2001). *Integrating educational research and practice.* Retrieved November 25, 2008, from http://www.virtual.gmu.edu/ss_research/cdpapers/policy.pdf

Senge, P. (1990). *The fifth discipline.* New York: Doubleday.

Stokes, D. (1997). *Pasteur's quadrant: Basic science and technological innovation.* Washington, DC: Brookings Institute Press.

Willis, J., & Wright, K. (2000, March/April). A general set of procedures for constructivist instructional design: The New R2D2. *Educational Technology, 40*(2), 5–20.

Willis, J. (2000). The maturing of constructivist instructional design: Some basic principles that can guide practice. *Educational Technology, 40*(1), 5–16.

Willis, J. (1995). A recursive, reflective instructional design model based on constructivist-interpretivist theory. *Educational Technology, 35*(6), 5–23.

Wilson, B. (1997). Reflections on Constructivism and Instructional Design. In C. R. Dills & A. A. Romiszowski (Eds.), *Instructional Development Paradigms* (pp. 63–80). Englewood Cliffs, NJ: Educational Technology Publications.

Winters, N., & Mor, Y. (2008). IDR: A participatory methodology for interdisciplinary design in technology-enhanced learning. *Computers & Education, 50*(2), 579–600.

CHAPTER 2

INSTRUCTIONAL DESIGN

Is it Time to Exchange Skinner's Teaching Machine for Dewey's Toolbox?

Karin Wiburg

Student learning is not accidental: it is the direct result of what has been designed, intentionally or unintentionally, by teachers, schools, curriculum developers and communities. Behind teaching and learning events are beliefs about learning which directly influence what students experience. This paper provides an historical perspective on current instructional design practices. While readers may find it easy to discount the inadequacy of traditional theories of instruction, an understanding of the historical evolution of these theories and their pervasive influence on K–12 education is necessary if we want to support teaching reform efforts. Without an understanding of the evolution of current teaching practices, change efforts may be unsuccessful.

Previously published as paper presented at the Computer Support for Collaborative Learning Conference (1995, Oct. 17–20), Indiana University, Bloomington.

THE EFFICIENCY MOVEMENT

Prior to 1900, educational practice possessed little in terms of a theoretical framework, was certainly not considered scientific or subject to scientific study, but was a holistic enterprise in which teachers were expected to teach facts while also shaping character. Popular curricula included the *McGuffy Readers*, a collection of moral tales designed to build character in American students.

In the second half of the 19th century, European scholars theorized that it was possible to develop methods which would make the study of human behavior more scientific. The ideas of Wilhelm Wundt in Germany and Francis Galton in England heavily influenced the development of a new American School of Psychology and along with it the emerging field of educational psychology. Stanley Hall and William James, leading American psychologists at the time, had as one of their students, Edward Thorndike, who enthusiastically applied the new scientific psychology to the control of learning. In his still influential book, *Principles of Learning* (1921) Thorndike suggested that learning would occur if subject matter were carefully refined and sequenced and students appropriately reinforced. His popular prescription for intense practice as a condition of learning remains popular today.

Educational theories advocating the scientific control of human behavior reflected larger reform efforts in society and business after the turn of the century. Frederick Taylor, an industrialist, had developed a method for studying the movements of workers on assembly lines and through a process of measurement and control, speed up production. Eliot Eisner writes about the scientific managerial approach as it relates to workers:

> What one sees here is a highly rationalized managerial approach to the production process. The worker's job is to follow the procedures prescribed. In this system, individual initiative and inventiveness by workers were regarded as sources of error, like sand in a motor, they impeded the operation of a smooth running machine that depended on adherence to formula. (Eisner, 1994, p. 10)

When schools faced criticism in the early 1900s related to "inefficient" practices and poor learning by students, they turned eagerly to scientific management as a way to improve. This standardization model remains attractive today and can be seen in the 1990's school reform efforts, such as the call for a standardized curriculum, *cultural literacy* (Hirsch, 1988), and specific, identical student outcomes.

AN ALTERNATIVE SCIENTIFIC VIEW

Like Thorndike, John Dewey, was also interested in the application of science to educational practice. However, unlike Thorndike, his scientific views were influenced not by connectivism but by an approach to the study of human organisms made popular by Darwin. As a result, Dewey saw learning as an activity driven, not by reinforcement, but by the learner's sense of disequilibrium when presented with new experiences and ideas. For Dewey if real growth was to occur the student must want to learn and be active in the learning process. He argued that the traditional reinforcement of information—given by the teacher, memorized and given back by the child—led only to superficial learning. The job of the teacher was to create a classroom in which the child would be presented with problematic situations which she/he would be motivated to resolve by learning. While Dewey's thinking did not influence the early development of instructional technology for a variety of reasons, his work served as an alternative framework for the study of learning throughout the twentieth century.

THE ORIGINS OF INSTRUCTIONAL TECHNOLOGY

Perhaps one of the reasons that Thorndike's views regarding learning came to dominate instructional design was the political situation in which the United States found itself as it faced two world wars. The country had a need to rapidly train military personnel. The training required was concrete and sequential, such as how to assemble a rifle, and could be easily described and arranged in step-by-step fashion. The rapid growth of audiovisual tools during the first half of the century also made it possible to tap the advantages of using sound and visualization for the design of the required training. The resulting training materials could then be used over and over, without the need for extensive teacher preparation.

Robert Gagne is his book on the foundations of instructional technology (1987) describes early industrial technology as the confluence of the scientific study of human learning practiced by Thorndike and his followers, and the availability of new technologies. The technologies of interest included both procedures and tools. New techniques, such as programmed learning tied to the use of audiovisual materials, were conceptualized as a way to increase the precision with which the learner is appropriately stimulated. Such materials were easily replicable and usable in far away locations without additional teacher training. Much of the training related to military objectives was accomplished with speed and precision using these methods.

Extensive research around these early audiovisual experiments helped to define instructional technology as an attempt to provide "conditions for effective learning" (Gagne, 1987, p. 3). Growth in the field came from work in two areas (1) research on human learning, most recently human information processing, and (2) research and development designed to increase the capacity of instructional technology to provide better strategies for learning, exemplified in intelligent tutoring systems (ITS). These two fields, the study of optimal conditions for human learning and the use of well-developed procedures and tools, combined into what is now generally accepted as "a systems approach for designing instruction" (Association for Educational Communications and Technology, 1977). Within this systems approach increased emphasis has been put on improving the accuracy of the procedures involved in defining learning tasks, learner characteristics and needs, conditions for optimal implementation, and the tools and procedures needed for managing the instructional design process. However, within this paradigm no questions are asked about the purpose of learning or the social context in which it is to occur.

The systems approach received additional support and funding from the federal government during the 1950's and 60s as a result of America's reaction to the Sputnik launch in 1957. In response to what was perceived as an educational crisis, Americans called for training and education which they believed to be scientific, systematic, rational, and reliable. Instructional technology had served the nation well during the war, why couldn't it also lessen what was perceived as a gap between American students as potential scientists and more scientifically skilled students in other countries, especially Russia.

INFLUENCE OF INSTRUCTIONAL DESIGN ON PUBLIC SCHOOL PRACTICE

A variety of instructional systems theories have had a profound and persistent influence on educational practice in K–12 settings. Several examples of these approaches will be discussed in the next section including programmed learning (Skinner, 1961), instructional objectives (Mager, 1962), conditions of learning (Gagne, 1965), mastery learning (Bloom, Madeus, & Hastings, 1981), and the work of Madeline Hunter (1967a 1967b) in popularizing these instructional theories.

Programmed Instruction

B. F. Skinner, the father of operant conditioning, is usually credited with the development of programmed instruction. In his classic 1954 article, *The Science of Learning and the Art of Teaching*, Skinner described the conditions of the typical classroom as particularly adverse to learning. A single teacher can not individually and appropriately reinforce 30 or more students at the same time.

In this article Skinner first conceptualized a teaching machine for the classroom for use by individual students. This machine could present information, reinforce appropriately and then branch to the next level of difficulty depending on the individual's performance. The roots of one form of computer-assisted instruction can be easily seen in Skinner's teaching machine.

Task Analysis and Behavioral Objectives

In order to fully implement programmed instruction two other areas of development were needed, task analysis and behavioral objectives. Task analysis is the process of identifying the tasks and subtasks that must be performed in order to complete a task or job. The concept of task analysis was applied to general education in early work by Frank and Lillian Gilbreth, expanded by Robert Miller (Miller, 1953) and utilized by Gagne (1987) as part of his description of the hierarchical nature of learning.

Methodologies associated with programmed instruction also required the identification of specific, observable behaviors that were to be performed by the learner. While objectives were advocated in teaching as early as the 1900's, Ralph Tyler (1975) has been called the father of behavioral objectives since he suggested as a result of his famous 8 year study of schools in the 1940s that many of the problems of instruction seem to be related to the fact that schools did not specify objectives, and that teachers and students were not aware of what they were supposed to be learning. However, the major implementation of behavioral objectives occurred only after Benjamin Bloom and his colleagues published the *Taxonomy of Educational Objectives* (1956). Even then, behavioral objectives were not widely used in practice until the publication of a small, humorous book by Robert Mager, *Preparing Instructional Objectives for Programmed Instruction*. This book has since been republished as *Preparing Instructional Objectives* (1962) and is still widely used today in both teacher education and training technologies. Robert Mager provides an excellent introduction to his book:

> Before you prepare instruction, before you choose materials, machine, or method, it is important to be able to state clearly what your goals are. This book is about instructional objectives. In it I will try to show how to state objectives that best succeed in communicating your intent to others. The book is NOT about the philosophy of education, nor is it about who should select objectives, nor about which objectives should be selected. (Mager, 1962, p. viii)

This is a very interesting little book which one reads in the manner of programmed instruction. Every aspect of the preparation of terminal behavioral objectives is well outlined and the reader is asked frequent questions, the answer to which determines one's path through the book. After completing this book, the learner will be able to develop complete and precise objectives which define clearly the *terminal behavior* to be displayed by the learner, the *criterion* or standard by which the behavior will be evaluated (e.g., 70% items correct on the test), and the *conditions* under which the behavior will be displayed.

One of the barriers to implementing programmed instruction and behavioral objectives in schools was the organization of schooling. Classrooms were not organized to support individualized learning. However, Benjamin Bloom developed a method for reorganizing instruction to allow for more individualized learning which became known as mastery learning.

Mastery Learning

Bloom's method is based on the idea that the learner will succeed in learning a task if given the exact amount of time he or she needs to learn the task. Bloom was a passionate opponent of the common educational practice of assuming that only about a third of the class will learn the material taught. He suggested "this set of expectations, which fixes the academic goals of teachers and students, is the most wasteful and destructive aspect of the present educational system" (Bloom, Madeaus, & Hastings, 1981, p. 51). Bloom suggested a variety of strategies that can be used in classrooms to provide conditions for mastery learning, including the use of tutors, small group study, peer tutoring, programmed instruction, audiovisual materials and games. Research by Bloom and others in many countries reported that slow learners can indeed achieve as much as faster learners when given the opportunity (Bloom, 1976; Block & Anderson, 1975; Yildiran, 1977)

The Conditions of Learning Model

Robert Gagnè is best know for his development of a model of instruction. Prior to Gagnè, learning was often conceptualized as a single, uniform concept. No distinction was made between learning to load a rifle and learning to solve a complex mathematical problem. Among Gagne's contributions was the notion that there are various types of human learning and that each of these types of learning require different kinds of instructional strategies. For example, while Thorndike advocated continuous practice as the key to learning, Gagne suggested practice was effective only for certain types of learning, such as typing or playing ball which involve kinesthetic learning. The learning of cognitive strategies for problem solving is a very different type of learning, requiring different instructional strategies and conditions. In order to learn cognitive strategies, the learner must be presented with, and assisted in solving, puzzling problems. For this type of learning, practice without a change in perception can be counter-productive.

Gagnè's development of a model of human learning foreshadowed later discoveries about human information processing and added significantly to our understanding of the stages in cognitive processing and their relationship to instruction. He argued that an understanding of the characteristics of, and functions of, short term and long-term memory were important for instructional designers. Students will not be able to retrieve learning from long-term memory for later use if they are not assisted in incoding new concepts in meaningful ways during the initial learning experience. These ideas influenced instructional designers who began to include the cognitive needs of the learner, but within the same top-down instructional approach.

Instructional Theory Into Practice

Many educators found in Gagnè's work a foundation for addressing the instructional problems found in schools, perhaps none more so than Madeline Hunter. She suggests that her strongest contribution to education was not additional theory, but the development of the technologies needed by teachers to implement new theories of learning. She describes her purpose in the opening to one of her many publications.

> Psychological knowledge that will result in significantly increased learning of students is now available for teachers. In most cases, this knowledge remains unused because it is written in language that takes an advanced statistician to decode, or is buried in research journals in university libraries. This book is one of a series written to make this important knowledge available to the classroom teachers. (Hunter, 1967a, forward)

In her years as a professor at UCLA and a prolific writer, she had a significant influence on educational practice. Students in teacher education programs over the last 40 years, studied her writings and videos and inevitably learned to write precise instructional objectives, engage in task analysis, design appropriate guided and independent practice and write for their student teaching supervisors six-step lesson plans. Hunter's work provided the basis for a popular approach to remediation of students who were not doing well in school. The model, diagnostic-prescriptive teaching was widely adopted by federally-funded programs designed to provide what became known as *compensatory education* for children from the lower socio-economic classes. It was theorized that basic skills teachers trained in Hunter's instructional technologies would be able to quickly and efficiently remediate the learning difficulties of those children who had received inequitable educational opportunities.

COMMONALITIES

Both traditional instructional design and design aided by insights from cognitive and humanistic pyschology had in common certain characterisitics which will be described in this section. It is these commonalities which prevade current instructional practice and which may limit the design of computer-based collaborative learning environments.

An Emphasis on Method or Technique

All of the designs for learning discussed thus far reflect a technological approach to learning. The aim is to avoid philosophy as Mager suggests when he writes that his book is "NOT about the philosophy of education, nor is it about who should select objectives" rather, one must learn the techniques required to make learning more efficient. The value or importance of what is to be learned is not considered. Questions are not asked, as Dewey might have asked, about the kinds of learning that would be useful to the larger society. As Dewey suggested, one can learn to be a better thief but such learning, however rapid, might be socially undesirable, an example of what he termed *miseducation*.

Psychology and the Individual Learner

In addition, the instructional design models described focus solely on the individual learner. None of the models consider the social or cultural context in which learning is to occur. The only relationships which are well-described are the relationship between the teacher or the teaching machine and the individual learner. This reflects the dominant influence of American individualistic psychology on both education and the development of instructional technology. Thorndike stressed the need for individual practice and reinforcement for learning to occur. Skinner proposed a system of individualized programmed instruction which would focus specifically on the reinforcement needs of each individual in order to increase learning in schools. Bloom with mastery learning provided an approach to teaching which would allow some reorganization of instruction in classrooms in order to increase the likelihood of individualized learning. Diagnostic-prescriptive teaching, task analysis, and individualized remediation were all based on the individual, much like a doctor's prescription for an individual patient.

Focus on Content

While larger questions about the social value of what should be taught were not asked, each of these models was designed around the assumption that the important act in school is the learning of content—facts, figures, and concepts. In terms of Dewey's original question in 1905 (Dewey, 1959) concerning the distinction the focus on the child versus the focus on content, traditional instructional technology is grounded in the content to be taught, not the needs of the learner and group, or the social context in which the learner is situated. Traditionally, instructional designers have asked themselves: What are the concepts to be learned? How should they be presented and sequenced? What ideas need to be taught prior to others? What media can be most useful in presenting each concept?

ALTERNATIVE CONCEPTIONS OF LEARNING RELEVANT TO DESIGN

Perhaps the most eloquent critic of the technological approach to education is Eliot Eisner (1982, 1994). Eisner, whose work is well-grounded in the work of Dewey with whom he agrees that there is an important distinction between education and learning. In his discussion of different philosophical orientations to education, Eisner identifies the technical approaches discussed in this paper as unconcerned with the larger purpose of education. He also describes the limitations of teaching to objectives which are an essential element of current instructional design practice. Eisner remains concerned with the result of an exclusive use of behavioral objectives which breaks learning into small, manageable pieces. These pieces of learning are easily reinforced and measured. The problem is that students while mastering each of these pieces, is unable to put the pieces together and apply them to new situations. For example, a student may be able to name each of the vowels with 100% accuracy but not be able to distinguish between vowels in new words. Eisner argues, in addition, that there may be some very important goals of education, such as an appreciation of the arts or the valuing of open-minded skepticism in science, that are not easily broken down into small units of behavior which can be taught singly and reinforced. Finally, he proposes that the evaluation of students has and always will drive the curriculum. So long as we evaluate students in terms of easily measured small units of behavior students will only learn such small units of behavior. If we want students to gain problem-solving skills we will have to evaluate and value problem-solving, a type of learning which, as Gagne pointed out, is not learned by rote practice.

TRAINING VERSUS TEACHING

It was desirable in many of the situations in which instructional technology evolved, such as the need to quickly train soldiers during World War II, that teaching and learning materials not require extensive training and involvement of on-site teachers. Yet there is a real danger in continuing to conceptualize teachers as factory workers. In fact, one can argue that it is this conceptualization that may be partially responsible for the fact that so few teachers have access to telephones in the classroom, much less Internet connections. After all, factory workers don't need telephones.

However, some current educational reform movements recognize the important role of the teacher in supporting changes in classroom instructional practice. Without changes in teacher behavior, the best possible curricula will remain unused and have little or no effect on student learning. Teachers have always had a significant influence on student learning. As the diversity of our population and the complexity of what students need to know increases, teachers who are capable of designing learning appropriate for their community and the students who will work in an information age become increasingly important. Any instructional design theory in which teachers do not have a central role as designers and facilitators of learning will become less and less useful. The tasks of the teacher have become increasingly complex and the staff development needed is no longer one shot training programs or providing teachers with so-called teacher-proof materials, but rather continous opportunities for professional development that occur in the context of the everyday work of the school.

THE SOCIAL CONTEXT FOR LEARNING

Traditional approaches to instructional design assume that knowledge is independent of the situations in which it is learned and used. More recent research on learning in everyday situations and different cultural settings suggests the knowledge is not an independent phenomena, but situated in the activity, context and culture in which it is developed (Brown, Collins, & Duguid, 1988). An understanding of the need to connect learning with doing leads to a fundamentally different view of teaching and learning than that advocated by current instructional design practices. Tharp (1989) and others have suggested that learning is a process which occurs in social interaction. For example, Tharp and Gallimore (1976) describe how Hawaiian children who are used to working with peers may be perceived by their teachers as possessing low academic motivation rather than as seeking their own preferred learning environment which stresses peer cooperation. Therefore an important task in the design for learning is to make the organizational structures of schools ones in which students from a variety of cultures will be productive and engaged participants.

The purpose of this chapter was not to develop a theoretical framework for such communities of learners. Work by Brown (1994), Roschelle (1994) and many others has already begun to accomplish that task. Rather, this chapter was written to provide insight into the paradigm within which instructional design operates. This paradigm is particularly pervasive in current school practice, especially as it applies to the implementation of educational technology. In much of the computer education community we are still building and selling Skinner's teaching machines. The book, *Computers as Cognitive Tools* provides what one reviewer describes as a "dialectic of instructional technology" (Koschmann, 1994) between those who would design better controls and models for learning, the roots of which can be found in Thorndike, and those who believe in a social constructivist approach, origins of which can be found in Dewey. An expanded concept of instructional design that includes the purpose of education, the need to teach the person as well as the content, and the importance of the social context of learning is required before we can implement computer-based collaborative learning for the children in our schools.

REFERENCES

AECT. (1977). *The definition of educational technology.* Washington, DC: Association of Educational Communication and Technology.

Block, J. H., & Anderson, L. W. (1975). *Mastery learning in classroom instruction.* New York: Macmillan.

Bloom, B. S. (1956). *The taxonomy of educational objectives.* Weston, FL: Susan Fauer.

Bloom. B. F. (1996). *Human characteristics and school learning.* New York: McGraw-Hill.

Bloom, B. S., George F. Madaus, & J. Thomas Hastings. (1981). *Evaluation to improve learning.* Columbus, OH: McGraw-Hill.

Brown, A. (1994). Advancement of learning. *Educational Researcher, 23*(8), 4–12

Brown, J. S., Collins, A., & Duguid, P. (1988). *Report No. 6886: Situated cognition and the culture of learning.* Palo Alto, CA: Institute for Research on Learning.

Dewey, J. (1959). *Dewey on education: Selections from the child and the curriculum.* New York: Teachers College Press.

Eisner, E. (1982). *Cognition and curriculum: A basis for deciding what to teach.* New York: Longman.

Eisner, E. (1994). *The educational imagination: On the design and evaluation of school programs* (3rd ed.). New York: Macmillan.

Gagne, R. M. (1965). *The conditions of learning.* New York: Holt.

Gagne, R. M. (1987). *Instructional technology: Foundations.* Hillsdale, NJ: Erlbaum.

Hirsch, E. D. (1988). *Cultural literacy: What every American needs to know.* New York: Vintage Books.

Hunter, M. (1967a). *Retention: Theory into practice.* El Segundo, CA: Tip. (Original work published 1970)

Hunter, M. (1967b). *Motivation: Theory for teachers.* El Segundo, CA: Tip. (Original work published 1970)

Koschmann, T. D. (1994). The dialectics of instructional technology. *Educational Researcher, 22*(7), 38

Mager, R. (1962). *Preparing instructional objectives.* Palo Alto, CA: Fearon.

Miller. R. B. (1953). *A method for man-machine task analysis* (Tech. Rep. No. 53–137). Wright-Patternson Air Force Base, OH: Wright Air Development Center.

Roschelle, J. (1994) Collaborative inquiry: Reflections on Dewey and learning technology. *The Computing Teacher: Journal of the International Society for Technology in Education, 21*(8), 6–9.

Skinner, B. F. (1961, Oct. 24). Teaching machines. Science, 128(3330), 969-977.

Tharp, R. G. (1989). Psychocultural variables and constants: effects on teaching and learning in schools. *American Psychologist, 44*(2), 349–359.

Tharp, R., & Gallimore, R. (1976). *The uses and limits of social reinforcement and industriousness for learning to read* (Tech. Report No. 60). Honolulu: Kamehamehameha Schools/Bishop Estate, Kamehameha Early Educaton Program.

Thorndike, E. L. (1921). *Principles of learning*. New York: Teachers College Press.

Tyler, R. W. (1975). Educational benchmarks in retrospect: Educational change since 1915. *Viewpoints, 51*(2), 11–31.

Yildiran, G. (1977). *The effects of level of cognitive achievement on selected learning criteria under mastery learning and normal classroom instruction*. Unpublished doctoral dissertation, University of Chicago.

CHAPTER 3

CONSTRUCTIVISM, INSTRUCTIONAL DESIGN, AND TECHNOLOGY

Implications for Transforming Distance Learning

Maureen Tam

This paper examines the characteristics and value of designed instruction grounded in constructivist theory. It also attempts to connect the theory to the prevailing technology paradigms to establish an alignment between pedagogical and technological considerations in support of the assumptions arising from constructivism. Distance learning provides a unique context in which to infuse constructivist principles where learners are expected to function as self-motivated, self-directed, interactive, collaborative participants in their learning experiences by virtue of their physical location. Hence, the aim of this paper is to provide a clear link between the theoretical principles of constructivism, the construction of technology-supported learning environments, and the practice of distance education. The

Previously published in *Educational Technology & Society* (2000), *3*(2), 50–60.

Constructivist Instructional Design (C-ID): Foundations, Models, and Examples, pp. 61–80
Copyright © 2009 by Information Age Publishing
All rights of reproduction in any form reserved.

questions driving the argument in this paper include: What do constructivist perspectives offer instructional design and practice? What do computing technologies offer? And what do the two afford in combination? In particular, how do the two combine to transform distance learning from a highly industrialized mass production model to one that emphasizes subjective construction of knowledge and meaning derived from individual experiences.

The chapter proceeds in five stages. First, it begins with a basic characterization of constructivism, identifying what is believed to be the central principles in learning and understanding. The philosophical assumptions of constructivism are contrasted with those of objectivism, which holds very different views and approaches to learning and knowing.

Second, the discussion ensues to identify and elaborate on those instructional principles for the design of a constructivist learning environment. Exemplars of a constructivist learning environment will be identified and used to illustrate the design process that is based on the epistemological frameworks of constructivism.

Third, the role of technology is examined for its support in the construction of constructivist learning environments. If aligned, possibilities and capabilities afforded by technology will help to influence how constructivist beliefs about learning and understanding ultimately become operational in any teaching and learning situation.

Fourth, a link is established with distance learning. To that end, this part of the paper is intended to illustrate the infusion of constructivist principles and computing technology in distance learning contexts to cause the transformation of the system from an industrial model to a post-industrial one, which is found congruent with the constructivist principles and developments in modern technology.

Finally, constructivism is critically appraised to identify some of its problems and limitations. The debate among theoreticians, researchers, and practitioners is essential to clarifying foundations and assumptions, and promoting understanding of the merits of different perspectives and methods. There is not a single way to conceptualize learning systems. The challenge is to understand and evaluate the worth of different perspectives and methods to guide the design of effective instruction for learners.

WHAT IS CONSTRUCTIVISM?

Constructivism is a fundamental departure in thought about the nature of knowing, hence of learning and thus of teaching. To facilitate understanding of the constructivist view and its implications, it is compared to a familiar model of learning held by many: the objectivist models of learning. Guiding the discussions on constructivism and its implications for teaching and learning are four main questions:

1. What is learning?
2. What is the learning process?
3. What is the teacher's primary role in the learning process?
4. What can the teacher do to carry out that role?

WHAT IS LEARNING?

The constructivist perspective describes learning as a change in meaning constructed from experience (Newby. Stepich, Lehman, & Russell, 1996). Constructivists believe that "knowledge and truth are constructed by people and do not exist outside the human mind" (Duffy & Jonassen, 1991, p. 9). This is radically different from what objectivism conceives learning to be. To the objectivists, "knowledge and truth exist outside the mind of the individual and are therefore objective" (Runes, 1962, p. 217). "Learners are told about the world and are expected to replicate its content and structure in their thinking" (Jonassen, 1991, p. 6). The role of education in the objectivist view is therefore to help students learn about the real world. It is asserted that there is a particular body of knowledge that needs to be transmitted to a learner. Learning is thus viewed as the acquisition and accumulation of a finite set of skills and facts.

Contrary to these notions about learning and knowing is the constructivist's view of learning being "personal" and not purely "objective" (Bodner, 1986). Von Glaserfeld (1984) has written

learners construct understanding. They do not simply mirror and reflect what they are told or what they read. Learners look for meaning and will try to find regularity and order in the events of the world even in the absence of full or complete information.

Constructivism emphasizes the construction of knowledge while objectivism concerns itself mainly with the *object* of knowing. It is the fundamental difference about knowledge and learning that distinguishes the two in terms of both philosophy and implications for the design of instruction.

WHAT IS THE LEARNING PROCESS?

Central to the tenet of constructivism is that learning is an active process. Information may be imposed, but understanding cannot be, for it must come from within. In her Educational Psychology textbook, Woolfolk

(1993, p. 485) describes the constructivist view of the learning process as follows:

> The key idea is that students actively construct their own knowledge: the mind of the student mediates input from the outside world to determine what the student will learn. Learning is active mental work, not passive reception of teaching.

During the process of learning, learners may conceive of the external reality somewhat differently, based on their unique set of experiences with the world and their beliefs about them (Jonassen, 1991). However, learners may discuss their understandings with others and thus develop shared understandings (Cognition and Technology Group, 1991). While different learners may arrive at different answers, it is not a matter of "anything goes" (Spiro, Feltovich, Jacobson, & Coulson, 1991). Learners must be able to justify their position to establish its viability (Cognition and Technology Group, 1991).

While the important point is that the learner is central to the learning process, as epitomized by the Piagetian individualistic approach to constructivism, it is the collaboration among learners that makes constructivism more than an example of solipsism (Jonassen, 1991). Rather, it encourages the construction of knowledge in a social context in which collaboration creates a sense of community, and where teachers and students are active participants in the learning process.

In a learning environment, there is always some stimulus or goal for learning. In Dewey's terms, it is the "problematic' that leads to and is the organizer for learning (Dewey, 1938). Savery and Duffy (1995) prefer to talk about the learner's "puzzlement" as being the stimulus and organizer for learning. The important point here is that it is the problematic situation or context that is central to the learning process in constructivism.

Hence, according to the constructivist perspective, learning is determined by the complex interplay among learners' existing knowledge, the social context, and the problem to be solved. Instruction, then involves providing learners with a collaborative situation in which they have both the means and the opportunity to construct "new and situationally-specific understandings by assembling prior knowledge from diverse sources" (Ertmer & Newby, 1993, p. 63). Various authors have described the characteristics of constructivist instruction (e.g., Brooks & Brooks, 1993; Cognition and Technology Group, 1993; Collins, Brown, & Holum, 1991; Honebein, Duffy, & Fishman, 1993). Two characteristics seem to be central to these constructivist descriptions of the learning process:

A. "Good" Problems

Constructivist instruction asks learners to use their knowledge to solve problems that are meaningful and realistically complex. The problems provide the context for the learners to apply their knowledge and to take ownership of their learning. Good problems are required to stimulate the exploration and reflection necessary for knowledge construction. According to Brooks and Brooks (1993), a good problem is one that requires students to make and test a prediction

- can be solved with inexpensive equipment
- is realistically complex
- benefits from group effort
- is seen as relevant and interesting by students.

B. Collaboration

The constructivist perspective supports the assertion that students learn through interaction with others. Learners work together as peers, applying their combined knowledge to the solution of the problem. The dialogue that results from this combined effort provides learners with the opportunity to test and refine their understanding in an ongoing process.

WHAT IS THE TEACHER'S PRIMARY ROLE IN THE LEARNING PROCESS?

Vygotsky's theory of social constructivism, as opposed to Piaget's individualistic approach to constructivism, emphasizes the interaction of learners with others in cognitive development. His concept of the zone of proximal development embodies his belief that learning is directly related to social development (Rice & Wilson, 1999). "The discrepancy between a child's actual mental age and the level he reaches in solving problems with assistance indicates the zone of his proximal development" (Vygotsky, 1986, p. 187). Vygotsky felt good instruction could be provided by determining where each child is in his or her development and building on that child's experiences.

This is congruent with what most constructivists advocate that instructional intervention should not only match but also accelerate students' cognitive development. According to Copley (1992), constructivism requires a teacher who acts as a facilitator "whose main function is to help students become active participants in their learning and make meaningful connections between prior knowledge, new knowledge, and the processes involved in learning." Omrod (1995) stated that teachers could encourage students' development by presenting tasks that "they can complete only with assistance—that is, within each student's zone of proximal development."

As Chung (1991) described, a constructivist learning environment is characterized by (1) shared knowledge among teachers and students; (2) shared authority and responsibility among teachers and students; (3) the teacher's new role as guide in instruction; and (4) heterogeneous and small groupings of students.

Resonant with the idea that the teacher is a guide instead of an expert, constructivism instruction has always been likened to an apprenticeship (e.g., Collins et al., 1991; Rogoff, 1990) in which teachers participate with students in the solution of meaningful and realistic problems. Here, the teachers serve as models and guides, showing students how to reflect on their evolving knowledge and providing direction when the students are having difficulty. Learning is shared and responsibility for the instruction is shared. The amount of guidance provided by the teacher will depend on the knowledge level and experience of the students (Newby et al., 1996).

WHAT CAN THE TEACHER DO TO CARRY OUT THAT ROLE?

Brooks and Brooks (1993) summarize a large segment of the literature on descriptions of "constructivist teachers." They conceive of a constructivist teacher as someone who will:

- encourage and accept student autonomy and initiative;
- use a wide variety of materials, including raw data, primary sources, and interactive materials and encourage students to use them;
- inquire about students' understandings of concepts before sharing his/her own understanding of those concepts;
- encourage students to engage in dialogue with the teacher and with one another;
- encourage student inquiry by asking thoughtful, open-ended questions and encourage students to ask questions to each other and seek elaboration of students' initial responses;
- engage students in experiences that show contradictions to initial understandings and then encourage discussion;
- provide time for students to construct relationships and create metaphors;

- assess students' understanding through application and performance of open-structured tasks.

Hence, from a constructivist perspective, the primary responsibility of the teacher is to create and maintain a collaborative problem-solving environment, where students are allowed to construct their own knowledge, and the teacher acts as a facilitator and guide.

CONSTRUCTIVISM AND INSTRUCTIONAL DESIGN

The constructivist propositions outlined above suggest a set of instructional principles that can guide the practice of teaching and the design of learning environments. It is important that design practices must do more than merely accommodate the constructivist perspectives, they should also support the creation of powerful learning environments that optimize the value of the underlying epistemological principles.

Bednar, Cunningham, Duffy, and Perry (1992) help to put things in perspective by saying:

> Instructional design and development must be based upon some theory of learning and/or cognition; effective design is possible only if the developer has developed reflexive awareness of the theoretical basis underlying the design.

Lebow (1993) has hit upon a strategy for summarizing the constructivist framework in a way that may help with the interpretation of the instructional strategies. He talks about the shift in values when one takes a constructive perspective. He notes that:

> Traditional educational technology values of replicability, reliability, communication, and control (Heinich, 1984) contrast sharply with the seven primary constructivist values of collaboration, personal autonomy, generativity, reflectivity, active engagement, personal relevance, and pluralism.

Such incompatibility between the traditional instructional design practice and the constructivist perspective of designing instruction is arising from the epistemological differences of the two contrasting theories of instruction.

In the traditional approach to instructional design, the developer analyzes the conditions which bear on the instructional system (such as content, the learner, and the instructional setting) in preparation for the specification of intended learning outcomes. Procedural ID models "describe how to perform a task and are formulated to simplify and explain a series of complex processes" (Bagdonis & Salisbury, 1994).

> Procedural models in the instructional systems design (ISD) field ... attempt to account for all relevant components using a systematic approach to designing instruction, from needs assessment through the development of material, implementation, and evaluation. Each component within the process builds upon the other.
>
> Traditional ISD models are generally viewed by individuals in the instructional design field as representing a linear process, a plan of separate steps that proceed in a linear sequence.
>
> A typical ISD model is divided into five stages: analysis, design, production/development, implementation, and maintenance/revision.
>
> The five stages consist of an integrated set of components that are sequenced so that each component within the process must be completed before continuing to the next. (Bagdonis & Salisbury, 1994)

To summarize, the traditional Objective-Rational Instructional Design model has the following eight characteristics (Willis, 1995):

- The process is sequential and linear
- Planning is top down and systematic
- Objectives guide development
- Experts, who have special knowledge, are critical to ID work
- Careful sequencing and the teaching of subskills are important
- The goal is delivery of preselected knowledge
- Summative evaluation is critical
- Objective data are critical.

Although the behavioral, objective-rational approach to instructional design is well entrenched in practice and has influenced teaching and learning in many ways, an alternative to this approach has emerged over the last decade. While constructivist approaches have not yet replaced behavioral approaches as the dominant theoretical framework, they have already made a significant impact on how learning should be conceived of and provide wide-ranging implications for instructional design derived from a constructivist view.

The constructivist view is very different from the objectivist approach to instructional design. It summons instructional designers to make a radical shift in their thinking and to develop rich learning environments that help to translate the philosophy of constructivism into actual practice. Reigeluth (1989) argues for a "new mindset" to combine constructivist elements in the instructional design models.

In a recent review of the literature on how instructional designing should respond to constructivism, Lebow (1993) proposes 'Five Principles toward a New Mindset' as constructivist values might influence instructional design.

Principle 1

Maintain a buffer between the learner and the potentially damaging effects of instructional practices by:

- Increasing emphasis on the affective domain of learning
- Making instruction personally relevant to the learner
- Helping learners develop skills, attitudes, and beliefs that support self-regulation of the learning process
- Balancing the tendency to control the learning situation with a desire to promote personal autonomy.

Principle 2

Provide a context for learning that supports both autonomy and relatedness.

Principle 3

Embed the reasons for learning into the learning activity itself.

Principle 4

Support self-regulated learning by promoting skills and attitudes that enable the learner to assume increasing responsibility for the developmental restructuring process.

Principle 5

Strengthen the learner's tendency to engage in intentional learning processes, especially by encouraging the strategic exploration of errors (Lebow, 1993, pp. 5–6).

These principles support many of the views of constructivism that objects and events have no absolute meaning; rather, the individual interprets each and constructs meaning based on individual experience and evolved beliefs. The design task, therefore, is one of providing a rich context within which meaning can be negotiated and ways of understanding can emerge and evolve (Hannafin, Hannafin, Land, & Oliver, 1997). Constructivist designers tend to avoid the breaking down of context into component parts as traditional instructional designers do, but are in favour of environments in which knowledge, skills, and complexity exist naturally. Hence, instead of adopting a linear and "building-blocks" approach to instructional design, constructivist designers need to develop procedures for situations in which the instructional context plays a dominant part, and the instructional goals evolve as learning progresses.

In his comprehensive review of the literature on instructional design models, Willis (1995) offers an alternative model to the traditional Objective-Rational ID model. Willis termed it the Constructivist-Interpretivist Instructional Design Model, which has the following characteristics:

1. The design process is recursive, non-linear, and sometimes chaotic.
2. Planning is organic, developmental, reflective, and collaborative.
3. Objectives emerge from design and development work.
4. General ID experts do not exist.
5. Instruction emphasizes learning in meaningful contexts (the goal is personal understanding within meaningful contexts).
6. Formative evaluation is critical.
7. Subjective data may be the most valuable.

The constructivist design principles, implemented within the framework of the values and procedures outlined in the above ID model, can lead to a variety of learning environments. Examples of these constructivist learning environments include situating cognition in real-world contexts, cognitive flexible learning, collaborative learning, and others.

In order to translate the philosophy of constructivism into actual practice, many instructional designers are working to develop more constructivistic environments and instructional prescriptions. One of the most important of these prescriptions is the provision of instruction in relevant contexts. Situated cognition (Brown, Collins, & Duguid, 1989) suggests that knowledge and the conditions of its use are inextricably linked. Learning occurs most effectively in context, and that context becomes an important part of the knowledge base associated with learning (Jonassen, 1991).

A related approach is anchored instruction (Cognition and Technology Group at Vanderbilt, 1992), which emphasizes embedding skills and knowledge in holistic and realistic contexts. Anchored contexts support complex and ill-structured problems wherein learners generate new knowledge and subproblems as they determine how and when knowledge is used (Hannafin et al., 1997). Apprenticeship models are similarly aligned as they promote scaffolding and coaching of knowledge, heuristics, and strategies, while students carry out authentic tasks (Collins, Brown, & Newman, 1989). Other related approaches include the problem based learning model (Barrows, 1985, 1992) and case-based learning environments in which students are engaged in solving authentic tasks.

Another important strategy is the presentation of multiple perspectives to learners. The constructivist view emphasizes that students should learn to construct multiple perspectives on an issue. They must attempt to see an issue from different vantage points. It is essential that students make the best case possible from each perspective; that is, that they truly try to understand the alternative views (Bednar et al., 1992).

Cognitive flexibility theory is a conceptual model for instruction that facilitates advanced acquisition of knowledge in ill-structured knowledge domains (Jonassen, 1991). Flexibility theory (Spiro, Coulson, Feltovich, & Anderson, 1988) avoids oversimplifying instruction by stressing conceptual interrelatedness, providing multiple representations or perspectives on the content because there is no single schema, and emphasizing case-based instruction that provides multiple perspectives or themes inherent in the cases.

A central strategy for building constructivist learning environments such as situated learning, multiple perspectives and flexible learning is to create a collaborative learning environment. Collaborative learning does not just entail sharing a workload or coming to a consensus. Rather, it is to allow learners to develop, compare, and understand multiple perspectives on an issue. It is the rigorous process of developing and evaluating the arguments that is the goal in collaborative learning (Bednar et al., 1992).

There must be other exemplars of a constructivist learning environment in addition to the few outlined in the previous paragraphs. All of them facilitate constructivist learning and have mirrored important underlying principles of constructivism towards learning and understanding. The task of the instructional designers is to assess and review instructional theories, tools and resources at their disposal, and to consider (if appropriate) how constructivist learning may be facilitated, and how instructional designing should respond to constructivism.

CONSTRUCTIVISM AND TECHNOLOGY

Instruction today faces two challenges. One challenge comes from the changing perceptions of what learning is all about. The second challenge comes from the new learning opportunities that technology now affords (Salomon, 1991). Constructivism has presented the first challenge of reconceptualizing learning as a constructive process whereby information is turned into knowledge by means of interpretation, by actively relating it to existing bodies of knowledge, by the generative creation of representations, and by processes of purposeful elaboration (e.g., Resnick, 1989).

Presenting the second challenge is the computer. Because of its versatility and accessibility, its use in education may help to shift the foci from knowledge-as-possession to knowledge-as-construction, and from learning as outside-guided to learning as self-guided. It also carries with it a renewed conception of instruction that shifts attention from instruction as the imparting of knowledge to instruction as the guidance of socially-based exploration in intellectually rich settings (Salomon, 1991).

It is no coincidence that these shifts, implied by the computer, happen to be highly congruent with the constructivist principles of learning and teaching. Constructivism and computing technology, separately and often together, have remade substantially the conception of the challenges of learning, and brought about new learning possibilities for almost all teaching and learning situations, including traditional classroom teaching, distance learning and self-learning.

Constructivism provides ideas and principles about learning that have important implications for the construction of technology-supported learning environments. One of these implications is the need to embed learning into authentic and meaningful contexts (e.g., Brown, Collins, & Duguid, 1989; Cognition and Technology Group at Vanderbilt, 1992). Here, students are required to engage actively in authentic problem-tackling or decision-making cases. There are numerous kinds of case-based learning environments. Among the better known forms are cognitive flexibility, hypertexts and anchored instruction (Jonassen, Davidson, Collins, Campbell, & Haag, 1995). The central tenet of these forms of learning is improvement of students' understanding and their transfer of information through exposure to the same material, at different times, in rearranged contexts for different purposes (Spiro et al., 1991).

Lancy (1990) reported that computers were effective in developing higher-order thinking skills, including defining problems, judging information, solving the problems, and drawing appropriate conclusions. The computer can serve in the process of information gathering, inquiry, and collaboration, not merely as a vestige of direct instruction with its reliance on integrating technology in the existing curriculum (Rice & Wilson,

1999). Technology tools that aid in case-based learning include various types of simulation and strategy software/CD-Roms, video discs, multimedia/hypermedia, and telecommunications (e-mail and Internet).

These tools present benefits including the ability to obtain relevant information in the form of documents, photographs, transcripts, video and audio clips; the capability of providing virtual experiences that otherwise would not be possible; and the opportunities for students to examine a variety of viewpoints so they can construct their own knowledge of various concepts (Rice & Wilson, 1995).

It is important that computer-supported constructivist environments should not involve the knowledge and intelligence to guide and structure learning processes, but rather should create situations and offer tools that stimulate students to make maximum use of their own cognitive potential (Scardamalia Bereiter, McLean, Swallow, & Woodruff, 1989).

Another important implication of constructivism for the construction of technology-supported learning environment is that learning is a personal, as well as a social activity. The penetration of technology into the learning process can have profound consequences for how learning takes place socially. On one side, one can see even more individual learning in a student sitting in front of his or her computer. But on the other hand, the technology allows for much more diversified and socially rich learning contexts; peer tutoring via computer; computer networks, e-mail, telecommunications.

Increased recognition of the potential of computer-mediated communications, computer-supported collaboration work has enabled building the kinds of more supportive, collaborative and social learning environments called for by the constructivist perspective. Recent growth in telecommunications has led to the use of online services, electronic networks, and the World Wide Web, readily accessible to both homes and schools. Telecommunications include e-mail and Internet access. E-mail makes online discussion groups, electronic pen pals, student-to-student projects, class-to-class projects possible. In addition, the Internet provides many resources, including text, pictures, video, sound, and downloadable software, and is an endless source of activities and information.

Telecommunication technologies easily lend themselves to constructivist principles by providing students with opportunities to communicate with people all over the world, conduct research, discuss issues and work cooperatively. The advent of computer-mediated communication (CMC), has permitted learners, in particular distance learners, to benefit from the shared experience of a group engaged in the same study and the opportunity to measure his or her ideas against those of others in the group. By way of CMC, the teaching and learning styles of both instructors and learners are transformed from information dissemination to critical inquiry and from instructor dominated to collaborative learning.

The potential of telecommunication technology lies in its ability to function as a gateway; a gateway to resources, collaborative learning and individual achievement. While it is true that telecommunication technology is not a necessary component to the development of a constructivist learning environment nor is it sufficient in and of itself to cause the emergence of such learning, it does provide a means which increases the possibility that constructivist learning can, in fact, take place.

IMPLICATIONS FOR TRANSFORMING DISTANCE EDUCATION

According to the constructivist views of learning, as individuals bring different background knowledge, experience, and interests to the learning situation, they make unique connections in building their knowledge. Students and teachers both play a role in facilitating and generating knowledge. Students are encouraged to question each other's understanding and explain their own perspectives. These opportunities help hand over responsibility for knowledge generation to the learners (Maxwell, 1995). Further, constructivism encourages active, rather than passive learning and the use of group-based cooperative learning activities, which can be best mediated through telecommunication technologies.

Such constructivist views of learning correlate nicely with the philosophy of open and distance learning. If learning truly depends on the unique base of experience and knowledge brought to the learning environment by the learner, the learner then certainly should play a role in determining the learning goals, strategies, and methods for building on his or her base of knowledge and understanding. The autonomy called for by open and distance learning advocates is reflected in the constructivist views to encourage active, collaborative and responsible learners.

Distance education provides a unique context in which to infuse constructivist principles. In distance education settings, where learners are not in close physical proximity to the instructors and where technology mediates the learning experience, there is a stronger need for the construction of technology-supported constructivist learning environments wherein students are required to work collaboratively with each other, and to move the teacher from podium to sideline, from leader to coach, from purveyor of knowledge to facilitator of personal meaning making (Romiszowski & de Haas, 1989).

Constructivist principles provide a set of guiding principles to help designers and instructors create learner-centred, technology-supported collaborative environments that support reflective and experiential processes. When applied to the distance learning context, there is no doubt that constructivism and the use of new technologies will help transform significantly the way distance education should be conducted.

To transform distance education, Jonassen et al. (1995) suggest using 'constructivist tools and learning environment that foster personal meaning-making and discourse among communities of learners (socially negotiating meaning) rather than by instructional interventions that control the sequence and content of instruction that seek to map a particular model of thinking onto the learners'.

If developers of distance education want to apply constructivist principles to their instructional environments, they need to implement fundamental structural changes. First, distance education should change from being a highly industrialized orientation to a post-industrial one which emphasizes learner's self determination, self-direction and self-control (Peters, 1994). Under the industrial model, distance instruction is perceived as a typical product of industrial society. "It pre-supposes that instruction can be planned, evaluated and improved considerably in the same way as the production of goods can be planned, evaluated and planned" (p. 237). It also ignores the role of the learners, their wish, needs and motivations, in the design process.

In the post-industrial perspective, the role of learners has become a more prominent one. It is no longer suffice to provide distance learners with pre-packaged self-instructional materials where there is very little opportunity for student choice and interaction. Instead, instructional designers and academics should allow distance learners to be more reflective, to give personal views on topics, to debate and argue their points of view, to question information given by the instructor and textbooks, based on personal observations and knowledge acquired elsewhere.

Second, distance education should exploit further the potentials and capabilities of information technologies to foster two-way, interactive communication and collaboration between the instructor and learners and among the learners themselves. Jonassen et al. (1995) believe that a constructivist approach to knowledge construction and learning can be well supported in distance education settings through a variety of technologies. Technology-supported environments—computer-mediated communication, computer-supported collaborative work, case-based learning environments, and computer-based cognitive tools, for example —can offer the field of distance education alternative approaches to facilitating learning. These constructivist environments and tools can replace the deterministic teacher-controlled model of distance instruction with contextualized work environments, thinking tools, and conversation media that support the knowledge constructions process in different settings.

CONCLUSION

Like other instructional theories, constructivism cannot be the panacea for all instructional problems. It has its own limitations and problems for learning situations that may mitigate against its application. Yet constructivism holds important lessons for how to interpret the results of learning and for how to design environments to support learning.

While a constructivist perspective makes perfect sense from a theoretical position, the notion of there not being "right" or "wrong" answers can easily strike fear in the heart of an educator responsible for demonstrating that his or her students are achieving "world class standards," have attained specific performance based outcomes, or mastered activities prescribed by national education goal (Wagner & McCombs, 1995).

The absence of specific learning objectives and outcomes has earned the criticism for constructivism as "inefficient and ineffective" (Dick, 1992). Furthermore, its lack of concern for the entry behaviors of students is being criticized for ignoring the gap between what a student must know or be able to do before beginning instruction.

Assessment may be a problem in a constructivist learning environment. Constructivists are concerned about context - but more for instruction than individual assessment. They have been accused of showing no concern for efficiency, and little apparent concern for certifying the competency level of individual students (Dick, 1992).

Constructivist instructional approaches in general are being criticized mainly for three things: (1) They are costly to develop (because of the lack of efficiency); (2) they require technology to implement; and (3) they are very difficult to evaluate (Dick, 1992). However, these allegations can be rectified by practitioners who are creative and innovative enough to introduce ways of measuring student learning and assessing individual progress. Constructivism can provide unique and exciting learning environments, it is the challenge for practitioners to engage the learners in authentic and meaningful tasks, and to evaluate learning using assessment methods that reflect the constructionist methods embedded in the learning environments.

Where the introduction of computers in constructivist learning involves providing students with greater autonomy as learners, this commonly conflicts with students' past educational experiences and can require a shift in their conception of what learning involves and what constitutes appropriate roles of students and teachers. Student resistance to the inevitable stress of such change is to be expected, irrespective of the potential learning benefits of introducing the technology (Akerlind & Trevin, 1995).

Despite these criticisms, constructivism does present an alternative view of learning other than the objectivistic conception of learning, and provides a set of design principles and strategies to create learning environments wherein learners are engaged in negotiating meaning and in socially constructing reality. However, it is not suggested here that all designers should adopt constructivism as the only solution to all instructional problems. Rather they should reflect upon and articulate their conceptions of knowing and learning and adapt their methodology accordingly. The challenge for the design community is to understand and evaluate the different perspectives, methods and assumptions appropriate to fundamentally different contexts. The possibility of different conditions for different outcomes is completely consistent with the long-standing notion in instructional design that different types of outcomes require different instructional conditions (Gagne, 1965; Reigeluth & Curtis, 1987).

REFERENCES

Akerlind, G., & Trevitt, C. (1995, December). Enhancing learning through technology: when students resist the change. *ASCILITE 95—Learning with Technology*, 3–7.

Bagdonis, A. & Salisbury, D. (1994). Development and validation of models in instructional design. *Educational Technology, 34*(4), 26–32.

Barrows, H. S. (1985). *How to design a problem based curriculum for the preclinical years*. New York: Springer.

Barrows, H. S. (1992). *The tutorial process*. Springfield, IL: Southern Illinois University School of Medicine.

Bednar, A. K., Cunningham, D., Duffy, T. M., & Perry, J. D. (1992). Theory into practice: How do we link? In T. M. Duffy & D. H. Jonassen (Eds.) *Constructivism and the technology of instruction: a conversation* (pp. 17–35). Hillsdale, NJ: Erlbaum.

Bodner, G. M. (1986). Constructivism: A theory of knowledge. *Journal of Chemical Education, 63*, 873–878.

Brooks, J. G., & Brooks, M. G. (1993). *In search of understanding: the case for constructivist classrooms*, Alexandria, VA: American Society for Curriculum Development.

Brown, J. S., Collins, A., & Duguid, P. (1989). Situated cognition and the culture of learning. *Educational Researcher, 18*(1), 32–41.

Chung, J. (1991). Collaborative learning strategies: The design of instructional environments for the emerging new school. *Educational Technology, 31*(6), 15–22.

Cognition and Technology Group at Vanderbilt. (1991). Some thoughts about constructivism and instructional design. In T. M. Duffy & D. H. Jonassen

(Eds.), *Constructivism and the technology of instruction: a conversation* (pp. 115–119. Hillsdale, NJ: Erlbaum.

Cognition and Technology Group at Vanderbilt. (1992). Emerging technologies, ISD, and learning environments: critical perspectives. *Educational Technology Research and Development, 40*(1), 65–80.

Cognition and Technology Group at Vanderbilt. (1993). Designing learning environments that support thinking: the Jasper series as a case study. In T. M. Duffy, J. Lowyeh, & D. H. Jonassen (Eds.), *Designing environments for constructive learning* (pp. 9–36), Berlin: Springer-Verlag.

Collins, A., Brown, J. S., & Newman, S. (1989). Cognitive apprenticeship: teaching the crafts of reading, writing, and mathematics. In L. Resnick (Ed.), *Knowledge, learning, and instruction* (pp. 453–494). Englewood Cliffs, NJ: Erlbaum.

Collins, A., Brown, J. S., & Holum, A. (1991). Cognitive apprenticeship: Making thinking visible. *American Educator, 15*(3), 6–11, 38–46.

Copley, J. (1992). The integration of teacher education and technology: a constructivist model. In D. Carey, R. Carey, D. Willis, & J. Willis (Eds.), *Technology and Teacher Education* (p. 681), Charlottesville, VA: AACE.

Dewey, J. (1938). *Logic: the theory of inquiry,* New York: Holt and Co.

Dick, W. (1992). An instructional designer's view of constructivism. In T. M. Duffy & D. H. Jonassen (Eds.), *Constructivism and the technology of instruction: A conversation* (pp. 91–98), Hillsdale, NJ: Erlbaum.

Duffy, T. M., & Jonassen, D. H. (1991). New implications for instructional technology? *Educational Technology, 31*(3), 7–12.

Ertmer, P. A., & Newby, T. J. (1993). Behaviorism, cognitivism, constructivism: comparing critical features from an instructional design perspective. *Performance Improvement Quarterly, 6*(4), 50–72.

Gagne, R. M. (1965). *The conditions of learning.* New York: Holt, Rinehart and Winston.

Hannafin, M. J., Hannafin, K. M., Land, S. M., & Oliver, K. (1997). Grounded practice and the design of constructivist learning environments. *Educational Technology Research and Development, 45*(3), 101–117.

Heinich, R. (1984). The proper study of instructional technology. *Educational Communications and Technology Journal, 32*(2), 67–87.

Honebein, P. C., Duffy, T. M., & Fishman, B. J. (1993). Constructivism and the design of learning environments: context and authentic activities for learning. In T. M. Duffy, J. Lowyck, & D. H. Jonassen (Eds.), *Designing environments for constructive learning* (pp. 87–108). Berlin: Springer-Verlag.

Jonassen, D. H. (1991). Objectivism versus constructivism: do we need a new philosophical paradigm? *Journal of Educational Research, 39*(3), 5–14.

Jonassen, D., Davidson, M., Collins, M., Campbell, J., & Haag, B. B. (1995). Constructivism and computer-mediated communication in distance education. *The American Journal of Distance Education, 9*(2), 17–25.

Lancy, D. (1990). Microcomputers and the social studies. *OCSS Review, 26,* 30–37.

Lebow, D. (1993). Constructivist values for systems design: five principles toward a new mindset. *Educational Technology Research and Development, 41,* 4–16.

Maxwell, L. (1995). Integrating open learning and distance education. *Educational Technology, 35*(6), 43–48.

Newby, T. J., Stepich, D. A., Lehman, J. D., & Russell, J. D. (1996). *Instructional technology for teaching and learning: Designing instruction, integrating computers, and using media*, NJ: Prentice Hall.

Omrod, J. (1995). *Educational Psychology: Principles and applications*, Englewood Cliffs, NJ: Prentice Hall.

Peters, O. (1994). Distance education in a post-industrial society. In D. Keegan (Ed.), *Otto Peters on distance education* (pp. 220–240), London: Routledge.

Reigeluth, C. M., & Curtis, R. V. (1987). Learning situations and instructional models. In R. M. Gagne (Ed.), *Instructional technology: Foundations*. Hillsdale, NJ: Erlbaum.

Reigeluth, C. M. (1989). Educational technology at the crossroads: new mindsets and new directions. *Educational Technology Research and Development, 37*(1), 1042–1629.

Resnick, L. B. (1989). Introduction. In L. B. Resnick (Ed.), *Knowing, learning, and instruction* (pp. 1–24), Hillsdale, NJ: Erlbaum.

Rice, M. L., & Wilson, E. K. (1999). *How technology aids constructivism in the social studies classroom*. Retrieved from http://global.umi.com/pqdweb

Rogoff, B. (1990). *Apprenticeship in thinking: Cognitive development in social context*. New York: Oxford University Press.

Romiszowski, A., & de Haas, J. (1989). Computer-mediated communication for instruction: using e-mail as a seminar. *Educational Technology, 29*(10), 7–14.

Runes, D. D. (1962). *Dictionary of philosophy* (15th ed.), Paterson, NJ: Littlefield, Adams & Co.

Salomon, G. (1991). From theory to practice: The international science classroom—a technology-intensive, exploratory, team-based and interdisciplinary high school project. *Educational Technology, 31*(3), 41–44.

Savery, J. R., & Duffy, T. M. (1995). Problem based learning: An instructional model and its constructivist framework. *Educational Technology, 35*(5), 31–38.

Scardamalia, M., Bereiter, C., McLean, R. S., Swallow, J., & Woodruff, E. (1989). *Computer-supported learning environments and problem-solving*, Berlin: Springer-Verlag.

Spiro, R. J., Coulson, R. L., Feltovich, P. J., & Anderson, D. K. (1988). *Cognitive flexibility theory: Advanced knowledge acquisition in ill-structured domains* (Technical Report no. 441), Champaign, IL: University of Illinois, Center for the Study of Reading.

Spiro, R. J., Feltovich, P. J., Jacobson, M. J., & Coulson, R. L. (1991). Knowledge representation, content specification, and the development of skill in situation-specific knowledge assembly: some constructivist issues as they relate to cognitive flexibility theory and hypertext. *Educational Technology, 31*(5), 22–25.

Von Glaserfeld, E. (1984). Radical constructivism. In P. Watzlawick (Ed.), *The invented reality* (pp. 17–40). Cambridge, MA: Harvard University Press.

Vygotsky, L. S. (1986). *Language and thought*, Cambridge, MA: MIT Press.

Wagner, E. D., & McCombs, B. L. (1995). Learner centered psychological principles in practice: designs for distance education. *Educational Technology, 35*(2), 32–35.

Willis, J. (1995). Recursive, reflective instructional design model based on constructivist-interpretist theory. *Educational Technology, 35*(6), 5–23.

Woolfolk, A. E. (1993). *Educational psychology.* Bosten: Allyn & Bacon.

CHAPTER 4

FOUNDATIONS OF INSTRUCTIONAL DESIGN

What's Worth Talking About and What Is Not

Jerry Willis

Instructional design has been in a period of relatively rapid change over the past 15 years. The dominant ID models in 1990 were base primarily on behavioral and information processing theories, but the emerging models in the first decade of the twenty-first century tend use constructivist theories. In this period of change it is important to focus on issues that advance the field and to avoid spending time on debates and issues that are more provincial and personal than systemic and conceptual.

The field of instructional design (ID) is in the middle of a significant paradigm debate. On one side, there are proponents of established ID models based on behavioral and cognitive science theories of learning (Braden, 1996; Dick, 1995, 1996; Merrill, 1996). These ID models have dominated the field for over 25 years and they have been taught to thou-

Parts of this chapter were previously published in *Educational Technology* (1998), *38*(3), 5-16.

Constructivist Instructional Design (C-ID): Foundations, Models, and Examples, pp. 81–108

sands of graduate students in educational technology programs. Many graduate students, in fact, learned only one ID model while they were in graduate school, and the vast majority of them were taught one of the traditional models. The general framework of the ID models in widespread use today was developed during the sixties and seventies when behavioral psychology still dominated not only educational technology but education in general. The ID models matured and evolved in the eighties and nineties, and, while their behavioral roots are still quite obvious, the influence of a number of other theories, such as information processing theory, general systems theory, cognitive science, and communications theory, is apparent in the current generation of ID models (Dick, 1996).

ID models have developed, changed, and matured in many ways over the past 25 years, but for most of that time the changes have been evolutionary. Existing models were adapted and revised as new theories of learning and instruction became popular and influential. Dick (1996), for example, commented that when the ISD (Instructional Systems Design) model was first developed in the 60s, the experience he used as the background for creating the model was "experience in developing Skinnerian programmed instruction and efforts to create CAI instruction for an IBM 1500 system" (p. 55). In another paper, Dick (1995) commented on "The historical roots of much of what today is referred to as instructional design was Skinnerian psychology, specifically as it was manifested in programmed instruction" (p. 5). As he traced the progression of ID during the 70s and 80s, Dick identified the major improvements on the original Skinnerian model. "Skinnerian ideas were supplemented with those of the three Roberts: Robert Mager on objectives, Robert Glasser on criterion-referenced testing, and perhaps most importantly, Robert M. Gagnè's events of instruction and conditions of learning" (p. 56).

In describing changes in a recent version of ISD, Dick commented that the model presented in the 1978 edition of his book (co-authored by Lou Carey) "remained essentially unchanged through the second and third editions" (Dick, 1996, p. 56), but he pointed out that the 1996 edition of the book takes into consideration several emerging concepts, including performance technology, context analysis, multi-level evaluation models, and total quality management. It is certainly true that the four influences on the 1996 version have had an impact, but it is also true that the changes in the model have been evolutionary. These new influences have been incorporated into the model without making any revolutionary changes. A behavioral psychologist interested in creating Skinnerian programmed instruction would be able to comfortably use the newest version of ISD to do that. On the other hand, a constructivist educator interested in developing a student-centered, problem-based learning environment for high school European history would still find the 1996 ISD model uncomfortable because it makes assumptions that are contradictory to the guiding principles that have been adopted by the constructivist educator. The same is true of the sixth edition of the Dick, Carey, and Carey (2004) book, *The Systematic Design of Instruction.*

Perhaps because most of the changes in the standard ID models have been evolutionary, there have been a number of attempts in the last few years to formulate models that represent a revolutionary departure from the standard models. Several authors, including Wilson, Teslow, and Osman-Jouchoux (1995) and Winn (1992), have written thoughtful papers on the general influence of constructivist theory on instructional design. Most of these papers, including those in the influential *Educational Technology* special issue which led to the book edited by Duffy and Jonassen (1992), have put an emphasis on the types of instructional strategies typically used by constructivist designers. That is, they have focused on the *pedagogical* aspects of design while paying less attention to the *process* of ID. A few, however, have dealt with the process of ID (Bednar, Cunningham, Duffy, & Perry, 1992; Wilson, 1997a, 1997b), but this type of paper is far less prevalent. And most of the papers that focus on the ID process have offered relatively general guidelines (Winn, 1992) rather than specific models, or examples, of how constructivist principles were applied in any particular design context. However, some recent papers and books have described ID models based on constructivst theory (Cennamo, Abell, & Chung, 1996; Cennamo & Kalk, 2004; Willis, 1995, 2000; Willis & Wright, 2000).

The emergence of alternative ID models has generated a surprising range of comment. Braden (1996) and Merrill, Drake, Lacy, Pratt, and the ID2 Research Group. (1996), for example, seem to have considered constructivist ID models a threat to the very survical of instructional design. Dick (1996), on the other hand, compared his ISD model to a constructivist ID model (my own R2D2 model) and concluded that there were not that many differences between them.

While I do not agree with Dick's assessment that there are not many differences between a constructivist ID model and the Dick and Carey ISD model, I do believe that the most important differences between them are at a conceptual and foundational level rather than at the level of techniques or strategies. Fortunately, we now have a significant and growing body of literature that discusses and debates the relative merits of instructional design procedures based on frameworks like the Dick and Carey ISD model and approaches based on alternative theoretical foundations such as constructivism. Reigeluth (1996a), for example, wrote a very informative article on the need for a new paradigm of instructional design. In his paper, he proposed an approach that would make design (1) less linear, (2) more iterative or recursive, (3) more attentive to the context for which instruction is to be developed, (4) more active in facilitating the participation of all stakeholder groups, and (5) more focused on the creation of materials that let users become "designers" of their own learning environments. In Figure 4.1 I organized his recommendations according to whether they relate primarily to the process of ID or to the pedagogies that should be used. Four of Reigeluth's five recommendations are about the processs of ID while only one, a recommendation that designers invest in helping learners design their own learning environments, deals primarily with pedagogy.

Although Reigeluth has been associated with traditional ID models for many years, and he has been quite critical of *radical constructivisism* (Reigeluth, 1992), his call for a new paradigm for ID includes a set of concepts that are quite *constructivist* in their formulation. They were so constructivist that Merrill (1996) wrote a stinging criticism of the proposal for a new ID paradigm. Among other things, Merrill called some of Reigeluth's proposals "a recipe for disaster in the real world of instructional development" (p. 58). This choice of words is particularly interesting in light of the recent book on Constructivist Instructional Design Models and theories that the authors (Cennamo & Kalk, 2004) called *Real World Instructional Design*. All of the changes that Reigeluth recommended are reflected in the ID model described by Cennamo and Kalk. Merrill also argued against involving students in decisions about how they learn because "Students are for the most part, lazy" (p. 58). Since that comment was written over a decade ago, considerable progress has been made toward involving students in the creation of their own learning experiences under a variety of terms and concepts including:

- (AILE) anchored interactive learning environments (Leonard, Davis, & Sidler, 2005),
- contextual design (Rodríguez, Diehl, & Christiaans, 2006),
- learner control,
- learner-centered design (Krajcik, Blumenfeld, Marx, & Soloway, 2000; Nesset & Large, 2004)
- user-centered design (Thursky & Mahemoff, 2007),
- collaborative learning,
- collaborative inquiry, (Verbaan, 2008), and
- participatory design (Brandt & Grunett, 2001).

Whether alternative ID models are a momentary blip on the screen or the beginning of a paradigm shift that will bring major changes in both how ID is taught and how it is done, is still open to question. Regardless of their future, they are now the subject of significant debate.

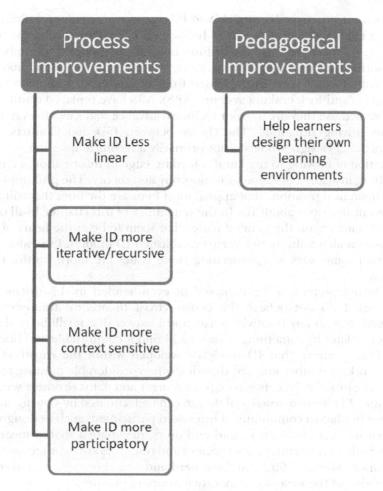

Figure 4.1. Reigeluth's recommendations for improving ID.

Although many designers feel very comfortable with their use of the Dick and Carey model, it is recognized that there are those who differ with the model on a philosophical level. They would argue that we are in a paradigm shift, and that the traditional model will soon be obsolete, if it isn't already. Most of the objectors would be classified as constructivists. They, in turn, would view the Dick and Carey model as an example of positivistic, objectivist thinking. (Dick, 1996, p. 60)

While I am one of those constructivists who thinks the Dick and Carey model is based on positivism and objectivism, I do not think the demise of that ID model is just around the corner. That is because the issues between constructivist ID models and ISD are not primarily technical. We are not talking about a new way of heating milk for hot chocolate without burning it, or a procedure for drilling holes in the wood of a fine bookcase without causing unsightly splinters. We are not even talking about a technical advance such as the move from traditional automobile brake systems to anti-lock braking systems (ABS). ABS have replaced traditional systems because they are a major technical advance that saves lives in hazardous driving situations. The choice between ISD and Constructivist Instructional Design Models is not primarily a technical question nor is it a question of moving to the latest, bleeding edge of information technology. In is, instead, a question of philosophy and theory. The philosophical and theoretical positions that appeal most to us are the ones that will also guide our decisions about ID. In the remainder of this chapter I will offer some comments on the general issues that seem to be at the heart of the discussions about alternative versus established ID models. I will also suggest that some ways of approaching those issues are more fruitful than others.

I do not pretend to be unbiased or even-handed in this discussion. Although I do not believe the constructivist theoretical framework is "proven" true in any positivist sense, and I expect that it will be replaced sooner or later by something better, I still find it comfortable and liberating. I also believe that ID models developed within the constructivist framework, and other kindred theories, offer considerable promise to the field. The promise I see, however, is the danger and damage others see. For example, I believe instructional design can be facilitated by viewing it as a process in which a community of interested participants such as designers, experts in a variety of areas, and end users, negotiate a shared meaning about both what learning opportunities and resources are to be created and supported. Merrill (1992), on the other hand, was very explicit in his condemnation of the idea of collaborative meaning making:

> If I hire a surgeon to do heart surgery PLEASE let me have one who has learned the trivial case and knows that my heart looks like every other human heart. Please don't let him negotiate new meanings and hook up my veins in some "self-chosen position" to which [she/he] can commit [herself/himself]. I want [her/him] committed to the standard objective view. (p. 108)

Merrill's concern reflects some fundamental misunderstandings of the concepts of collaborative learning, negotiated meaning, and multiple perspectives as those ideas are used in contemporary constructivist theory, but perhaps more importantly he seems to believe that all this constructivist stuff has not reared its ugly head in important areas like medical education and practice. But it has, both in terms of how the creation and evolution of medical knowledge is characterized and how physicans are taught. For example, in a paper titled "Constructing Knowledge Within a Medical Doman: A Cognitive Perspective," Bouchard, Lajoie, and Fleiszer (1997) discussed the growing dissatisfaction among medical school students and medical educators with the dominant instructor-centered approach to teaching in most medical schools. The McGill University authors then describe a constructivist approach to medical education that makes heavy use of multimedia and communications technology.

> Our ongoing research and development activities reflect our assumptions that cognitive technology applied to the medical curriculum will foster the development of new skills, behaviors, and attitudes in the construction of knowledge. Our underlying premise is that learning can be facilitated by situating the learner within an authentic context where students learn to do things that they will do in the real world.... For medical students this means learning basic sciences in the context of how such knowledge will be used in the clinical setting. (p. 46)

The paper then goes on to describe a collaborative group problem-based learning approach which "encourages the creation of authentic settings and real world contexts leading to a social construction of knowledge" (Bouchard et al., p. 46). The authors take the position that "we construct meaning in the context of learning situations, where knowledge is negotiated from multiple perspectives. We impose meaning on the world, rather than meaning being imposed on us" (p. 48). The work at McGill is only one example of many efforts at curricular reform in medical education that is being approached from a constructivist perspective.

Thus, while Merrill may want to believe his heart surgeon was taught the *one right way* to do heart surgery, a recent graduate may well have had learning experiences that have all three of the things Merrill was concerned about—collaborative learning, negotiated (or constructed) meaning, and multiple perspectives. Further, the physician may not have as much confidence in the "rightness" of the decisions he or she makes as Merrill would like. Physicians know that the "right" way to treat a particular medical problem today may be grounds for a malpractice lawsuit next month. "The standard objective view" as Merrill characterizes it, is not a static, permanent, universal entity. It is, instead, something groups of experts in the field agree on at the moment. Different medical specialists may actually take contradictory positions on some issues and the same group of specialists may drastically shift their views over time.

A good example of this in the area of heart surgery is the radical new form of heart surgery that was the topic of a PBS special in 1997. Dr. Randas Batista of Brazil developed a form of surgery that actually reduces the size of the enlarged heart in some very serious forms of heart disease. His work was originally criticized as dangerous and ineffective by some American heart specialists. Today, however, what is now referred to as the Batista Procedure is being carefully studied at a number of heart research centers, and his work has inspired others to develop what they hope are more effective procedures to accomplish the same thing. My point is that even in a highly empirical, research-based field like medical research, there is often considerable debate over just what "the standard objective view" is and whether it has changed recently.

Merrill's worries about constructivism as a foundation for instructional design will be addressed again later in this chapter, but his concerns are typical of one group with a considerable investment in the standard approaches to ID that emerged from behavioral psychology in the sixties and were then influenced by the cognitive turn in psychology during the seventies. As noted earlier, a somewhat opposite perspective is presented by Dick (1995), who seems to feel there is very little "new" in constructivist ID. I do not believe either extreme—that constructivism is a danger to the field or that it does not point to much that is different from what is being done now—is correct. As a field, we need to continue the dialog about the influence of alternative theories like constructivism and chaos theory on instructional design. We have begun that process of dialog, and there are aspects of the literature to date that are promising. There are also aspects that are disappointing. In chapter 1 I have already discussed one of my disappointments—the tendency to equate theories and discussions about pedagogy with ID rather than acknowledging the important role the process of ID plays in creating excellent educational resources. I think there are also some other core issues that deserve more debate, and some approaches to the dialog that seem less useful.

IS THE HEART OF THE DEBATE OVER STRATEGIES OR PRINCIPLES?

Behavioral, information processing and cognitive science theories can lead us rather quickly to very precise prescriptive instructions about what instructional strategies to use. Merrill et al. (1996), for example, make this point when they say:

> There are known instructional strategies. The acquisition of different types of knowledge and skill require different conditions for learning.... If an instructional experience or environment does not include the instructional strategies required for the acquisition of the desired knowledge or skill, ... the desired outcome will not occur. (p. 6)

Quoting Merrill et al. again (1996):

In our view, which learning strategy to use for a particular instructional goal is not a matter of preference, it is a matter of science. There are learning strategies (conditions of learning) that do promote particular learning outcomes. If the conditions of learning do not match the desired learning outcome, then the outcome will not be accomplished. (p. 6)

They are saying that we have a solid foundation of empirical research upon which we can base decisions about what strategies to use. If the type of learning you desire students to accomplish is X, then you should use instructional strategy Y. X is produced by Y. However, if you're interested in another type of learning, say Z, then instructional strategy B may be preferred.

There is thus a tendency on the part of proponents of ID models based on behavioral and cognitive science theories to see the discussion about constructivist ID as one of whether you use this or that instructional strategy. This focus on Pedagogical ID is understandable, however. It is a natural thing to do when your theory leads you so quickly and so directly to a set of clear cut guidelines that link types of content and types of learning to specific instructional strategies. Dick (1995) took that position when he graciously suggested that his model could be used to create instruction that includes constructivist strategies.

There are three problems with situating the conversation about constructivism and ID at the instructional strategy level. The first is the lack of agreement on what an instructional strategy is. When Reigeluth (1996b) argued that we should consider a much wider range of instructional strategies when designing instruction (e.g., apprenticeships, debates, games, group discussions, team projects, simulations, case studies, role playing, and so on) Merrill (1996) argued that "the different approaches to instruction depicted in the paper are in fact not instructional strategies but different social environments" (p. 57). Merrill then argued that there are scientifically discovered and validated conditions of learning that must be present for successful learning to occur and

these conditions or learning strategies are not determined by the social environment. A learner can learn in any of these settings. It is not the setting but the learning strategy (learning conditions) employed within the setting that will be the more important determinant of effective learning. (p. 57)

This position is again understandable given that of the social sciences, it is psychology that has dominated both education and educational theory for the past 100 years or so. Had American sociology, or anthropology, had as much control over the thinking in education as American psychology did during the twentieth century, I doubt the social and contextual setting within which learning occurs would occupy such a small and infrequently visited basement room in the edifice of influence implied by Merrill. Constructivist theories are an antidote to the much too limiting perspective of behavioral learning theory because they generally emphasize the crucially important role that social factors play in all forms of learning. We learn everything, from our language to our cultural values, social history, and professional practices through the social process of interaction.

Merrill's position on what an instructional strategy is and is not, is a convenient one because it eliminates by executive order many of the "instructional strategies" that are preferred by proponents of opposing theories of teaching and learning. And, at the same time, his narrowing of the focus to certain "conditions," forces us to think about learning in the way his preferred theory proposes.

In a rebuttal, Reigeluth (1996b) pointed out that the instructional strategies Merrill was trying to eliminate met the criteria specified by Merrill and Reigeluth in papers they had previously coauthored. Reigeluth (1996b) concluded that "social environments, even more than the instructional media Dave compares them to, can indeed influence learning outcomes – quite dramatically for some students" (p. 59).

Clearly, there is no general agreement even on what is or is not an instructional strategy. But even the discussion of that question is a call for us to move up to a higher level of generalization and to discuss both the increasingly important philosophical issues that are the foundations of different types of ID and the issues embedded in the theories of learning that are both a source of guidance for both pedagogy and the process of instructional design.

A second problem with the focus on instructional strategies is that traditional design models, by their very nature, push designers toward the use of some instructional strategies and away from others. The Dick and Carey ISD model, for example, is based on Gagnè's events of instruction. Gagnè's approach is quite compatible with most forms of direct or "teacher-centered" instruction (what many call instructionism), but it can actually get in the way of designing instruction based on other approaches to teaching and learning such as the use of case studies, anchored instruction, or microworlds.

I think discussions of ID models that focus on which instructional strategies the designer will select miss the point that there are issues at higher and more general levels that help determine what strategies will be considered. The previous chapters in this section have made the point over and over that keeping the discussion on the strategy level would mean we do not address some of the more important points of departure between established and emerging ID models.

When I wrote the original version of this chapter (Willis, 1998) almost a decade ago I said that

> I find myself in strong agreement with Bednar, Cunningham, Duffy, and Perry (1992) when they argue for a theory-based approach to instructional design. We do not have one large bowl of instructional strategies from which to make our selections each time we design instructional materials. We have many bowls, and they contain families of instructional strategies that are based on different theories of learning and instruction. There is a behavioral family of instructional strategies, a cognitive science family, and at least two constructivist families. The selection of an instructional strategy should be made from within a theoretical framework that guides decision making. (p. 8)

While I still find the paper by Bednar and her colleagues stirring and inspirational, I no longer share their enthusiasm for what I referred to as pedagogical fundamentalism in chapter 1. Fundamentalism is only desirable if you know without a doubt that the theory you advocate is completely and unquestionably Right in every sense of its meaning. (Even then, I can think of situations where it still might not be desirable.) However, a basic guiding principle of constructivism is that you can never be confident the theory that currently appeals to you is Right. Further, the history of education and the social sciences is a history of one certain theory after another being overturned by another that, at least for the moment, seems to offer us more. I thus believe now that designers who find constructivist theory, or other theories such as nonlinear systems dynamics, appealing as a foundation for their design practice, should use it and incorporate it into their decision making process. However, I think behaviorists make a mistake when they decide the empirical research tells them that virtually every new learning task calls for some form of direct instruction. And, I think constructivists make a mistake when they select instructional strategies only from the bowl of constructivist pedagogies. In my own teaching I often use what I call "just in time" direct instruction, which usually involves stopping a class engaged in some project and giving a short lecture on a topic that most students in the course seem to need. Constructivists do use direct instruction methods, a fact which is often ignored by critics. Braden (1996), for example, made the comment that "systematic ID … and structured instruction are not going to go away no matter how much the constructivists and their bedfellows might yearn for that result" (p. 18). In general, constructivists don't have such yearnings. What they do want to change is the way we think about teaching and learning. They have a set of guidelines or concepts that can guide our thinking, but they do not advocate completely eliminating any particular instructional strategy.

That leads me to the third problem with concentrating on instructional strategies when comparing ID models based on different paradigms. Constructivist theory, while it does point us to some instructional strategies, such as anchored instruction and collaborative learning, is less focused on particular strategies than it is a set of principles. As Wilson, Teslow, and Osman-Jouchoux (1995) put it when discussing the influence of constructivism on ID, the ID concepts and models produced by constructivists "are still somewhat vague about actual design practices. A certain fuzziness may be inevitable since constructivism is a broad theoretical framework, not a specific model of design. Moreover, constructivism tends to celebrate complexity and multiple perspectives" (p. 137). Like some graduate students, some instructional designers, particularly novices, are more comfortable with concrete, detailed directives on what to do first, second, third, and so on. Some ID models provide this, just as some course outlines do. However, the end result of such detail in an ID model is reductionistic. It reduces the process of designing instruction to what Donald Schön called a "technical-rational" process. I still find considerable artistry in the ID process and like most constructivists I see it as a social and participatory process when practiced at its best. Add a few more terms like holistic, contextual, and inclusive, and you have a process that cannot be reduced to concrete, detailed directives because too much must emerge from the design process itself instead of be imposed on it by preexisting rules and laws of pedagogy and design. Constructivist-ID calls not for laws and rules but for principles and guidelines.

Constructivist theory yields principles that are the framework for thinking about teaching and learning as well as instructional design. Those principles, along with the theories of learning and epistemology they are based on, should be the focus for discussions about ID much more often than instructional strategies. The late William Winn (1997) made this point eloquently. He took a position that is in opposition to that of Merrill discussed earlier. Winn did not believe we will ever have a set of research-based formulas that link a particular type of instructional objective to a particular instructional strategy.

> A theory of instruction that contains a prescription for every combination of learning outcome and student characteristic will collapse under its own weight, as aptitude-treatment interaction theory ... did when people tried to use it as a framework for instruction rather than as a framework for research. (p. 35)

Learning is too complex for this reductionist approach. Teachers and designers will always need to creatively use what Winn calls "first principles" to make many of the day to day professional practice decisions.

Thus, a significant part of instructional design is rational and reflective rather than empirical; it requires thoughtful consideration of the complex factors that face the designer. We do not have, nor are will we probably ever have, a "completely knowledgeable prescriptive instructional theory (Winn 1997, p. 36). Winn's first principles constitute a framework within which we can think about design. This is where the action is, and that is at a higher level of abstraction than instructional strategy.

The approach of moving the discussion to a more abstract level is sometimes equated with the idea that designers can make choices based solely on their personal whims. This is a charge made by designers who believe there is, or will be in the future, an empirical research base for our field that makes decision making as simple as matching the new teaching task with what the research says about how it should be done. As Braden (1996) put it, "If we cannot progress beyond intuition and personal preference in the way that we put instruction together, then the legitimacy and viability of the entire profession are open to questions none of us wants to answer" (p. 14). I think this is simply wrongheaded. The members of most, if not all, professions cannot practice in a completely technical way, "given conditions Y, treatment X is the obvious choice." Many, many professional decisions are fuzzy and difficult. Schön's (1987) ideas about reflection in action as the foundation for professional practices seems to me to be more realistic than the technical-rational approach embodied in the framework proposed by Braden and others. If that is true, we should, as Winn suggests, learn more about the theories and frameworks that underlie all aspects of instructional design because they are the source of the "first principles" we will use as designers.

OLD TERMS, NEW MEANINGS AND THE PROBLEM OF CROSS PARADIGMATIC COMMUNICATION

An important implication of Thomas Kuhn's (1970) sociological theory of scientific revolutions is that shifts in the underlying paradigm of a field often involve giving new meaning to old terms, or creating new terms that supplant or replace existing terms in the field. There are, for example, many terms that seem to mean something very similar to Piaget's concept of *schema*. As new theories of cognitive development have emerged, theorists have the option of redefining existing words like schema, assimilation, and accommodation. Many, however, prefer to coin new terms because they feel the existing terms are already too heavily laden with with excess meaning that is associated with another theory or paradigm. Dills and Romiszowski (1997), in their discussion of whether PSI (Personalized System of Instruction) is dead, comment that many of the elements of PSI have been incorporated into new plans of instruction. But "new generations of researchers and practitioners feel they need to improve and modify the plans; having done so, they tend to also change their names" (p. xxi). New terms thus come into vogue that replace well-used existing terms that may have had similar meanings.

Coining new terms, however, is not without its problems. A theory with many new terms may be harder to understand and grasp than one that recycles terms already in use in the field. Rorty's (1991) use of the term *truth* is a good example of recycling. Rorty does not accept the idea that the scientific method can lead us to a clear and unambiguous understanding of reality, and Rorty proposes a worldview that is similar in many ways to Kuhn's—groups socially construct meaning. Thus, what is true, and what is not, is negotiated among members of the group that is deciding what is true and what is not. Rorty could have coined a new term for this new meaning but he did not. He used the word truth, but his *truth* is quite different from the truth of an empiricist, although it serves similar purposes. Theory builders thus always face the dilemma of deciding whether to coin new terms or recycle old ones. The word *constructivism* is itself an example of this. Constructivism has much in common with John Dewey's progressive education, but that old term was not resurrected and used.

One danger of using new terms is that it makes it more difficult for interested parties to understand new concepts and theories. On the other hand, if you use existing terms in new ways, the reader may well miss the point that some terms have quite different meanings in the new context. In developing the R2D2 ID model, I initially decided to use existing ID terms in different ways. I believe my recycling of existing terms is one reason why Dick (1996) concluded there was not very much difference between his ID model and R2D2. Recycling leads to assumptions that terms have the same meaning in the old and new contexts, but that is often not true. For example, I used terms like front-end analysis, learner analysis, and task analysis in early descriptions of the R2D2 design model. My use of terms that were originally developed within ISD-style models let Dick to conclude that "It seems that when constructivist models are proceduralized, they look very much like traditional design models" (p. 62).

Appearances can be deceiving, since my use of the term learner analysis, for example, is quite different from the use of that term in ISD models. For me, learner analysis means learners participate in the process of deciding what is to be learned and how. Learners are not objectives to be studied, they are participants in the design process.

I do not see an ideal solution to the problem of whether different theoretical and conceptual frameworks can or cannot share specialized vocabulary. This may make sense in some situations and be disastrous in others. However, what does seem conclusive is that whether old or new terms are used, they should be discussed in enough detail that newcomers have no trouble understanding them. And, f there are nuances of difference between a new and old term, they should be made clear to readers.

DO STRAW MEN (OR WOMEN) HAVE A PLACE IN THE DISCUSSION?

One of the easiest levels of argument to handle is bashing straw people, First, you take a perspective (say Poisition K) you disagree with, then you state clearly that anyone who proposes Position K also advocates X, which virtually no one thinks is a good idea. X is the "straw man" of this attack. Then you use the "fact" that X is part of Position K to bash the position. Of course, it matters little that few, if any, people who advocate Position K also support X, but that is not the point. The point is to "win" the argument.

I write this at the height of the 2008 Presidential election campaign in America and there are so many examples of this in the political arena that one wonders whether there could be a campaign without straw men and women, and the responses that carefully explain that an opponent or his/her surrogate has used a straw man argument.

Straw man arguments are not limited to politics, however. They are commonplace in scholarly and professional discourse as well. For example, Dick (1996) has argued that I have used straw man arguments in my discussion of his ID model. There are, I think, four ways we can end up making straw man arguments, and only one can be justified.

Failure to Understand the Position. A common reason for making straw man arguments is that the author is strongly opposed to a general position but does not actually understand the position well enough to see that the straw man argument he or she uses is inaccurate or used inappropriately. This is not an acceptable reason for using a straw man argument. Acquiring a thorough understanding of a position before you oppose it will make your arguments both more relevant and more influential. That may also lead to a change in your position. For example, over the last 15 to 20 years I have read the philosophical papers and books of Stanford University professor D. C. Phillips. I have always found his strong defense of positivist, empiricist foundations for educational research to be glib, sometimes arrogant, and occasionally illogical. However, I continued to read his work because he is one of the best-known advocates of postpositivism as the foundation for research in American education. Then I read his 2000 book with Nicholas Burbules, *Postpositivism and Educational Research*. In his effort to address some of the more problematic issues of a postpositivist perspective, I felt Phillips has modified postpositivism so much that there is now a much smaller distance between it and my own preferred research paradigm, interpretivism. I do not write the same about Phillips today because I my current understanding of his views have changed considerably, and my view of Phillips' kinder and gentler form of positivism is considerably better.

Making Straw Men in Full Awareness of the Distortion. This is another unacceptable reason for using this type of argument. Scholarly and professional discourse is hindered by disingenuous straw man arguments.

Out of Context Explanations. A third, somewhat understandable, path to a straw man argument involves taking a particular aspect of a theory out of context and arguing against it. In his defense of the ISD model, Dick (1996) discussed one possible future in which people use that model when they want to "design and develop efficient and effective instructional solutions to human performance problems" (p. 62) while public education adopts a constructivist approach because you have to motivate and engage students and "learning seems sometimes to be a distant second priority" (p. 62). Dick (1995) made this same point in another paper when he worried about whether "creativity will replace effectiveness, that having fun will replace learning something, and that we won't know the difference.... We could easily find ourselves creating mere educatainment or infotainment" (p. 10). Dick seems to be saying here that characteristics such as motivation, engagement, creativity, fun, and learner appeal have no place in efforts to design instruction. They seem somehow to be presented as the opposite of "efficient and effective instructional solutions." It is true that constructivist-ID models do respect and value things like creativity, motivation, engagement, and fun, but Dick seems to assume you cannot design instruction that is both engaging and efficient, or appealing and effective. I would argue that not only can you do both, it is essential that you try to do just that. This is simply another version of the argument that the better the medicine is for you the worse it tastes. That doesn't work for medicine and it certainly does not work when we are designing instructional materials. It is true that constructivist theory does put more emphasis on the engagement of students in active learning while the building blocks of instruction in behavioral learning theory put little if any emphasis on those values. The difference between ISD and C-ID is not that one wants to produce good instruction and the other, like Madonna, just wants to have fun. No C-ID model proposes that we concentrate on making learning experiences fun while ignoring whether they are strong learning experiences or not. Many constructivists do argue, however, that one potential reason for poor learning is boring, out-of-context instruction that does not capture the interest of students, and instruction that, because the content is taught out of context, produces inert knowledge that is difficult for learners to apply or transfer to meaningful contexts. Boiling the differences between constructivist ID and ISD down to whether we want to develop efficient and effective instruction or entertain students is simply wrong. It takes desirable features of C-ID models (such as emphasizing enhancing the engagement and participation of students) out of context and gives the impression that either constructivists just wanna have fun or real learning must be difficult and boring.

Arguments Against One Theory From Within Another. This type of straw man argument is fully justified so long as it is clearly labeled as a critique that is not based on the internal logic and belief system of the paradigm or theory being criticized. For example, Braden (1996) created a straw person argument when he characterized new approaches to ID as "calling for intuitive, artistic ID with no structure, no rules, no constraints —instructional design of an unspecified nature" (p. 20). While this seems to be a common conception of constructivist ID (partly because people like Braden keep saying that is the case), the detailed descriptions of ID models in Section III of this book make it clear there is structure, there are rules or guidelines, and there are constraints. They are simply not the structures, rules or guidelines, and constraints that Braden's preferred paradigm would produce. Braden's mistake in making this sort of argument is not that he compares what "intuitive, artistic ID" does with a set of standards he believes in. That is reasonable—to state your preferred source of standards and then to use standards derived from that source for evaluation. His mistake was in taking the position his standards are absolute standards that are more like laws than rules or guidelines. Braden's positions come from a variety of theoretical perspectives including positivism and hard systems theory. Even if you believe a particular theory and its standards are True, stating the paradigmatic and theoretical foundations for your position is still incumbent upon all of us.

ONE SET OF EXPERT-DETERMINED GOALS OR MANY GOALS?

There is a general issue so deeply embedded in some of the ID models that it is virtually hidden. Many of the dominant ID models are based on the assumption that a common set of instructional objectives will be developed at the beginning of a design project and that the goal of the instructional package created will be to get as many students as possible to achieve those objectives. This perspective is sometimes taken to be the only viable option when developing instruction. Braden (1996), for example, says that "A basic premise of linear instructional design is that it is driven by performance goals." And "whenever instructional designers encourage a learning need that cannot be rendered into a sensible, observable, achievable student performance, they should abandon attempts to meet that need through linear instructional design techniques" (p. 9). For Braden, it is not instructional design (or at least not linear ID) if specific, observable objectives are not developed early in the process. Further, he argues that the objectives created are used to create a matching set of test items, and then "instruction should be designed which 'teaches to the test'" (p. 12).

Predetermined objectives selected by designers or teachers or school districts or state and federal governments, are not the only option, however. The constructivist literature has made the point over and over again that a viable alternative is to help students develop the skills and interests to develop some or all of the objectives of a learning experience. The natural result of involving students in the selection of objectives is, of course, the fact that different students will pursue different objectives. And, if the alternative is pushed further, the differences in perspective on this issue become even more dramatic. Giving students the opportunity to participate in the process of creating goals and objectives for a learning experience means that specific objectives may not be defined, or negotiated, in the beginning of an instructional design process. Instead, they may emerge as students explore and learn about a new topic. The perspective taken on this issue—should objectives be selected in advance by experts or can they sometimes be negotiated in context with significant student participation—has major implications for instructional design theory that need extended discussion.

Much of the linear nature of standard ID models derives from the assumption that where we are headed (e.g., accomplishing the objectives) must be decided before material can be developed. That approach, however, does not fit all types of instructional materials. Perkins (1992) describes five aspects of a learning environment—information banks, symbol pads, construction kids, phenomenaria, and task managers. The design of these types of instructional resources is not well served by creating specific behavioral objectives at the beginning of ID.

The "one set of objectives fits all" approach is more suited to behavioral instruction strategies, such as drill and practice programs, tutorials, and other forms of direct instruction. However, the alternative perspective on objectives—that they may be developed in collaboration with students, that different students may have different goals and objectives, and that they may be changed and revised throughout the learning experience—is a much better fit for all five items in Perkins' list. Creating an information bank, such as a DVD of material on a particular topic—industrial pollution in rivers and bays, for example, would not normally begin with the creation of a large number of specific objectives, while creating a math drill and practice program might. I thus strongly disagree with Braden (1996b, p 17) when he says "This is where the line in the sand must be drawn. Objectives have been under fire for years but have outlasted all detractors. Without learning objectives ID would be a farce."

"RESEARCH-BASED" VERSUS BRAND X MODELS OF INSTRUCTIONAL DESIGN

Proponents of some ID models emphasize that the way they have come to do things is the right way because it is based on "research." Dick and Carey (1996), for example, say "Systems approach models are an out-growth of more than twenty-five years of research into the learning process. Each component of the model is based upon theory and in most instances research that demonstrates the effectiveness of that component" (p. 5). But to their credit, Dick and Carey (1996) also acknowledge that their model is based on several sources of knowledge. "the model is based not only upon theory and research but also on a considerable amount of practical experience in its application" (p. 5). And, in reviewing the research on "whether it makes any difference if objectives are used" they acknowledge that "the results of this type of research have been ambiguous" (p. 118).

I do not want to belabor the point, but all claims, yes, ALL claims, that a certain ID model is superior because it is based on "research" are simply false. Dills and Romiszowski (1997) made this point when they suggested that

> some action research on the real-world effects of adopting specific ID approaches should be done in order to determine their real-world effects, and so their real-world values. This has been tried in the past, but only in a few cases, because of the cost required to carry out such studies and the number of people, both students and designers, that are required to fill all the required experimental and control blocks involved. (p. xiii)

I agree that the research has not been done to date. I don't agree that it needs to be done. I don't believe it can be done in any way that would "settle" the issue. When authors make a claim that their ID model is based on research, they are always talking about a certain body of related research literature that is appealing to them. Authors who support an opposing view can also cite related research literature that supports their own perspective. ID models, ALL ID models, are based on unproven, and probably unprovable, paradigms and theories, and on professional practice knowledge. All three of these sources of influence—general paradigms, more detailed and specific theories, and professional practice knowledge—are more influential in the shaping and development of ID models than research in the empirical scientific tradition. Supporters of no approach—not behaviorism, not information processing theory, not cognitive science, and not constructivism—can claim their ID model is better supported by research. There is very little research of any sort on ID models and their impact. The claim of being "based on research" is more Madison Avenue hype than it is a realistic representation of the origins and foundations of an ID model. It is generally made by those who adhere to a positivist or postpositivist philosophy of science because both of these foundations that insist the scientific method is the only source of "truth." Thus, if you don't have research to support your viewpoint, you have no foundation for advocating it. But simply saying there is research to support your viewpoint is not the same as actually having the research.

A side effect of claiming that your model is based on research is a reluctance to change your approach unless you are presented with research that supports the proposed alternative. Braden (1996) takes this approach when he argues that all steps in an ID model should be taken rather than, as Wedman and Tessmer (1993) suggest, letting designers decide what steps are necessary for a particular project and which ones are not. Braden argues that Wedman and Tessmer base their suggestions on a set of assumptions and "until much more evidence is available, those assumptions will not be accepted here" (p. 15). The problem with this sort of dismissal of Wedman and Tessmer's approach is that there is no empirical evidence for the opposite assumptions, which Braden supports.

Further, replacing the word "research" with "scientific" does not improve the situation. In their paper titled "Reclaiming Instructional Design" Merrill, Drake, Lacy, Pratt, and the ID2 Research Group (1996) assert that

> Education and its related disciplines continue to flutter this way and that by every philosophical wind that blows. In an uncertain science and technology, unscientific theories flourish.... When answers are slow in coming, uncertain in statement and difficult to find; then the void is filled with wild speculation and philosophical extremism. (p. 5)

The extremism the authors appear to be criticizing is constructivism and they go on to take the position that "there is a scientific discipline of instruction and a technology of instructional design founded on this science" (p. 5). The paper makes the case for an approach to design that is based on the "right" ways to teach different types of content. Those right ways are, of course, discovered by conducting scientific research, and they are universal.

The authors compare the right ways to instruct with the aerodynamic principles of flight: lift and drag, and they point out that "it was not until the Wright brothers discovered the correct principles of aerodynamics (science), that they could invent an airplane that would sustain powered flight (technology)" (p. 5). They then comment that "it is not until we discover the correct instructional strategies that we can invent instructional design procedures and tools that will promote student learning" (p. 5).

There are two problems with this analogy. First, even after almost a hundred years of intensive scientific research, the discovery and validation of the "correct instructional strategies" remains a holy grail that is yet to be obtained. It is a hope of the empiricist, not an accomplishment. Second, even if we accept that there are empirically validated instructional strategies, any standard we use to make that assertion would admit a wide range of strategies not commonly used by Merrill and other traditional designers. There is thus not a single "instructional science" and no single "technology of instructional design" based on that science. There are many instructional sciences, and they have different theoretical foundations, different bodies of research supporting them, and different approaches to instructional design associated with them. What they all have in common is that not a single one can offer a solid body of empirical or scientific research that definitively shows that it is better or more desirable than the others. We can adopt a behavioral, a cognitive, a critical, or a constructivist framework for design, but whatever framework we adopt, it will be on the basis of something other than a convincing body of research that demonstrates that framework is stronger than the others.

It therefore makes no sense to assert, as Merrill et al. (1996) do, that

> those persons who claim that knowledge is founded on collaboration rather than empirical science, or who claim that all truth is relative, are not instructional designers. They have disassociated themselves from the technology of instructional design. We don't want to cast anyone out of the discipline of instructional science or the technology of instructional design; however, those who decry scientific method, and who deride instructional strategies, don't need to be cast off; they have exited on their own. (p. 1)

While not common in our field, such statements are still too frequent in the ID literature. We simply don't know enough about how people learn and how to design good instruction to be so arrogant. Membership in the sorority of instructional design cannot be based on whether someone adheres to a particular paradigm or theory of learning. There should be no secret handshake that is required of all members. There are a number of competing theories of learning today—information processing theory, cognitive science, chaos theory, Piagetian constructivism, Vygotskian constructivism, to name a few—and all of them have associated instructional science theories as well as implications for the process of instructional design. There are thoughtful, reasonable people who support these viewpoints and it makes little sense for us to set up a theoretical litmus test for membership in the ID fraternity. We need all the help and ideas we can get!

THE PERSONALIZING OF INTELLECTUAL ISSUES

I must admit that I was offended by Braden's (1996) personal characterization of me as "an arrogant adolescent with a chip on his shoulder" (p. 19) [although at my advanced age perhaps I ought to take any comment about being a teenager as a compliment] and my writing as "a threat to a valid understanding of ID" (p. 19). His complaint was with the ID paradigm I support and the constructivist ID model I developed (Willis, 1995). Attributing to opponents unscholarly and unprofessional reasons (e.g., acting like an arrogant adolescent) for taking a position you oppose is commonplace in political life, but we should probably be more watchful about it in our field. My concerns about Braden's comments are similar to Reigeluth's (1996b) objection to Merrill's (1996) depictions of Reigeluth's views. Merrill attributed unsavory reasons for statements of Reigeluth that Merrill considered unacceptable. However, it is clear that Reigeluth believed our society has changed so significantly that a different way of thinking about ID is called for. His was a bold proposal that deserves serious attention. There are thoughtful, reasonable people who support a range of diverse viewpoints, and it makes little sense for us to either set up theoretical litmus tests for membership in the ID sorority or attribute other than honorable and scholarly reasons for someone taking a different position than the one we support. That applies to constructivist and critical theorists as well as positivists!

The same is true of inflammatory language. It is one thing to say "I believe the future of ID will be brighter if we base our work on a scientific, empirical approach" and another thing to say "Too much of the structure of educational technology is built upon the sand of relativism, rather than the rock of science. When winds of new paradigms blow and the sands of old paradigms shift; then the structure of educational technology slides toward the sea of pseudo-science and mythology. We stand firm against the shifting sands of new paradigms and 'realities.' We have drawn a line in the sand" (Merrill et al., 1996, p. 7).

It is perhaps ironic that much (but not all) of this type of rhetoric comes from the proponents of the scientific approach to ID, but all sides practice it to one degree or another. The language is that of battle and verbal warfare, however, instead of communication and collaboration. The less we have of this, the better we probably are as a field.

The sharp and warlike rhetoric of Merrill and his colleagues is in contrast to the way Dick (1996) frames the differences between constructivists and ISD proponents.

> In recent years, constructivists have written of the subjectivity of language and the arbitrariness of descriptions of reality. They object to pre-specifying objectives and criterion-referenced evaluations for all learners. They endorse contextualized learning environments in which learners can explore and set their own goals, and be assessed via an examination of portfolios and other idiosyncratic accomplishments. (p. 60)

Dick has as many, if not more, disagreements with constructivists as Merrill, but his approach to framing the differences between the paradigms is much more likely to encourage inter-paradigm dialog and discussion.

WILL THIS END WESTERN CIVILIZATION AS WE KNOW IT?

The final general point I would like to make relates to the worry and concern some scholars have about the impact on the field of adopting a different paradigm. Merrill (1996), for example, uses terms like "recipe for disaster" when discussing some of Reigeluth's recommendations for a new paradigm for instructional design. Similarly, Braden (1996) called the idea of considering design an art rather than a procedural process "dangerous" because accepting the idea that ID is an art "has opened a crack. Others have squeezed into and expanded the crack. The path is an easy one. Begin with art. Next comes artistic license. Soon there is just license—total license.... Art in the design of instruction, when applied at the expense of craft and/or science, would be disastrous" (p. 15–16). Braden worries that constructivism, if accepted, could end ID as we know it.

> Do constructivism's basic tenets spell doom for instruction as we have known it? If so, widespread acceptance of radical constructivism would mean the demise of instructional design and development. Neanderthal instructional designers like this author would find that hard to accept. (p. 17)

This worry about the effects of constructivist ID can also be seen in much of Dills and Romiszowski's (1997) rhetoric on the place of alternative paradigms for ID.

I simply can't see it. If ID, as it is generally practiced today, were the foundation for a thriving educational system, or a world beating business and industrial base, then I would have more sympathy for these worries. But the fact of the matter (remember that my "facts" may not be the same as yours) is that ID has thus far had very little impact on education, and it is not a significant part of business and industry, either. Many successful, and unsuccessful, curricula in schools were created with little attention to the general procedures of traditional ID. Many training programs in business and industry are based on theories and models created outside the field of ID, and many very successful educational software programs have been brought to market after being developed via models that are quite different from traditional ID.

If ID, as it is expressed in real-world instructional design, were to disappear off the face of the earth, no great calamity would befall the world. We must simply admit that while we can point to a few "successes," the field has thus far failed to have any major and long-lasting impact. We mainly discuss, argue, and debate among ourselves, and very few people outside our small group give our work the slightest attention. I, therefore, am less worried about a calamity befalling the field if we change paradigms, and am more worried about it dying a slow death from being unused and ignored if we continue to practice the way we have.

HOW SHOULD WE APPROACH THE DISCUSSION?

In this chapter, I have suggested that there are some critical issues at the heart of the paradigm dialog as it relates to instructional design. A particularly crucial one is whether this discussion is about what strategies to use in instructional materials we create, or should the discussion be at a higher and more general or conceptual level? I believe the dialog must occur at the higher level for two reasons. First, instructional strategies are embedded in and based on theoretical philosophies of science. If we keep the debate at the strategy level, we will overlook much of the reason why there are differences on issues, such as whether end users should participate fully in the design process, whether objectives should be prespecified by experts or allowed to emerge across the design process, and whether the creation of assessment instruments is a crucial precursor to development, or sometimes an intrusion on the creative design process.

Simply put, we cannot understand why there is strong disagreement on some issues related to ID unless we understand the theoretical and epistemological framework being use by proponents of different viewpoints. A good example of this is the question of whether the social context can constitute an instructional strategy. Is group problem solving an instructional strategy or not? Merrill's emphatic No can only be understood if we have some awareness of the theory of learning he is using. And, when we understand Vygotskian learning theory with its emphasis on the role that cultures, as well as other humans, play in a child's learning, we can predict that constructivists would argue strongly that collaborative learning is a viable and important instructional strategy. Similarly, the question of whether objectives are created early, and in some detail, or emerge from the work on the instructional material is best understood as an implication of the particular philosophy of science and epistemology the design team has adopted. If you adopt an epistemology that posits a knowable external reality, then it makes sense for experts to preselect objectives. On the other hand, if you believe the reality we deal with is socially constructed and therefore subjective, then it makes more sense for objectives to emerge across the design and development process as a collaborative team works on the instructional package.

I am not suggesting that everyone who advocates prespecifying objectives before doing development work sat down and thought for hours on the nature of reality. In many cases, we do what generally feels comfortable to us and we also do what we were taught to do. I suggest, however, that there is an implicit assumption about what reality is in the way we approach objectives, and that any discussion of the role objectives should play in the ID process must deal with the assumptions and theories that are the foundation for the different perspectives.

I have also suggested that some ways of framing the issues are not very productive or helpful. Personalizing the debate, predicting doom for the field, making unfounded claims of superior research supporting a particular view, and setting up straw people to knock down are common but relatively unhelpful approaches.

Much is uncertain, fuzzy, and unclear about instructional design theory today, but one thing that is clear is that different paradigms are being used today to guide ID, and those paradigms lead to quite different answers to very basic questions. Thus the differences between the paradigms are not cosmetic. Their origins are deep in the core of the paradigms. What does that mean for the individual designer? For the ID theorist? Should all of us make a choice about what paradigm I we "believe" and then practice totally within that paradigm? Some authors have certainly taken that position, and there is logic to support such a decision. In a way, it does not make sense to do some design work within a behavioral paradigm and then to do other work within a constructivist paradigm. These two paradigms are based on opposing answers to some fundamental questions about teaching and learning.

There are, however, two other ways of thinking about this question, and both lead us to different conclusions. The first involves the ability to separate core paradigm issues from instructional strategy issues. It is clear that proponents of ISD such as Walter Dick support the use of "constructivist" instructional strategies under some circumstances, just as proponents of constructivist ID support the use of direct instruction strategies under some circumstances. Thus, strategies developed within one paradigm are used by those who support another. Using an instructional strategy from another paradigm may involve considerable adaptation, but borrowing from a different paradigm is already common. We thus need to be aware of developments in other paradigms and be willing to consider their use in our own work.

Yet another way of thinking about this issue is to consider both (1) what we believe to be right as well as (2) our degree of confidence in the rightness of our beliefs. I have argued in this chapter that none of the ID models available today have such firm foundations that they can be considered infallible. In such a situation, Rorty's (1991) democratic pragmatism seems very appropriate as a framework for design and development. Do ID guided by your preferred paradigm but remain open to methods and results from other paradigms. Make an effort to understand what others are saying and be willing to change your paradigm, or even give it up for another, if proponents of opposing views can convince you they have merit. This approach—open democratic pragmatism—is the one I recommend to the reader.

REFERENCES

Bednar, A., Cunningham, D., Duffy, T., & Perry, D. (1992). Theory into practice: How do we link. In T. Duffy & D. Jonassen (Eds.), *Constructivism and the technology of instruction: A conversation* (pp. 17–34). Hillsdale, NJ: Erlbaum.

Bouchard, R., Lajoie, S., & Fleiszer, D. (1997). Constructing knowledge within the medical domain: A cognitive perspective. *Proceedings of the First International Cognitive Technology Conference, Hong Kong*, 45–56.

Braden, R. (1996, March/April). The case for linear instructional design and development: A commentary on models, challenges, and myths. *Educational Technology, 35*(2), 5–24.

Brandt, E., & Grunnet, C. (2000). Evoking the future: Drama and props in user-centered design. In T. Cherkasky, J. Greenboum, P. Mambrey, & J. K. Pors (Eds.). *PDC 2000 Proceedings of the Participatory Design Conference.* New York, NY, USA, 28. November – 1 December 2000. Retrieved from http://intranet.dkds.dk/postnukephoenix/images/imageUpload/305/files/PDC00_evoking_the_future_FINAL.pdf

Cennamo, K., & Kalk, D. (2004). *Real world instructional design.* Belmond, CA: Wadsworth.

Cennamo, K., Abell, S., & Chung, M. (1996). A "layers of negotiation" model for designing constructivist learning materials. *Educational Technology, 36*(4), 39–48.

Dick. W. (1995, July/August). Instructional design and creativity: A response to the critics. *Educational Technology, 35*(4), 5–23.

Dick, W. (1996). The Dick and Carey model: Will it survive the decade? *Educational Technology Research and Development, 44*(3), 55–63.

Dick, W., & Carey, L. (1996). *The systematic design of instruction* (4th ed.). New York: HarperCollins.Dick, W., Carey, L., & Carey, J. (2004). *The systematic design of instruction* (6th ed.) Boston: Allyn & Bacon.

Dills, C., & Romiszowski, A. (1997). Preface and overview. In C. Dills & A. Romiszowski (Eds.), *Instructional development paradigms* (pp. ix–xxxi). Englewood Cliffs, NJ: Educational Technology Publications.

Duffy, T., & Jonassen, D. (Eds.). (1992). *Constructivism and the technology of instruction: A conversation.* Hillsdale, NJ: Erlbaum.

Krajcik, J., Blumenfeld, P., Marx, R., & Soloway, E. (2000). Instructional, curricular, and technological supports for inquiry in science classrooms. In J. Minstrell & E. H. V. Zee (Eds.), *Inquiring into inquiry learning and teaching in science* (pp. 283–315). Washington, DC: American Association for the Advancement of Science.

Kuhn, T. (1970). The structure of scientific revolutions (2nd ed.). Chicago: University of Chicago Press.

Leonard, J., Davis, J., & Sidler, J. (2005, Spring). Cultural relevance and computer-assisted instruction. *Journal of Research on Technology in Education, 37*(3), 63–284.

Merrill, M. (1992). Constructivism and instructional design. In T. Duffy & D. lonassen (Eds.), *Constructivism and the technology of instruction: A conversation* (pp. 99–114). Hillsdale, NJ: Erlbaum.

Merrill, M. (1996, July/August). What new paradigm of ISD? *Educational Technology, 36*(4), 57–58.

Merrill, M., Drake, L., Lacy, M., Pratt, L., & The ID2 Research Group. (1996, September/October). Reclaiming instructional design. *Educational Technology, 3*(6), S–7.

Nesset, V., & Large, A. (2004, Spring). Children in the information technology design process: A review of theories and their applications. *Library & Information Science Research, 26*(2), 140–161.

Perkins, D. (1992). Technology meets constructivism: Do they make a marriage? In T. Duffy & D. jonassen (Eds.), Constructivism and the technology of instruction: A conversation (pp. 45-56). Hillsdale, N]: Erlbaum.

Phillips, D. C., & Burbules, N. (2000). *Postpositivism and educational research.* Lanham, MD: Rowman & Littlefield.

Reigeluth, C. (1992). Reflections on the implications of constructivism for educational technology. In T. Duffy & D. Jonassen (Eds.), *Constructivism and the technology of instruction: A conversation* (pp. 149–156). Hillsdale, NJ: Erlbaum.

Reigeluth, C. (1996a, May/June). A new paradigm of ISD? *Educational Technology, 6*(3), 13–20.

Reigeluth, C. (1996b, July/August). Of paradigms lost and gained. *Educational Technology, 36*(4), 58–59.

Rodríguez, J., Diehl, J. C., & Christiaans, H. (2006, September). Gaining insight into unfamiliar contexts: A design toolbox as input for using role-play techniques. *Interacting with Computers, 18*(5), 956–976.

Rorty, R. (1991). *Objectivity, relativism, and truth: Philosophical papers* (Vol. 1.). New York: Cambridge University Press.

Schön, D. (1987). *Educating the reflective practitioner.* San Francisco: Jossey-Bass.

Thursky, K., & Mahemoff, M. (October, 2007). User-centered *design* techniques for a computerised antibiotic decision support system in an intensive care unit. *International Journal of Medical Informatics, 76*(10), 760–768.

Verbaan, E. (2008, August). The multicultural society in the Netherlands: technology-supported inquiry-based learning in an inter-institutional context. *Teaching in Higher Education, 13*(4), 437–447.

Wedman, J., & Tessmer, M. (1993). *The intelligent design of computer—assisted instruction.* New York: Longman.

Willis, J. (2000). The maturing of constructivist instructional design: Some basic principles that can guide practice. *Educational Technology, 40*(1), 5–16.

Willis, J., & Wright, K. (2000, March/April). A general set of procedures for constructivist instructional design: The New R2D2. *Educational Technology, 40*(2), 5–20.

Willis, J. (1998). Alternative instructional design paradigms: What's worth discussing and what isn't? *Educational Technology, 38*(3), 5–16.

Willis, J. (1995, November/December). A recursive, reflective instructional design model based on constructivist-interpretivist theory. *Educational Technology, 35*(6), 5–23.

Wilson, B. (1997a). Reflections on constructivism and instructional design. In C. Dills & A. Romoszowski (Eds.), *Instructional development paradigms* (pp. 63–80). Englewood Cliffs, NJ: Educational Technology Publications.

Wilson, B. (1997b). The postmodern paradigm. In C. Dills & A. Romoszowski (Eds.), *Instructional development paradigms* (pp. 297–310). Englewood Cliffs, NJ: Educational Technology Publications.

Wilson, B., Teslow, K., & Osman-Louchoux, R. (1995). The impact of constructivism (and postmodernism) on ID fundamentals. In B. Seels (Ed.), *Instructional design fundamentals: A review and reconsideration* (pp. 137–157). Englewood Cliffs, NJ: Educational Technology Publications.

Winn, W. (1992). The assumptions of constructivism and instructional design. In T. Duffy & D. Jonassen (Eds.), *Constructivism and the technology of instruction: A conversation* (pp. 177–182). Hillsdale, NJ: Erlbaum.

Winn, W. (1997, January/February). Advantages of a theory-based curriculum in instructional technology. *Educational Technology, 37*(1), 34–11.

CHAPTER 5

CONSTRUCTIVISM IN INSTRUCTIONAL DESIGN THEORY

Frank Dinter

In this chapter, I will argue in favor of a basic constructivist foundation for instructional design theory by putting forward common sense arguments. By doing so, I intend to clarify some issues of the ongoing discussion on constructivism among instructional theorists and instructional designers. This chapter is neither criticizing any specific viewpoint held by some instructional designers nor is it advertising for a particular brand of constructivism. My concern is with epistemological arguments which are relevant to every scientific discipline related to the field of instructional design. An additional concern is methodological consideration on the status and function of epistemological arguments. Methodological and epistemological considerations should go hand in hand for the simple reason that all participants in the discussion can recognize immediately which kind of argument is presented and which kind of argument would be an adequate counter argument.

Previously published in *Journal of Structural Learning & Intelligent Systems* (1998), *13*(2), 71–89

The contemporary discussion on constructivism in the field of instructional design is similar to theoretical discussions in other scientific disciplines. The discussion on constructivism concerns the fundamental question of whether scientific work needs an explicit epistemological foundation or can do without it. Interestingly, the recent discussion did not originate in philosophy, but in the workings of the neurobiologist Maturana and his collaborators. Scientific research on the functioning of the nervous system and of cognition as a biological phenomenon led to the fundamental epistemological question of how living organisms (and humans among them) acquire a representation of their environment. Maturana's answer is: "The organism in its operation does not act upon an environment, nor does the nervous system operate with a representation of one in the generation of the adequate behavior of the organism" (Maturana, 1990, p. 103). At first glance, this assumption appears nonsensical but within the framework of Maturana's elaborated theory of cognition (and autopoiesis) and the epistemological conclusions Maturana derived from it, the proposition cited above makes perfect sense. Highly sophisticated epistemological assumptions (of more or less radicality) are hard to grasp without the theoretical context from which they are derived. Unfortunately, in the field of instructional design the discussion on constructivism has developed (almost) irrespectively of Maturana's theoretical framework. However, I do not want to expound Maturana's theory. I only mention it because, historically, it was one of the main reasons for the discussion on constructivism in so many scientific disciplines in the past two decades. Perhaps because the results of empirical scientific research led to epistemological questions, scientists cannot refuse the related discussion as a "mere philosophical" one (in the sense of "without any interest for science").

Thus, we are confronted with a specific situation *in instructional design*: on the one hand some curtailed, simplified radical constructivist assumptions entered the discussion (like, "There is no external world."). Considered *in* isolation and without further explanation, such assumptions weaken every reasonable constructivist standpoint, because they lead to the wrong impression that a refutation of these assumptions means a refutation of constructivist epistemology. On the other hand the neglect of the theoretical background of radical *constructivism* simplifies the discussion *in* a certain way. It spares us some detailed interpretations of radical constructivist perspectives. Instead, I will argue *in* favor of a "basic" *constructivism* by putting forward common sense arguments. By doing so, I intend to clarify some issues of the ongoing discussion on *constructivism* among *instructional* theorists and *instructional* designers.

CURRENT STATUS OF EPISTEMOLOGICAL AND METHODOLOGICAL CONSIDERATIONS

Epistemology and methodology serve the function of setting the foundations of a scientific discipline. In contrast to some philosophers of science (e.g., Salmon, 1992) who think that methodology includes heuristic considerations or other aspects of the context of discovery, I think that methodology concerns the logic of science (especially the logic of scientific method). Methodology concerns the context of justification. Epistemology concerns questions on how knowledge is acquired and what knowledge represents. In history, epistemology focused on human perceptual abilities. In our century (since the so-called "linguistic turn"), philosophers recognized language—or more precisely: human conceptual abilities—as an important factor for our epistemic access to reality. Thus, epistemological and methodological considerations always include conceptual considerations.

For reasons of space, the scope of this chapter is restricted to conceptual arguments. Empirically validated facts concerning the functioning of the brain and the mechanisms of perception can justify severe objections against the realist idea of a "mapping" of the external world into the brain/mind. These objections will not be further discussed but taken for granted. Many scientists, including *instructional* theorists and designers, are not concerned with epistemological questions. Therefore, questions like, "Why do we discuss such questions at all?" "Why don't we leave epistemological issues to the philosophers and return to really important problems?" sometimes are raised among *instructional* designers, who are bored with the more or less philosophical nature of the *constructivism* debate. Such questions are intended to end the discussion and to return to an epistemologically unreflected view that we are used to *in* our everyday lives.

In defense of the subject this chapter is dealing with, I can reply:

- Every scientific discipline must reflect its own foundations; epistemology and methodology are an important part of the foundations of a scientific discipline. This applies to many disciplines which refer to the scientific investigation of human cognitive abilities. Epistemology has impact on our models of human cognition.
- Every scientific theory reflects implicitly the epistemological viewpoint of the theorist. I argue in favor of the demand to make it explicit in order to make it discussable for others.
- Actually, there is a discussion about epistemological questions and it seems to me that this discussion is going wrong in certain aspects.

The discussion itself reveals where the field of instructional design and the field of philosophy were at odds with one another.

My intention is to abbreviate this discussion, but not by simply dropping the questions! This chapter is neither criticizing any specific viewpoint held by some instructional designers, nor is it advertising for a particular brand of constructivism. My concern is with epistemological arguments which are relevant to every scientific discipline related to the field of instructional design. An additional concern is methodological consideration of the status and function of epistemological arguments. Methodological and epistemological considerations should go hand in hand for the simple reason that all participants in the discussion can recognize immediately which kind of argument is presented and which kind of argument would be an adequate counter argument.

A simple and crude scheme (see Table 5.1.) should illustrate how to distinguish relevant questions in the following discussion at four levels.

1. Philosophical level	Epistemology, methodology, philosophy of science
2. Cognitive science level	implications of the results of level 1 (for neurology, philosophy of mind, instructional, cognitive, etc.) psychology, educational science ...
3. ID theory level	Implications of the results of level 1 (via level 2) and of level 2 for a theory of instructional design (which we do not have today)
4. Instructional design level	Implications of the results of level 3 (and levels 1 and 2 via level 3) for the design of learning environments

The "deeper" levels provide the meta-*theories* for the higher levels. To a great extent this chapter deals with level 1 considerations.

SOME CONFUSIONS AND MISAPPREHENSIONS IN THE CONSTRUCTIVISM DEBATE

Is There an Epistemological Continuum?

An observer of the constructivism debate in the field of instructional design may get the impression that some of the participants think of a sort of "epistemological continuum" with realism at one end and constructivism at the other. Such a continuum might be suggested, for example, by Rescher's (1992, p. 189) description: "the thesis of naive (or commonsense) realism that 'External things exist exactly as we know them' sounds realistic or idealistic according as one stresses the first three words of the dictum or the last four." But we have to keep in mind that two opposite points of view must not be linked by a continuum that provides a half-way compromise between one extreme and the other. If we affiliate with constructivism, we are excluded from a realist standpoint and vice versa.

I think the picture and the corresponding idea of a continuum between realism and constructivism is misleading because there is a strict distinction between the variety of realist positions on the one hand and the variety of constructivist positions on the other hand. As I will show, the distinction concerns the concept of knowledge. In fact, some realist epistemologies are incompatible with one another and we cannot even talk of a continuum from moderate to extreme among them. Because there is no such continuum among realistic epistemologies, the idea is apparently misleading that a conception of learning and instruction based on a realist viewpoint can easily be transformed to a constructivist (i.e., anti-realist) conception by simply changing some terms. Perhaps, especially for instructional designers, it must be a tempting idea to assume such a continuum.

This might become clear by some rough-and-ready remarks on the history of instructional design. It is generally accepted that the instructional design field has its historical roots in behaviorist psychology. Behaviorism is an obvious example of a realist scientific paradigm. In the early 1960s, behaviorism went out of fashion and cognitivist psychology became mainstream. Instructional designers hesitated to follow the new cognitive movement for a quite understandable reason: instructional design theories and instructional design models which were founded on behaviorist psychology had been rather successful. Probably, they thought: "Never change a winning team!" Anyway, sooner or later they gave up their resistance and began to restructure piece by piece their behaviorist theories and models and transformed them into cognitive theories and models. The smooth transition from behaviorist to cognitivist orientation might suggest the idea that there is a similar transition from realist to constructivist epistemology (i.e., that there is an epistemological continuum). Or it might suggest an even more attractive idea that the transition from behaviorism to cognitivism already includes the transition from realist to constructivist epistemology. In the latter case the misapprehension stems from a careless identification of psychological cognitivism and epistemological constructivism. In spite of the inflationary usage of the adjective "constructive" by cognitive psychologists, almost all of their theories are epistemologically realistic.[1] The empirical findings which demonstrate the constructive nature of the cognitive processes of the persons under examination do not necessarily lead the examiner to the conclusion that his/her own cognitive processes are constructive as well. Therefore, they do not necessarily lead to further considerations on the epistemological implications of empirical findings.

The objection against the assumption of an epistemological continuum excludes neither the idea of a continuum among cognitive theories as proposed by Lowyck and Elen (1993) nor a continuum among instructional designs which are based upon such theories. In either case, instructional design theories or models are not positioned with regard to an epistemological criterion but with regard to implications of epistemology, for example, the theoretical modelling of the learner.

Is There a Problem With the External World?

Certainly, the question is not whether there is an external world; this much is taken for granted. Questions concerning the very existence of an external world are unavoidably of a metaphysical nature. Even among philosophers there is no general agreement about the conceptual prerequisites necessary for reasonably raising such questions. Try to imagine a discourse in which one of the participants claims that an external world does not exist. Such a claim leads to a version of the solipsist's paradox. The situation is paradoxical because the only consistent reaction is ignoring the claim. From the perspective of everybody else the herald of the non-existence of the external world himself is part of the external world. But if the external world does not exist, it is quite unjust to charge constructivism with an antipathy to reality. For it is not the existence but the nature of reality that the constructivist puts in question. Instead, the problem, in rough approximation, can be formulated by asking why a person gains knowledge of the external world. So understood, the problem seems to admit an easy solution. There is knowledge of the external world which people acquire primarily by perceiving objects and events which make it up. However, this easy solution seems problematic because we have to clarify the concept of knowledge and the concept of perception. And that is the reason why we are discussing epistemological questions.

The debate on epistemology in the field of instructional design is a heterogeneous terrain. Some of the main terms occurring frequently in the discussion seem to be without any precisely established usage. Unfortunately, this especially holds for the terms "realism" and "constructivism." Therefore, I will try to give a short description of the epistemological points of view related to these terms.

EPISTEMOLOGICAL REALISM

Usually, the realist is the player in the theatre of epistemology who thinks himself/herself to hold the normal, the plausible or even the self-evident point of view. Accordingly, the constructivist plays the role of the challenger, who is charged with the burden of proof for his/her alternative viewpoint.

Realism claims that there are such-and-such entities (e.g., trees, people, mental states, electrons) that are indeed real (cp., Pettit, 1992, p. 420). Realism requires two things (cp., Westphal, 1992):

- the characteristics of real entities do not depend upon our cognitive capacities or mental processes like perception, reasoning, thoughts or language; and,
- the entities are knowable, that is, under ordinary circumstances our knowledge of reality is reliable.

A version of realism without frills is direct realism. Direct realism holds that the objects of the external world exist independently of any mind that might perceive them. For a direct realist there is no need of some non-physical intermediary (like a sense datum) or any mental go-betweens in order to explain our perception of the world. For him/her perception provides a one-to-one mapping of the external world like it is "in itself." As far as I know this seemingly crude version of realism plays no role in the discussion on constructivism. It is mentioned here only because it is easy to see that direct realism rules out other realist views. For example, representative realism, which seems to be an attractive version of realism among scientists, assumes sense data and other mental entities in order to explain our epistemic access to the external world (for the representational aspects, see e.g., Putnam, 1988; Stich & Warfield, 1994). That is, representative realism assumes what direct realism denies. Even among philosophical realist viewpoints the idea of an epistemological continuum seems to be inadequate. Representative realism holds that:

1. There is a world whose existence and nature is independent of us and of our perceptual experience of it;
2. perceiving an object necessarily involves causally interacting with it (causal *theory* of perception); and,
3. the information acquired *in* perceiving an object is indirect.

Clause (1) makes it a species of realism, clause (3) puts it in opposition to other species of realism like direct realism. Anyway, there is one central feature common to all species of realism. Subtleties aside, it is this: there is no metaphysical distinction between appearance (the world as it is perceived by us) and reality (the world "in itself") which blocks knowledge of reality (Westphal, 1992). Therefore, the world "in itself" is the ultimate reference point for every realist conception of knowledge. True knowledge tells us how the world is in itself. The realist conception of knowledge usually is combined with a correspondence theory of truth. In short: when we are to decide the truth or falsity of a proposition, we must look at whether the proposition corresponds to reality or not. If the proposition corresponds to reality, it is true; otherwise it is false.

This standpoint faces serious problems. One of them concerns the question of how we can explicate the relation of correspondence between a linguistic entity (= the proposition) and a non-linguistic entity (= the world). But we shall not linger on this problem.

"If realism itself can be given a fairly quick characterization, it is more difficult to chart the various forms of opposition, for they are legion" (Pettit, 1992, p. 420). The form of opposition I wish to support in this chapter insists that the entities in question, "are tailored to our human capacities and interests and, to that extent, are as much a product of invention as a matter of discovery" (Pettit, 1992, p. 420). A standard example for such opposition is constructivism.

EPISTEMOLOGICAL CONSTRUCTIVISM: A STANDARD CONSTRUCTIVIST ARGUMENT AGAINST REALISM

From the realist point of view, a person obtains knowledge if she or he succeeds in (mentally and/or verbally) "representing" a segment of the given world "in itself," that is to say, independent of any human cognitive capacities. That is the essential premise.

A usable conception of knowledge must provide procedures that allow distinguishing in an intersubjectively controllable way true knowledge from pretended knowledge. This is a methodological requirement that every conception of knowledge should meet. Evidently, for a realist conception of knowledge there is no such procedure. If somebody wants to verify a representation of a segment of the world as true knowledge, she or he must compare the representation to (the segment of) the world "in itself." That is impossible because as human beings our only access to the world is necessarily mediated by our cognitive processes. Obviously, an epistemological position based on a more sophisticated conception of knowledge—the realist would say: a weaker conception of knowledge—not "realist" in the recent sense of the word.

CHARACTERISTICS OF CONSTRUCTIVIST EPISTEMOLOGY

For a German speaking scientist it is an astonishing fact that "constructivism" is not a major heading in English encyclopedias or dictionaries of epistemology. "The term constructivism does not occur with great frequency in these other bodies of literature—for example, the recent encyclopedic volume edited by Dancy and Sosa (1992), *A Companion to Epistemology*, gives it only three passing references—but nevertheless closely related ideas are the subject of vigorous debate" (Phillips, 1995, p. 5).

It looks like "constructivism" is subsumed to the headword "idealism." Actually, constructivism and idealism have several general features in common. Therefore, I can use a part of the description philosophers give of idealism for a description of constructivism as well. Under constructivism/constructivist epistemology we understand the

> philosophical doctrine that reality is somehow mind-correlative or mind coordinated—that the real objects comprising the "external world" are not independent of cognizing minds, but only exist as in some way correlative to the mental operations. The doctrine centres on the conception that reality as we understand it reflects the workings of mind. And it construes this as meaning that the inquiring mind itself makes a formative contribution not merely to our understanding of the nature of the real but even to the resulting character we attribute to it. (Rescher, 1992, p. 187)

Constructivism maintains that all of the characterizing properties of physical entities have no standing at all without reference to minds. Weaker still seems to be an explanatory constructivism which merely holds that an adequate explanation of the real always requires some recourse to the operations of mind (cp., Rescher, 1992). But this only seems to be a weaker standpoint. Explanatory constructivism emphasizes the essential methodological requirement that every epistemology must meet: an adequate explanation must be given, how we can act in and talk about a shared reality on the basis of acquired knowledge. It is the fundamental premise common to all brands of constructivism that such an explanation must mention the workings of the mind in order to be an adequate explanation. I do not accept an explanation that ignores one or another constitutive factor of our efforts to understand the world in which we live. The assumption of mental constructions as mediators of human epistemic access to reality necessarily puts into question what we call our knowledge of the real along with other important terms like "reality," "objectivity," "truth," and so on. As scientists we cannot, and we do not want, to do without these terms.

On What is Not Constructed: Problems With Some Radical Constructivist Assumptions[2]

Unfortunately, the philosophical discussion on constructivism provides several statements that may cause serious confusions. Some of these statements suggest a construction of concrete single entities/ objects/ things, in short—a construction of matter. And in some cases it is stated that the world, including the constructing subject, is constructed. Not taken as a metaphor but seriously and quite literally, those radical constructivist assumptions lead to evident absurdities, if we presuppose our normal-language concept of "construction." On the one hand some radical constructivists are to be blamed for the related misapprehensions. In fact, some radical constructivists fancy themselves in the role of a troublemaker. On the other hand misunderstandings are unavoidable when a single proposition is cited without the necessary context or explanatory background (like the citation of Maturana in the introductory sentences of this article).

"To construct" means "to create," "to engender," "to bring forth" something that is not available in the time before the process of construction. And "to construct" means to generate something intentionally and according to a plan.

Some radical constructivists assume that "the world," "the experiential world," or "the reality" is constructed. This manner of speaking leads to a quite trivial paradox.[3]

1. If one asks where the activity of constructing takes place, the answer surely is: "In the world."
2. If the world itself is the object or result of construction, it cannot be in existence during the activity of constructing. Thus, the activity of constructing must take place outside the world.

To avoid this simple paradox radical constructivists make a terminological distinction between

- the ontic world (the world "in itself" or the world "beyond the phenomenal world"), and
- the constructed reality (the world "for us").

But this distinction is of no advantage for the constructivist position. Due to their own premises radical constructivists cannot pick out as a central theme, the part of their distinction, which is supposed to be the interesting one, namely, the ontic world. If there is nothing to be said about the ontic world except in terms of speculation and mere fantasy—as constructivists assume—then the essential point of constructivism, that is to say, construction, is shrouded in mystery.

Constructivism need certainly not go so far as to affirm that mind makes or constitutes matter. From a logical point of view it is absurd to deny the existence of the real world. From a consistent constructivist point of view, one cannot say anything about the world "in itself." The world "in itself" is the world as it is intrinsically, that is, the character of the world apart from any relations in which it happens to stand. The notion of the "world-in-itself" is a metaphysical abstraction which we can generate effortlessly with our conceptual abilities. As a conceptual construction this notion may serve certain functions—but that holds for other conceptual constructions like "unicorn" or "absolute velocity" as well.

The constructivist's point is the relation between the inquiring mind and the external world, that is, reality as it is experienced by us. Valid and true knowledge about the world "in itself" is the realist's pretended privilege. The constructivist does not deny the world "in itself." He/she denies that we can refer to the world "in itself" for a justification of knowledge claims, that is, he/she denies that the world "in itself" is a usable criterion for the verification or falsification of propositions.

So far we have dealt with the refutation of the realist epistemology and the realist conception of knowledge. This is the destructive part of a constructivist line of argumentation. Now we have to sketch an alternative based on the principles of constructivism.

On What May Be a Matter of Construction

The history of epistemology was dominated by the discussion of the role and function of perception, especially visual perception. Concepts as an important means for human knowledge acquisition have been neglected for a long time. Perception provides sensory input which is processed automatically. But perception alone neither leads to a representation of the external world nor to knowledge. For the higher levels of processing, concepts are necessary. In the philosophical tradition concepts are considered certain abilities to identify, to classify, and to describe entities. Concepts constitute the ability to act upon objects qua objects of a certain kind. Due to our concepts, we structure our experiences of the world. The world appears in the way it is mediated by concepts. This implies that we would experience a completely different world if we depend on a different system of concepts.

This concept of a concept should not be understood to imply:

- that there are at first objects of a certain kind for a person (without a concept); and subsequent to a sort of "direct understanding" of these objects the person develops the ability to behave more or less adequately with regard to those objects.

- that every kind of ability to act upon objects qua objects is a concept. For example: the ability of a single-celled organism to respond to stimuli is the ability to behave towards something qua something. But nobody would say that such organisms have concepts because it would extend our normal use of the term "concept" in an inappropriate manner.

In addition, for the purpose of this article, I restrict the term "concept" to the meanings of words shared by a language community. By doing so, I avoid the difficulties with concepts that may—or may not—constitute a private language as discussed and rejected by Wittgenstein (1953).

Usually it is supposed that concepts are connected with the faculty of speech and the ability to carry out linguistic acts like the classification of certain objects.[4] A person is endowed with a concept when she or he is able to apply correctly the corresponding general term. Concept use is rule-based. But to prove the person's conceptual ability it is not enough to observe a certain regularity with regard to the person's linguistic behavior. Wittgenstein (1953) has shown in his inquiries on the logic of rule-following that we cannot find out by mere observation whether a person follows a specific rule (the rule hypothesized by the observer's interpretation) or intends to follow a completely different rule (a rule unknown to the observer) or does not intend to follow a rule at all, but acts purely by chance (cp., Kripke, 1982).

I draw the conclusion from this fact that the best method to reach a well-grounded assessment of an individual's faculty of speech consists in an interaction with the individual. Within this interaction the person can explain her/his view concerning the correct use of some linguistic expressions in question.

For the proposed conception of a concept, I conclude that concepts are not only the abilities to use correctly general terms by identifying and describing objects. Additionally, a person must be able to explain (herself or himself or another individual) her or his application of general terms (or equivalent terms).

Now, the crucial question is, where do concepts come from? Most scientists and philosophers abandoned the idea that concepts (or ideas) are innate. This old idea faces too many logical problems and empirical counter-evidence.[5] Today's general conviction is that concepts are acquired. But it is still an open question how the process of acquisition takes place.

One realist's thesis might say that the individual (somehow) discovers prebuilt concepts in the external world. The advantage of this thesis consists in the idea that concepts which are part of the external world are identical for everybody. Therefore, they can guarantee intersubjective communication. Additionally, by this thesis the realist might try to preserve his/her conception of knowledge. He/she can admit that our epistemic access to the world is mediated by concepts; and he/she might conclude that if we know our concepts, we know part of the world "in itself."

The drawback of this line of thought is twofold. First, the ontological status of concepts is unclear. If concepts are part of the external world, they cannot be cognitive or mental entities. If concepts are cognitive entities, they are not independent of the human mind. Secondly, if concepts are a decisive means for us to act in the world, and if the discovery of concepts is some kind of action, how can we discover them without having already acquired (some) concepts? In the case of concepts the point is that we cannot start searching without having already available what we are supposed to search for.

Another realist thesis says that concepts are real entities but not part of the external world. This thesis leads to a version of Platonism. Frege (1975) took this view. He has to postulate a "third realm" of "concepts/meanings" and "eternal truths." Again, intersubjective communication would be guaranteed and the realist conception of knowledge would be preserved. However, Frege cannot explain how humans have access to this "third realm," and how concepts refer to entities of the first realm (= the physical world) and the second realm (= the mental world).

A third realist thesis might assume that concepts must automatically fit and reflect the structure of the world because reality directly determines human concept formation. The specific human way of structuring and ordering the world is determined by the structure of the world. In the strongest sense the notion "determination" means "causation." In fact, if reality causes us to form concepts in one and only one way, we cannot miss the true recognition of the world. For example, misconceptions simply would not be possible. Unfortunately, a huge amount of empirical counter-evidence contradicts this thesis. Additionally, it is hard to conceive how physical causes can effect cognitive phenomena like concepts. If "determination" is not meant in the strong sense of "causation," that is, if the human mind has some "degrees of freedom" to form concepts in one way or another, then nothing guarantees that our concepts reflect reality "in itself."

By these considerations I only want to illustrate some general problems a realist conception of concepts has to face. The problems are due to the realist conception of knowledge. In order to avoid these difficulties the constructivist assumes that concepts are not found in the external world but are mentally constructed by the individual.

Advantages and Problems of a Constructed Concepts' Approach

In fact, with the assumption that concepts are the constructed entities, I seek to avoid several problems of the radical constructivist's assumptions that either concrete entities (things) or the world as a whole (matter) are constructed. I do not need the distinction between "ontic world" and "constructed reality." When we use our common concepts, we have no difficulties in saying that they were constructed in the world;—in the world as we recognize it on the basis of our use of the concepts at hand.

The brain (or the mind or whoever is the constructor) generates the concepts which enable us to ask about the existence of certain entities and to describe ourselves as persons who have acquired the ability to generate concepts. Questions about the existence of things are treated as semantic questions, not as ontological or metaphysical ones. (That's what the so-called "linguistic turn" in philosophy is all about!) The question of how mentally constructed concepts fit the structure of the world turns either trivial or meaningless. The trivial version is: Concepts are constructed in order to structure our experiences of the world. Therefore, concepts perfectly fit the structure of the world because they generate this structure. The meaningless version is: How do concepts fit the structure of the world "in itself?" For the principal reason that we have no access to it independent of our cognitive capacities, we cannot know the structure of the world "in itself." Therefore, we cannot compare it to our system of concepts. Certainly, the external world determines mental constructions because the external world embodies all conditions, requirements, and constraints which such constructions must meet. The attempt to examine these conditions, requirements, and constraints leads straightforwardly into an epistemological circle: we must use concepts for the examination, concepts, which have been constructed in dependence of the matter under examination.

There remains a conceptual and/or terminological difficulty. Concerning our explicated concept of construction, we want to apply this concept when an acting subject is generating something intentionally and according to a plan. Undoubtedly, it is possible that a person is endowed with a concept—that is, the person is able to use a general term correctly and to explain her or his use—without knowing anything about concepts in principle. Strictly speaking, in this case we cannot say that the person has constructed the concept. This looks bad, but things get worse! It follows that no concept can be constructed in the strict sense of the word because nobody can know how a particular concept will look like before the concept is generated.

Thus, we can speak about the construction of concepts only in a "diminished" sense. Human history shows that preconceptions are transformed, revised, and refined piece by piece in a "semi-controlled," co-operative procedure that is similar to a process of construction. Obviously humans construct preconceptions without having prior knowledge of how the future concept will look. We must assume that preconceptions are less specific compared to concepts and less clarified in their relations to other concepts. To illustrate the "semi-controlled" constructional process, we can draw a comparison to the work of an artisan or craftsman who, first of all, has a rough idea with regard to the thing he wants to create. His idea slowly takes shape during his creative activity.

The construction of concepts which is described as a co-operative activity within the social framework of a language community, disposes the question of solipsism. There is an old belief which is hard to eliminate: it is the opinion that constructivism leads to solipsism sooner or later. Sometimes realists like to attack constructivism with this reproach. They try to suggest that an epistemology which leads to solipsism is totally unusable as an epistemological foundation of science. In defense of constructivism, I can say: in a co-operative way, constructed concepts provide shared meanings for the members of a language community. Furnished with an adequately explicated concept of construction, the realist's reproach of solipsism as well as the problem of the external world become straw men. Both problems miss the basic point at issue because the realist confuses construction and arbitrary, random, or accidental generation.

HOW TO GET ALONG

Due to methodological considerations the realist conception of knowledge and truth is rejected. I have tried to explicate the term "construction" and I have answered the question concerning which entities are a matter of construction.

A constructivist approach that takes into account the construction of concepts avoids most of the problems realism has to face. But methodological considerations are only a preliminary step towards a constructive theory of instruction. Most theoretical and empirical questions concerning mental construction are still to be asked and answered.

As I have argued elsewhere (cp., Dinter & Seel, 1994), there are no direct implications derivable from epistemological considerations for instructional design theories and models. The impact of epistemological considerations is on the construction of psychological theories of learning and instruction which serve as meta-theories for instructional design. Some supporters of constructivism among the instructional designers disagree with this hypothesis. They cannot resist the appealing idea to demand new instructional design models based on constructivist epistemology and constructivist principles. For example, Knuth and Cunningham (1993, p. 164) draw upon the writing of Maturana and their purpose is "to assess the implications of this view for existing and possible new instructional systems. In other words, would instructional design and the use of technology look different based upon Maturana's notion of autopoiesis?" But such models would be ill-founded if they are developed independently of an epistemologically refined psychological model of the learner. Otherwise, so-called constructivist principles of designing instruction appear to be arbitrary. To give an example: It is popular among constructivist designers to demand "authentic learning tasks" and "authentic learning environments." Such a demand is sound and reasonable, but it is not derived from any particular constructivist designing principle. It is consistent with every designing principle perhaps even behaviorist principles.

Anyway, from the view I have reached in my discussion, at least one point should be clear: Knowledge (based on the meaning of linguistic expressions) cannot be instilled in the learner. The process of knowledge acquisition is a process of constructing meaning. The design of learning environments should serve this function among other functions. Until today most learning environments probably were designed according to the designers' idea that knowledge can be transmitted into the learners head. Some of these environments are pretty good, that is, they lead to knowledge acquisition on the part of the learner. Does this fact disprove the constructivist point of view? Surely not! It demonstrates that learners cannot be prevented from constructing knowledge, even when the designer holds an inadequate conception of learning. The really interesting question is whether we can optimize constructive learning processes by designing environments according to a constructivist conception of learning. But such a conception—along with other important psychological concepts—remains to be developed.

ACKNOWLEDGMENT

I am indebted to Michael Spector who generously took the time to provide very constructive and patient discussions on an earlier draft of this paper. The remaining errors are solely the responsibility of the author.

NOTES

1. It is my personal conviction that a deeper examination would show that modern cognitive psychology does not decisively differ from neo-behaviorism if one disregards the changes in psychological terminology.
2. The following argumentation refers to the critical analysis of the philosopher A. Ros; for a sophisticated and detailed discussion see Ros (1994a, 1994b).
3. Only to call it to mind: A paradox is a contradiction between two (or more) propositions that are regarded as true each one alone.
4. By this we do not assume that, for example, deaf-mutes do not use concepts! The faculty of speech is not a necessary condition of concept use. In general, we can prove our correct usage of a concept by adequate behavior (whether verbal or non-verbal), e.g., by pointing at an object *in* question. Speech only makes it easier to control our ordinary concept use in our language community.
5. The philosopher J. A. Fodor (1975, 1980, 1983) tried to renew the idea of innate concepts; he speaks of "semantic representations" which are innate, and he assumes, that our concepts are decomposable into such representations. Putnam (1988) demonstrated the unavoidable problems of Fordor's approach.

REFERENCES

Dancy, J., & Sosa, E. (Eds.). (1992). *A companion to epistemology.* Oxford, England: Blackwell.

Dinter, F. R., & Seel, N. M. (1994). What does it mean to be a constructivist. In J. Lowyck & J. Elen (Eds.), *Modelling I.D. research. Proceedings of the first workshop of the Special Interest Group on Instructional Design of EARLI* (pp. 49–66).

Fodor, J. A. (1975). *The language of thought.* New York: Crowell.

Fodor, J. A. (1980). Fixation of belief and concept acquisition. In M. Piattelli-Palmarini (Ed.), *Language and learning: The debate between Jean Piaget and Noam Chomsky* (pp. 143–149). Cambridge, MA: Harvard University Press.

Fodor, J. A. (1983). *The modularity of mind.* Cambridge, MA: MIT Press.

Frege, G. (1975). *Funktion, Begriff, Bedeutung. Funf logische Studien.* (4. erg. Aufl.). Gottingen: Vandenhoeck & Ruprecht.

Knuth, R. A., & Cunningham, D. J. (1993). Tools for constructivism. In T. M. Duffy, J. Lowyck, & D. H. Jonassen (Eds.), *Designing environments for constructive learning* (pp. 163–188). Berlin, Germany: Springer.

Kripke, S. A. (1982). *Wittgenstein on rules and private language: An elementary exposition.* Oxford. England: Blackwell.

Lowyck, J., & Elen, J. (1993). Transitions in the theoretical foundation of *instructional design.* In T. Duffy, J. Lowyck, & D. Jonassen (Eds.), *Designing environments for constructive learning* (pp. 213–229). Berlin, Germany: Springer.

Maturana, H. (1990). The biological foundations of self consciousness and the physical domain of existence. In N. Luhmann (Ed.), *Beobachter: Konvergenz der Erkenntnistheorien?* (pp. 47–118). Munchen: Fink.

Pettit, P. (1992). Realism. In J. Dancy & E. Sosa (Eds.), *A companion to epistemology* (pp. 420-424). Oxford, England: Blackwell.

Phillips, D.C. (1995). The good, the bad, and the ugly: The many faces of constructivism. *Educational Researcher, 24,* 5–12.

Putnam, H. (1988). *Representation and reality.* Cambridge, MA: MIT Press.

Rescher, N. (1992). Idealism. In J. Dancy & E. Sosa (Eds.), *A companion to epistemology* (pp. 187–191). Oxford, England: Blackwell.

Ros, A. (1994a). "Wirklichkeit" und "Konstruktion." Der Status der Wirklichkeit in der Genese kognitiver Strukturen bei Jean Piaget. In G. Rusch & S. J. Schmidt (Eds.), *Piaget und der radikale Konstruktivismus* (pp. 139–175). Frankfurt/M.: Suhrkamp.

Ros, A. (1994b). "Konstruktion" und "Wirklichkeit." Bemerkungen zu den erkenntnistheoretischen Grundannahmen des Radikalen Konstruktivismus. In G. Rusch & S. J. Schmidt (Eds.), *Piaget und der radikale Konstruktivismus* (pp. 176–213). Frankfurt, Germany: Suhrkamp.

Salmon, W. C. (1992). Methodology. In J. Dancy & E. Sosa (Eds.), *A companion to epistemology* (p. 279). Oxford, England: Blackwell.

Stich, S. P., & Warfield, T. A. (Eds.). (1994). *Mental representation. A reader.* Cambridge, MA: Blackwell.

Westphal, K. R. (1992). Hegel. In J. Dancy & E. Sosa (Eds.), *A companion to epistemology* (pp. 167–170). Oxford, England: Blackwell.

Wittgenstein, L. (1953). *Philosophical investigations.* Oxford, England: Blackwell.

CHAPTER 6

CONSIDERING THE PHILOSOPHIES OF WITTGENSTEIN, DEWEY, AND RORTY AS POTENTIAL FOUNDATIONS FOR C-ID

Jerry Willis

Four social science theories—behaviorism (Skinner), cognitive constructivism (Piaget), social constructivism (Vygotsky), and critical theory—are commonly used as foundations for development and practice in the field of instructional technology. Different developers and educators, however, have taken diverse views on theory, from ignorant or informed ignoring to tenacious adherence to the tenets of a single theory. The chapter will suggest that theories of teaching and learning are actually expressions of more basic philosophical and epistemological positions. The chapter will develop a conceptual scaffolding for relating practice and learning theory to "bedrock" philosophy and epistemology. It suggests that Rorty's version of democratic

Previously published (n.d.) as paper presented at the First Annual Research Conference of the Association for Information Technology in Teacher Education, Cambridge, England, Cambridge University.

pragmatism, which is based on the work of Wittgenstein and Dewey, offers a sound foundation, and justification, for an integrative polytheorism that can guide research and practice in IT and the practice of instructional design (ID).

For much of the short history of computer-assisted learning, the great majority of both researchers and practitioners have been much too busy "doing" to be seriously concerned about theoretical foundations, much less the philosophical underpinnings of research and practice. The pressure to do, to act, to create, and to use, has been intensive since Patrick Suppes created the first drill and practice program for a large, time-shared computer system and helped launch the field of computer-assisted learning as well as modern instructional technology (IT). Lack of attention to foundations does not, however, imply there are no foundations, no givens, that guide research, development, and practice. The choice is actually not one of whether you have foundations or not; it is between having implicit, perhaps unacknowledged and vaguely understood foundations, and foundations that are explicit, understood, and thoughtfully applied to the common scholarly and professional tasks of IT and ID. One very good example of work in an area of educational technology that uses an unacknowledged foundation is Schaefermeyer's (1990) effort to outline a set of "general standards" everyone can use for evaluating educational software. For example, Schaefrmeyer argued that "behaviorally—based objectives ... should always be clearly stated in the documentation" (p. 9), and she mentioned only five types of software—drill and practice, tutorial, simulation, games, and problem solving. Although problem solving is listed, few of the standards proposed fit this type of software. In addition, anchored instruction software, microworlds such as Logo, tool software such as children's writing and publishing programs, collaborative writing programs, electronic encyclopedias, and hypermedia are only a few of the types of instructional software that does not "fit" her categories or her criteria. The language of the standards—branching, cues and/or prompts, feedback,—is distinctly behavioral. In her conclusion, Schaefermeyer argued that standards such as the ones she proposes are derived from "research-based principles" and that there is a "paramount need to develop at least 'minimum standards' for educational software" (p. 15). She suggested the problem could be approached by creating a checklist. Her approach, the standards proposed, and the solution are all expressions of a behavioral approach that would be vigorously opposed by IT experts who have adopted alternative theoretical foundations.

The twentieth century ended with four theories from the social sciences that were very active in the field of instructional technology. They were behaviorism, cognitive constructivism, social constructivism, and critical theory. All four were commonly used as foundations for instructional design (ID) and practice in IT. The current IT literature contains papers about the use of these theories. For example, the classic book, *Constructivism and the Technology of Instruction: A Conversation* (Duffy & Jonassen, 1992) contains chapters on constructivist approaches as well as critiques of the approach from authors more commonly associated with behavioral approaches. This book, which was adapted from a special issue of *Educational Technology* (September, 1991) is one of several that relate a social science theory to IT practice. More recently, books by Jonassen, Howland, Moore, and Marra (2002) and Wegeriff (2007) have focused on how constructivist theory and methods can guide the integration of technology into learning environments.

While behavioral theories (including cognitive theories based on an information processing model and cognitive science theories) have been discussed in the literature for over 40 years, both cognitive and social constructivism have only recently become "popular" perspectives in the IT literature. And, while critical theory has been a framework used to censure the ways technology is allocated and used in education for decades, the century ended with very little development work and even less instructional design practice based on this theory. (In the broader field of education, however, there is an active and engergetic critical pedagogy practice, often guided by the theoretical and practical work of Paulo Freire.) However, in IT critical theory has tended to emphasize the negative role of technology in either depowering and deskilling teachers, or the positive potential for empowering groups that are not a part of the mainstream power structure (Bigum & Green, 1992). Recently, however, a growing literature on how critical theory can be put to work as a framework for using educational technology has begun to appear (Evans, 2005; Feenberg, 2002).

Thus, as a new century begins, there is one "established" theory that may be on the wane—behaviorism—and three alternative theories that are enjoying increasing popularity—cognitive constructivism based on the work of Jean Piaget and Jerome Bruner, social constructivism based on the work of Lev Vygotsky and American supporters like Luis Moll, and critical theory with a sizable group of active American, European, and Australian theorists and practitioners. The influence of these three alternative theoretical foundations have already had a significant impact on education. One example of this is Greg Goodman's (2008) book, *Educational Psychology: An Application of Critical Constructivism*, which is based on what I am calling critical theory here.

As a field IT is currently in a state of flux with regard to theoretical foundations. In that context a comparison of these four theories, accompanied by a discussion of implications for IT, would be beneficial. There is, however, a growing body of literature on the four approaches. The 21st century literature is based on work that began in the 20th century. The book edited by Forman and Pufall (1988) has many examples of the use of Piagetian-style cognitive constructivism in IT. Moll's (1992) book has several chapters on uses of technology in education that are based on Vygotskian social constructivism. Critical theory perspectives on IT were presented in books edited by Muffoletto & Knupfer (1993) in the United States, Bigum and Green, (1992) in Australia, and in a special issue of *Educational Technology* edited by Andrew Yeaman (the February, 1994 issue). Also, much of the general IT literature in the 1960s to 1990s expresses a behavioral approach (broadly defined to include information processing and cognitive science theories that tend to "break down" content to be taught into smaller units

Au: Muffoletto 1992 publication year does not match year in refs.

and to support direct instruction strategies) to both research and practice. And, a number of books (Alessi & Trollip, 1991; Criswell, 1989; Fleming & Levie, 1993; Kemp, Morrison, & Ross, 1994) as well as special issues of journals (see for example, *Educational Technology*, October, 1993) were devoted to IT from a behavioral perspective.

That scholarship has grown and expanded in the first decade of the 21st century and there is a healthy and growing literature on the implications of social science theories for IT. There is far less, however, on the philosophical foundations that underlie those theories. Each of the four theories mentioned earlier is founded on some "bedrock" philosophical assumptions that must be accepted if the theory is to be meaningful. It is this second type of foundation that will be addressed in this chapter. I will explore some "bedrock" assumptions of each of the four social science theories most influential in IT today.

A FRAMEWORK FOR DISCUSSION

Even the way the topic is discussed has theoretical implications. It would not be difficult, for example, to list the four social science theories in one column and then list the philosophies of science normally associated with them in another column, as is done in Figure 6.1. Then, each of the bedrock assumptions of the supporting philosophies of science (PoS) could be summarized along with their implications for IT. In Figure 6.1, for example, the type of reality each PoS assumes is listed in the third column. Empiricism and postempiricism, for example, define reality as external to the individual. Reality is not only "out there," it can be understood by researchers who use proper methods and it can then be communicated to practitioners who base their practice on the knowledge of reality developed by researchers. Critical theorists also take this type of reality for granted, but, because other assumptions are different, their reality is frequently not the reality of the empiricists and their "proper methods" are quite different from the proper methods of empiricists.

The type of reality assumed by each PoS is but one of the many aspects of a PoS. Empiricists, who consider truth to be a match between what is said and what is "out there in the real world" are comfortable listing the assumptions of a particular approach (such as empiricism, postempiricism, interpretivism, or critical theory) and then empirically testing each assumption separately to determine if it is "true." Empiricists, and scholars who use related philosophies of science, believe humans can learn more and more about the "real world," which is a material or physical world, through the careful application of the scientific method. In contrast, interpretists and fellow travellers insist that humans have no direct

Social Science Theory	Philosophy of Science	Type of Reality
•Behaviorism	•Empiricism, Positivism or Postempiricism, Postpositivism	•"Out there" Realism, Materialism, Physicalism
•Cognitive Constructivism	•Interpretivism	•Constructed Reality, Consensus Reality, Antifoundationalist Reality, Nonrealism*
•Social Constructivism	•Critical theory or interpretivism	•Same as either cognitive constructivism or critical theory
•Critical Theory	•Critical theory	•Ideological (Marxist/NeoMarxist Realism, Socio-cultural Realism, Historical Determinism

Note: The term "nonrealism" is used by Smith (1993) to indicate interpretivists do not find arguments about an external reality particularly useful or important.

Figure 6.1.   Philosophies of science normally associated with each of the four social science theories and their definitions of reality.

contact with reality; their experiences and the experiments are always filtered through their beliefs, their prior experiences, and their expectations. Thus, what is "real" or true is, for interpretivists, what a particular group agrees upon.

Critical theorists have adopted another, and very different, view of reality. They have accepted modern versions of Marxist theories that have two foundational assumptions. One is that the core or critical issue in human relations is that of power. Traditional Marxism framed discussions of power around the tension between who owns the means of production (capitalists) and those whose labor is used to produce. The capitalist versus laborer conflict, and the domination of capitalists over labor, was the core of classical Marxism. The views of modern Marxists or critical theorists remain focused on the question of power, who dominates who, but the focus has expanded to include power relationships based on ethnic, class, gender, age, culture, and much more. And the focus on means of production, which might have been appropriate in an industrial age, has been expanded to include control over the media and other aspects of a culture. Critical theorists, for example, would be very critical and suspicious of the right wing politician Silvio Berlusconi's frequent and successful election campaigns for the position of Premier of Italy while controlling half the national television news producers and news magazines.

This chapter could be organized around the four groups of philosophies of science listed in Figure 6.1, with each bedrock assumption of a PoS dealt with in turn. The idea would be to reject or accept the bedrock assumptions and thus determine which of the PoS has the most support. Empiricists might find this acceptable because they use a correspondence theory of truth that allows them to make separate judgments on each aspect of "truth." A statement, assumption, or prediction can be tested by determining if it corresponds to external reality. On the other hand, postempiricists in the tradition of Lakatos (1974), would have difficulty with that approach since they believe it is much harder to get in touch with the "out there" reality than empiricists normally admit, and they thus are more willing to doubt the results of any particular study. For the postempiricist there are never enough studies to allow you to be completely sure your theory or assumption is True. And, a few studies with negative results don't justify abandoning a theory. The result may be due to factors such as poor instrumentation, improper interpretation, or problems with the subject selection rather than an inaccurate theory. The PoS developed by Lakatos proposes that social science theories be evaluated as a whole, rather than piece by piece. A paper organized according to this approach would be different from one based on the empiricist model.

If the seemingly simple issue of how a paper should be organized is a theory-laden process, then foundational issues certainly impinge on the important decisions ID researchers and practitioners make daily. A major goal of this chapter is to explore that linkage—the connection between foundational assumptions and action in three areas:

1. Applied IT research
2. Instructional design and development work
3. IT professional practice

The discussion will thus be anchored on one end by the actions of IT work. On the other end are the philosophies of science that underlie current social science theories. I will not explore the PoS in detail, but there are a number of excellent books for non-philosophers that introduce the current crop of philosophies of science (see for example Chalmers, 1982, 1990; Rosenberg, 2008; and Delnty & Strydom, 2003). These generally divide the alternatives into four or five general categories such as empiricist, postempiricist, critical theory, sociological (e.g., Lakatos, 1974; and Kuhn, 1970), and anarchistic or interpretive (e.g., Feyerabend, 1975). More specialized books written specifically for readers in the social sciences and education have generally covered the same ground, but with more attention to the implications for both research and practice in the areas of human and cultural endeavor. For example, a book by Phillips and Burbules (2000) is a well reasoned and thoughtful defense of the use of postpositivism as the preferred philosophy of science of education scholars.

In a 1989 book, Smith roughly divided the options into two categories: empiricism and interpretivism. After a bit more thought, in a book titled *After the Demise of Empiricism: The Problem of Judging Social and Educational Inquiry*, Smith (1993) discussed four alternatives: empiricism, postempiricism, critical theory, and interpretivism. In *The Paradigm Dialog* (Guba, 1990), the alternatives were conventional positivism (or empiricism, which is generally agreed to be outmoded), postpositivism (or postempiricism), critical theory, and constructivism. Those same four *paradigms* were used by Guba and Lincoln (1994) in the *Handbook of Qualitative Research* (Denzin & Lincoln, 1994), which extensively covers the large and growing family of qualitative research models and paradigms.

Unfortunately, very little of the ongoing developments reported in publications like the various editions of the *Handbook of Qualitative Research* have been carried over into discussions of IT research, development, and practice. IT, perhaps less than many other fields in the social sciences and education, seems more inclined to act than reflect and analyze. Several books could be written on the implications of competing philosophies of science for IT as well as on the relationship of emerging qualitative research paradigms to IT research and practice methods. However, in the space available for this chapter, an alternative approach seems more suited to the task. It involves identifying critical or "core" issues on which the competing PoS take distinct and different positions, and looking at the implications of those positions for IT.

THREE CORE ISSUES

Much of the debate about appropriate theoretical foundations in the social sciences and education emphasizes "grand theories" or paradigms such as critical theory, constructivism, or interpretivism on the one hand and the various forms of positivism or empiricism on the other. Each of these *isms* is really a prepackaged collection of positions on fundamental issues. Accepting one of the current isms as a foundation for research or practice is convenient—much like buying the lunch special at a restaurant. There is only one, rather than many, decisions to make. Though convenient, this approach leaves unexamined many of the fundamental issues that undergird research and practice in the social sciences and education. The remainder of this chapter revolves around three fundamental or "foundation" issues that have major implications for both research and practice: the nature of truth or reality, the role of language in human cognition, and the question of what constitutes "research."

The Role of Language in Human Cognition

For logical positivists, empiricists, and postempiricists, language is a means of describing the real world. It is thus a passive vessel for communicating that, when properly used, does nothing more than transfer information. For empiricists and postempiricists it is essential that language play a neutral role in the knowledge transmission process because the theory of knowledge they use requires several types of separation—separation of the knower from what we wish to know, separation of fact from value, separation of objective data from subjective data, separation of what you "want" reality to be from what the data shows it to be. Since language is the medium of knowledge transmission, if it cannot be made "neutral" then dichotomies such as those noted above collapse. If language itself helps to determine reality, or influences how we define it, then there is no possibility of a purely objective or rational approach.

Although the role of language in different PoS varies widely, many begin with the assumption that ordinary, everyday language is too imprecise, too subjective, to be useful in professional and scientific discourse. Empiricists and postempiricists generally invest a great deal of effort in creating specialized, objective language for both research and practice. Constructivists and interpretivists, on the other hand, often contend that language is a vehicle for constructing reality rather than the medium through which knowledge about reality is transmitted. Critical theorists, by contrast, often treat language as a weapon that is wielded by the power elite. Some of the work of critical theorists in IT a relates to the identification of biases, including gender bias, in educational software (DeVaney, 1993).

One theory of language that has had a significant impact on IT is the later Wittgenstein's concept of language games (Wittgenstein, 1953). The term "later" is used because early in his career Wittgenstein developed a picture theory of language that significantly influenced the development of logical positivism, an extreme form of empiricism that is still influential today (Monk, 1990). Monk described Wittgenstein's picture theory of language this way:

> Just as a drawing or a painting portrays pictorially, so, he came to think, a proposition portrays logically. That is to say, there is—and must be—a logical structure in common between a proposition ("The grass is green") and a

state of affairs (the grass being green), and it is this commonality of structure which enables language to represent reality. (p. 118)

This approach to language is a *vessel model* which means language is simply the container for meaning. However, as Wittgenstein continued his analysis of language, his theory changed drastically. He concluded that efforts to explain language as a logical reflection of "out there" reality was futile. He proposed instead, the *language game* concept. Games, in fact, are central to several of Wittgenstein's positions. He argued that a language is like a game. You learn the rules of a game by playing the game, not by studying a rulebook. Similarly, you learn the game of language by playing that game. Language acquires meaning through use, not study of a dictionary.

> For not only do we not think of the rules of usage - of definitions, etc. - while using language, but when we are asked to give such rules, in most cases we aren't able to do so. We are unable clearly to circumscribe the concepts we use; not because we don't know their real definition, but because there is no real "definition" to them. To suppose that there must be would be like supposing that whenever children play with a ball they play a game according to strict rules. (Wittgenstein, 1969, p. 25)

The core of the early Wittgenstein's view of language was a relationship between language and external reality. The core of the later Wittgenstein's view was the relationship between language and groups of humans.

> In a word: "To imagine a language means to imagine a form of life." (Philosophical Investigations, #19)

> It is clear from his examples that by "a form of life" Wittgenstein means the entirety of the practices of a linguistic community.... A form of life, like a language, may be compared to an old city, with its "maze of alleys and plazas, old and new houses, and houses with additions from various periods; all this surrounded by a number of new suburbs with straight, regular streets and uniform houses." (Philosophical Investigations, # 18)

> What seems strange or familiar to us depends on our form of life.... "The term 'language game' is meant to emphasize that the speaking of a language is part of an activity or a form of life" (PI, # 23).... Wittgenstein's point is not that our experiences and views about the world play no role. His point is that there is not just one group of factors that make linguistic communication possible. (Schulte, 1992, pp. 108–110)

Once he had at least partially disconnected language from an external reality, Wittgenstein went one step further and argued against his "picture theory" of language on another front. When striving to understand a concept, traditional logic calls for definitions that prescribe the essential elements or characteristics of something. An apple, for example, has much in common with a pear, but distinguishing between them could be approached by creating a list of characteristics of all apples that are not characteristics of pears. Wittgenstein found this approach less than satisfactory. "According to Wittgenstein, this striving to formulate the 'essence' of a thing—that is, common characteristics deemed to be necessary and sufficient for its existence—has led us astray again and again" (Schulte, 1992, p. iii).

Wittgenstein proposed instead the idea of *family resemblances*. Members of the same family are often recognizable as kin even though no single feature is the same across all family members. He illustrates his point in a famous comment about the meaning of the word "game."

> Consider, for example, the proceedings we call "games." I mean board games, card games, ball games, athletic games, etc. What is common to all of these?—Don't say: "They have to have something in common or they would not be called 'games'—but rather look and see if there is something common to all of them.—For if you look at them, you will not see something common to all; you will see similarities and relationships—a whole series of them. As I said: don't think, look!—Look, for example at the board games with their variety of related features. Now go on to the card games: here you will find many things that correspond to the first group, but many common traits disappear while others appear for the first time. If we move on to ball games, much that is in common is retained, but much is lost.... And we can proceed through the many, many other groups of games in the same way, seeing similarities surface and disappear. And the result of this investigation is like this: We see a complicated network of similarities intersecting and overlapping one another—similarities large and small. I can give no better characterization of these similarities than "family resemblances"; for it is in just this way that the various resemblances to be found among members of a family overlap and intersect: build, facial features, eye color, gait, temperament, etc., etc.—And I would say: "games" for a family." (Wittgenstein, 1953, # 66f)

The later Wittgenstein thus moves well away from the view of language as something to be improved to the point it is a transparent medium for communicating information about reality (*language as a calculus*)—an important goal for empiricists and postempericists, but something Wittgenstein thought less than useful. Language for Wittgenstein is a living, changing entity whose meaning develops through use.

Language Roles and Their Consequences in IT Practice

Discussions of the role of language in philosophies of science may seem far too abstract and ephemeral to have an influence on everyday practice, but they, in fact, have a phenomenal impact. For example, Wittgenstein's view of language is central to Spiro's cognitive flexibility hypertext model of instruction (Antonenko, Toy, & Niederhauser, 2004; Spiro, Feltovich, Jacobson, & Coulson, 1992), a popular framework for developing instructional multimedia for complex, ill-structured knowledge domains. Spiro based his concept of criss—crossed information landscapes on Wittgenstein's discussion of language, especially the concept of family resemblances.

Wittgenstein, while at Cambridge University in 1939, was also involved in an indirect way with the development of computers themselves. Alan Turing, one of the founders of modem electronic computers, regularly attended Wittgenstein's lectures on the foundations of mathematics and just as regularly argued with Wittgenstein about those foundations. The excerpt below is from an interchange about the role of language, particularly the question of whether we must have a precise, clear-cut, language that has no logical contradictions, as the empiricists require, or whether we must live with the fuzzy, inexact language we actually have:

Turing: You have a system of calculations, which you use to build bridges. You give this system to your clerks, and they build a bridge with it and the bridge falls down. You then find a contradiction in the system.—Or suppose two systems, one of which has been used satisfactorily for building bridges. The other system is used and the bridge falls down.

Wittgenstein: I am a general. I tell Rhees to be at Trumpington at 3:00 and at Grantchester at 3:30, and I tell Turing to join Rhees at Grantchester at 3:00. They compare orders and find "That is quite impossible." Given a certain training, if I give you a contradiction, you do not know what to do. In logic, we think of calculations and ways of thinking we do in fact have the technique of language we all know. Here contradictions do not normally occur—or at least occur in such restricted fields (e.g., the Liar) that we may say: If that is logic, it does not contain any contradictions worth talking about.

Turing:	If one takes Frege's symbolism and gives someone the technique of multiplying in it, then by using a Russell paradox, he could get a wrong multiplication.
Wittgenstein:	Suppose I convince Rhees of the paradox of the Liar, and he says, "I lie, therefore I do not lie, therefore I lie and I do not lie, therefore we have a contradiction, therefore 2 X 2 = 369." Well, we should not call this "multiplication," that is all.
Turing:	If there are contradictions, it will go wrong.
Wittgenstein:	Nothing has gone wrong that way yet. Why not? (Quoted in Leiber, 1991, p. 89).

Turing, who believed we would build a thinking machine by the end of the century that could not be distinguished from a human, was concerned about contradictions in human language and logic because such contradictions could bring a computer to a screeching halt. And, since Turing saw human and machine thinking as similar, he was concerned about the contradictions in human language as well as in computer languages. Wittgenstein, who was primarily concerned with human thought and language, argued that the problems predicted by Turing simply did not happen. Humans seem to be inoculated against such problems with the language and, when faced with contradictions "do not rush about every which way, generating general anarchy—no, they will simply ask for further orders. We must see, Wittgenstein stresses, that 'a contradiction is not a germ which shows general illness'. We are inoculated somehow in a way that computers sometimes are not" (Lieber, 1991, p. 90). For much of the 20th century philosophers were concerned with contradictions in the language that could not be directly resolved. The Liar's Paradox, for example, involves someone saying to you, "I am lying." Is that statement true or false? Those who follow Turing's path often spend a great deal of time trying to create "metalanguages" in which such contradictions and paradoxes are not possible. Wittgenstein added a third category to the list of True and False. He said things could also be *Nonsense* or *Meaningless*. The Liar's Paradox is meaningless and need not be pursued further. There are other, more important, things to explore.

Views on the role of language continue to influence IT research and practice in many ways. Language as a precise mirror of external reality, for example, is a foundation for many types of educational software as well as some approaches to expert systems. Wittgenstein's approach to language, on the other hand, is a foundation for constructivist uses of technology and supports the fuzzy logic approach to expert systems. Table 6.1 lists some of the implications of the two approaches to language in three areas of IT.

Table 6.1. Potential Roles of Language and Their Influence on IT Practice Role Influence

---

Empiricist/Objectivist Language Perspectives

Language is a Neutral Vessel for Carrying Information About the Real World

---

Applied Research

Calls for a specialized language that is different from both ordinary language and the language of practice.

Often creates "hypothetical constructs" such as "simulations" or "direct instruction" and conducts research as if these constructs are "reality."

Goal is discovery of "universals" that apply across contexts.

Proposes that all problems are clear-cut problems, when properly framed.

Proposes that solutions to problems are clear-cut, when properly framed.

Instructional Development

Precise behavioral objectives are essential.

Experts, who have special knowledge, are critical to ID work.

Emphasis on delivery of "facts" selected by experts, favors drill and practice, tutorial, and other direct instruction methods. Computer takes the roles of a traditional teacher—information deliverer, evaluator, record keeper.

Professional Practice

Use objective tests to determine effect.

Problems, when properly defined, have straightforward solutions. If the situation seems fuzzy it is because we have not thought it through properly.

Wittgenstein's "Language Games" View of Language

Meaning is Defined by "use" and is Constructed Socially by Groups of Humans

Applied Research

Researchers should become involved in the context of practice since "playing the game" is the only way to understand the meaning of the language of that context.

Types of pedagogy such as direct instruction or student-centered learning cannot be precisely defined and categorized. Categories will share family resemblances but not necessarily a "core" of common characteristics. Also, there may be considerable overlap between software with different labels—e.g., tutorial and simulation, hypermedia and drill and practice, multimedia and information resources.

Meaning comes from practice, and is often limited to the context of practice. Thus, even common instructional terms like "simulation" or "drill and practice" may have different meanings across different application contexts, different groups of educators, and different educational systems.

Researchers should be "insiders" rather than "outsiders;" otherwise they will not understand.

Findings are local and may or may not have meaning across contexts.

Table continues on next page.

Table 6.1.   Continued

Wittgenstein's "Language Games" View of Language

Meaning is Defined by "use" and is Constructed Socially by Groups of Humans

Instructional Development

General ID specialists, who can work with subject matter experts from any discipline are a myth. You must understand the "game" being played before you can help develop instruction.

For IT development, the "citizen legislator" model, as opposed to the "professional politician" model is preferred. That is, those who will actually use the instructional materials being developed should be heavily involved in the ID process. This approach is preferred to a model that puts primary responsibility for ID in the hands of specialists who take the role of general experts in ID and who, therefore, can "design anything."

Development should be collaborative, the design group must work together to create a shared vision. That vision may emerge over the process of instructional development. It cannot be "established" at the beginning through a decision of the designer. Shared visions emerge from communication within the design team.

The instructional emphasis is on developing meaning in context. That favors strategies such as anchored instruction, collaborative learning, problem-based learning, situated cognition, cognitive apprenticeships, and cognitive flexibility hypertext.

The emphasis is also on access to knowledge in context: favors development of hypermedia and multimedia information resources, electronic encyclopedias, and a wide range of accessible, navigable electronic information resources.

Professional Practice

Meaning and understanding develops in context. Objective tests, especially those created outside the context of practice, have little meaning. Portfolio assessment and similar approaches are preferred.

Professional preparation and development are best acquired in context. For example, university - based teacher education programs are less powerful than university/school collaboratives such as professional development schools that develop a shared language and culture.

"In context" instructional approaches are preferred: authentic instruction, language experience approaches, situated cognition, anchored instruction, cognitive flexibility hypertext.

Many professional practice problems are fuzzy and ill-structured. They often have complex solutions or no solutions at all. Solutions are, however, learned by playing the game, not just by reading the research literature.

The Nature of "Truth:" Alternative Conceptions of Reality

What do we mean by the word "truth?" There are several approaches to the concept of truth and how it is discovered. Foundationalist approaches such as positivism, objectivism, and empiricism use an "out there" approach that assumes what we believe to be true can be tested by comparing it to the real world. An external world, separate from, and independent of, human consciousness, thus serves as a foundation for any claims to truth—from the claim that students learn to write better when they use collaborative writing systems such as Wikis to the assertion that Internet-based instruction is effective and should be used more often in schools. Asserting that there is a real world out there is not the major point of disagreement, however, between positivists and those who use other theoretical foundations. Positivism proposes not only that there is an external reality independent of the human mind, it also assumes that through the thoughtful use of proper methods—such as the scientific method—humans can come to know what that external reality is. Experimentation and systematic observation of the "real world" are thus the source of knowledge about reality in the foundationalist approaches such as rationalism and empiricism. Logical positivism, (also known as logical empiricism or the Vienna Circle because the founders met regularly during the 1920s in Vienna), is perhaps the best known and most extreme branch of empiricism. Logical positivism was the philosophical foundation for radical behaviorism, and while the influence of behaviorism has decreased over the past 40 years, logical positivism is still very influential. Each year the structure of hypotheses, statements of problems, definitions of terms, experimental design, independent variables, and dependent variables in the dissertations of hundreds of social science and education graduate students are shaped and controlled by the tenets of logical positivism. That is in spite of the fact that the flaws and contradictions of this approach have been acknowledged for at least 50 years. Less extreme forms, often called postempiricism or postpositivism, were proposed by several people including Popper (1972). Postempiricism is a less confident form of empiricism since the approach to both research and practice is similar but postempiricists acknowledge that we cannot be as confident of having found the Truth as empiricists believed.

Although the foundationalist approaches of empiricism, postempiricism, and rationalism are still dominant, a number of other alternatives are also influential in IT today. For example, aspects of cognitive constructivism derived from Piaget's theory of cognitive development reflects an idealism that harks back to Plato. Plato's idealism is the view "that true, absolute reality is the realm of the perfect, independently existing, unchanging, timeless forms (ideas), and the true object of all knowledge" (Angeles, 1992). Plato's concept of innate truths or ideal forms that are "revealed" through analysis is illustrated by the story of the slave boy in the dialog *Meno,* who, when asked a series of questions, came to the Pythagorean theorem—thus demonstrating that knowledge of geometry is innate and in all of us. Critics might argue that knowledge of geometry was in the questions asked rather than in the boy, but this concept of innate ideas is an important part of Piaget's theory of cognitive development. Piaget's "mother structures" have much in common with Plato's ideal forms because they are innate. They are also similar to Jung's *archetypes*. Idealism, like objectivism, is foundationalist. That is, both posit an extemal reality. They are not the same realities however. For objectivists, reality is the physical world around us that we can sense. Thus "sense data" such as the results of research studies are a primary way of coming to understand reality. Aristotle is one of the earliest philosophers who proposed this approach. For idealists in the tradition of Plato, reality is less concrete. It is the "ideal form" that is imperfectly reflected in the examples of that form that we see around us. A horse we can see and touch is simply an imperfect representation of a "real" horse—the ideal form that exists only as an idea. Idealists tend to emphasize careful thought and analysis over experimentation as a means of coming closer to reality. In educational computing the best known application of idealism is probably some aspects of the programming language Logo and the theories of instruction that emerged from the development of Logo as a learning tool. Idealism and empiricism-rationalism are often presented as opposites. However, if we substitute "tutorial software" for "horse" in the example above, you can see how an empiricist's hypothetical construct "tutorial software" has much in common with an idealist's ideal form of tutorial software.

Rationalism underlies behavioral theory, and idealism is implied in some of the cognitive constructivist work based on Piaget's theories. Both have foundational frameworks, something to believe in and rely on. A third, and quite different, answer to the question "what is reality" is the foundation for a social constructivist approach to IT that is based on the theories of Vygotsky, the philosophy of the later Wittgenstein, and the democratic pragmatism of Richard Rorty. In this section the views of Rorty (1982, 1991, 2000) will be emphasized because they reflect the current perspective on reality from an interpretivist viewpoint. Central to much of Rorty's philosophy is his definition of *objective* and *subjective*. In empiricist and postempiricist philosophies of science, "objective" refers to statements that can be empirically demonstrated to correspond to an "out there" reality while subjective statements are influenced by background, experience, opinion, values, morals, feelings, and other non-empirical influences. Rorty's definition of objective and subjective is quite different. Rorty (1979) considers objective to be what a group agrees on and subjective to be what a group considers "strange" or on which there is not agreement within the group. Thus truth or reality is found by a group rationally discussing an issue until the group reaches consensus (Smith, 1993).

Much of IT work conducted on an empiricist or postempiricist foundation is based on the core assumption that objective can be distinguished

from subjective and objective is always better. Many instructional development models, for example, call for the creation of specific, detailed
instructional objectives before developing instruction. Dick and Carey
(1985), for example, consider behavioral objectives very important:

> Perhaps the best-known component of the instructional design model is the
> writing of performance objectives, or, as they are more commonly called,
> behavioral objectives. (p. 97)

> The objectives guide the designer in selecting content and developing the
> instructional strategy. (p. 98)

Objectives also lead to increased use of objective tests.

> In recent years, classroom testing has taken a very different turn. Much of
> this change can be attributed to the impact of behavioral objectives. As
> more and more emphasis has been placed on statements of explicit behav-
> iors that students must demonstrate, it has been increasingly obvious that a
> fair and equitable evaluation system is one that measures those specific
> behaviors. (Dick & Carey, 1985, p. 106).

Objectives (and objective assessment) which are supposed to support
and enhance instruction, often seems to overwhelm it. In the 1985 edition
of their book, *The Systematic Design of Instruction*, Dick and Carey devoted
nine chapters to goals, subgoals, objectives, and assessment (entry
behaviors, exit behaviors, grading, formative evaluation, and summative
evaluation) and only three to the design of instructional materials. Some
(including this author) would argue this is the tail wagging the dog.

There is, of course, a large and growing literature that is opposed to
this type of objective instruction for many reasons. One objection is that a
heavy emphasis on creating and teaching to detailed behavioral objectives
that can be easily measured encourages teachers to concentrate on simple, discrete "facts" rather than higher order thinking skills. However, the
important point here is that these objective approaches are based on a
value-based dichotomy—objective versus subjective—that has been collapsed by Rorty and other interpretivists. Even if we do not take up the
point that the objectivist's belief that "objective" as better than "subjective" is a value judgment and therefore subjective, the interpretivist viewpoint is very troubling to empiricists. Interpretivists deny the possibility
that humans can take a "Gods-eye view" and make objective decisions.

Objective, in the empiricist sense, simply does not exist. Even if it is
desirable and preferable, it cannot be attained by humans. If the interpretivist perspective is accepted, many commonly accepted aspects of IT
practice must also be questioned: heavy reliance on behavioral objectives,

hours of instructional time spent on the objective assessment of learning objectives, and the traditional, linear sequence of activities in an instructional design model that begins with learner analysis, and task and concept analysis, followed by design, development, and summative evaluation. All these activities are based on an assumption that objective can be distinguished from subjective and that objective is better. Many of the "givens" of IT are not so easily accepted as givens if this objective-subjective dichotomy is collapsed. Interpretivists are antifoundationalists; they believe "there is no particular right or correct path to knowledge, no special method that automatically leads to intellectual progress" (Smith, 1993, p. 120), a position that is particularly troubling to empiricists and postempiricists who believe no progress can be made unless "standards" are upheld. Interpretivists don't have a problem with standards—in research, in the process of designing instructional materials, or in professional practice. It is just that they do not believe those standards are in any way universal or special. They are, instead, the products of a particular group or culture. Interpretivists don't abandon standards, they simply accept that whatever standards are used are subjective, and therefore fallible, rather than objective and universal.

The collapse of the empiricist's objective-subjective dichotomy has many implications for IT practice. Many fear the loss of an objective foundation for research and practice puts us straight on the road to hell - in this case, a wimpy relativism that treats every opinion and theory equally since there is no way to make objective decisions. Rorty's brand of interpretivism goes beyond the simple assertion that everything is subjective. Since we cannot escape our own backgrounds and experiences, he believes we must practice, conduct research, and develop theories within that framework—it is our only option. We cannot work in a vacuum and we cannot shed our background. We need not remain totally within our own group, however. If we are behaviorists who develop tutorial and drill and practice software, we need not remain completely ignorant of the work of other groups such as constructivists who create software to support anchored instruction. As Rorty (1991) put it:

> I urge that whatever good the ideas of "objectivity" and "transcendence" have done for our culture can be attained equally well by the idea of a community which strives after both intersubjective agreement and novelty—a democratic, progressive, pluralist community of the sort which Dewey dreamt. If one reinterprets objectivity as intersubjectivity, or as solidarity ... then one will drop the question of how to get in touch with "mind-independent and language-independent reality." One will replace it with questions like "What are the limits of our community? Are our encounters sufficiently free and open? Has what we have recently gained in solidarity cost us our ability to listen to outsiders who are suffering? To outsiders who have new

ideas?" These are political questions rather than metaphysical or epistemological questions. Dewey seems to me to have given us the right lead when he viewed pragmatism not as grounding, but as clearing the ground for, democratic politics. (p. 13)

Rorty's democratic pragmatism thus accepts the subjective nature of both research and professional practice, but it argues that we can reduce the danger of what he calls "ethnocentrism" by encouraging free and open discussion within our own group and by making a special effort to seek out and understand the truths of other groups.

One consequence of antirepresentationalism is the recognition that no description of how things are from a God's-eye point of view, no skyhook provided by some contemporary or yet-to-be-developed science, is going to free us from the contingency of having been acculturated as we were. Our acculturation is what makes certain options live, or momentous, or forced, while leaving others dead, or trivial, or optional. We can only hope to transcend our acculturation if our culture contains (or, thanks to disruptions from outside or internal revolt, comes to contain) splits which supply toeholds for new initiatives. Without such splits—without tensions which make people listen to unfamiliar ideas in the hope of finding means of overcoming those tensions—there is no such hope. The systematic elimination of such tensions, or the awareness of them, is what is so frightening about Brave New World and 1984. So our best chance for transcending our acculturation is to be brought up in a culture which prides itself on not being monolithic - on its tolerance for a plurality of subcultures and its willingness to listen to neighboring cultures. This is the connection which Dewey saw between antirepresentationalism and democracy. We should not look for skyhooks, but only for toeholds. (Rorty, 1991, pp. 13–14).

Table 6.2 lists some of the implications of adopting an objectivist versus interpretivist/subjectivist definition of reality.

Paths to Truth: What Constitutes "Research?"

The question of what constitutes research has already been discussed. It is inextricably linked to our views of the role of language and our definition of reality. The traditional empirical approach, which is based on a transparent, neutral language and an objectivist reality, seeks to support theory and to discover law like generalizations that can be used to guide practice. This approach has a long history in both computer science and instructional technology. For example, the intellectual ancestry of Herbert Simon, who won a Nobel prize for his work, can be traced back to the

Table 6.1.   Types of Reality and Their Influence on IT Practice

Empiricist/Objectivist

*Reality is External To Human Cognition and Can be Understood
By the Proper Application of Scientific Method*

Applied Research

There is one source of Truth: good, empirical research. Practice can provide hunches, and suggestions for research, but we must rely on research to tell us what the truth is.

Good research must meet objective standards in terms of design and analysis. There are universal rules, developed primarily in the natural sciences, for those standards.

Technical standards come first, then the importance of research is considered.

Disagreements are settled by more research.

Good research calls for predictions to be made before conducting the research. Ad hoc conclusions are viewed with suspicion.

Quantitative methods and experimentation are preferred.
The greatest sin is to be subjective.

The goal of research is to discover lawlike generalizations about external reality.

Instructional Development

Begin with a precise plan of action including clear behavioral objectives.

Considerable effort should be invested in creating instructional objectives and objective assessment instruments.

Emphasize the delivery of "facts" selected by experts: favors direct instruction methods.

The design process is sequential, objective, and focused on experts.

Invest the most assessment effort in the summative evaluation because it will prove whether the material is effective or not.

Professional Practice

Objective tests are used to determine success.

The appropriate approach to practice is technical-rational: look to the research for universal solutions that can be applied to your particular problem.

Rorty's "Nonfoundational Interpretivist Approach to Reality"

*Objective Truth is What Your Group Can Agree on, Subjectivity is When You Don't Agree.
Reduce Ethnocentrism by Seeking out Alternative Truths Created by Other Groups.*

Applied Research

There are many sources of truth—one of which is empirical research. It has no special place, however. It is just one more voice in the conversation.

Professional practice knowledge, for example, has as much right to be considered as empirical research.

Disagreements are settled by free and open exchange that allows the parties to reach the point of sharing similar values and interests.

Truths can be discovered by "wandering around."

There are no universal standards for good research. Each community of scholars must agree among themselves what they will listen to and what they will ignore. This is the "Consensus theory of truth."

In determining what is good and bad research, importance and relevance come first, with technical standards playing a less important role.

No method is necessarily preferred over another, but because so many important topics are amenable only to qualitative approaches, they will be used more often.

There are two great sins: narrowness in method and approach, and ethnocentrism. The cure is to be open to alternative perspectives including those from outside your group, and to look beyond your own type of research at least as a consumer.

The goal of research is to understand in context.

Instructional Development

Development should be collaborative, the group must work together to create a shared vision. That may emerge over the process of development. It cannot be "established" at the beginning. Participatory approaches to ID are essential.

The emphasis is on developing meaning in context: favors strategies such as anchored instruction, cognitive flexibility hypertext and student-centered/constructivist instructional methods.

Begin with a vague plan of action and fill in the details as you progress.

Development is recursive, not linear. Some problems, improvements, or changes will only be discovered in the context of use. Plan for recursive evaluations by users and by experts.

Invest the most assessment effort in the formative evaluations because they are the ones that provide feedback you can use to improve the product. Summative evaluation does nothing to help you improve the product.

Professional Practice

Measures of success should be multifaceted, including the opinions and evaluations of consumers and professionals. Objective tests are one of many ways to evaluate success.

The appropriate approach to practice is reflective: reflect on actions and events, revise practice based on that reflection; share and invite sharing with others.

logical positivism of the Vienna Circle. When Nazi terror threatened many members of the Vienna Circle, one of the leaders, Rudolf Carnap, came to the University of Chicago where Simon was a student. Simon mentions in his autobiography that his view of thinking and problem solving was "influenced by Rudolf Carnap's lectures at the University of Chicago and his book, and by my study of Whitehead and Russell's *Principia Mathematica*" (Simon, 1991, p. 193). Simon mentions other influences such as Allan Turing, Alfred Tarski, and Claude Shannon, who used the binary logic of Boolean algebra, which has only two values—1 and 0—to design electrical switching circuits. Shannon's work is also one of the foundations of modem digital computers. Virtually all of the influences Simon mentions are representatives of the objectivist approach. Many, for example, were concerned with creating a special, objective language that would avoid the subjective, fuzzy nature of ordinary language. Many, including Turing, also felt the research on computers would help us understand human thinking. Simon, as Turing did before him, continues to believe that computers will be able to think in the ways that humans do. Systems theory, artificial intelligence, means-ends analysis, the general problem-solving model, the use of computers to study human thinking— these are all powerful influences on computer science today as well as on instructional technology. They were all created or heavily influenced by the Newell-Simon collaboration that is grounded in the empiricist— objectivist tradition.

With effort, much of the field of instructional technology—from educational AI to intelligent tutoring systems, from meta-analyses of studies to determine whether computers improve school performance to research on feedback mechanisms in tutorial software—can be traced back to the logical positivism of the Vienna Circle. That intellectual tradition includes the work of behavioral psychologists and empiricist philosophers. Modern expressions of this tradition often acknowledge the flaws and fallacies of earlier work, but the world view and the goals remain essentially unchanged.

That is one tradition. Another tradition is interpretivism. Although this approach has a long history in philosophy and psychology, it has not been as influential in the areas of computer science or instructional technology. Behaviorism and logical positivism after all, dominated the American social sciences and education when electronic, digital computers were invented. There is good reason to think that interpretivists like Rorty, a philosopher, and computer scientists like Terry Winograd and Fernando Flores (1987) may have pointed the way to a future in IT that is quite different from the one created by objectivism and behaviorism. [It is an-interesting side note that in his autobiography Simion publishes a letter to his daughter explaining the heated debate between himself and Joseph Weizenbaum on the possibilities of AI and the role of computers in simulating human thought while the cover of Winograd and Flores' book, has a very positive note of support from Weizenbaum.]

Changing the foundations of our field from objectivism/empiricism (and the behavioral family of theories based on that philosophical base) to interpretivism (and the constructivist theories based on interpertivist philosophies) would have a number of significant implications. Some of them are enumerated in the next section.

The Implications of an Interpretivist Approach to IT Include

Computer Science is *Naturwissenschaften* while studies of human behavior and thinking are *Geistweswissenschaften*. While much of the American thrust in both the philosophy of science and the development of the social sciences has placed the social sciences in a position of emulating big brother natural science, there are traditions in German philosophy that distinguish between natural sciences (Naturwissenschaften) and the "cultural sciences" (Geistweissenschaften). Research in natural and cultural or social sciences may have different purposes, different methods, and different frames of reference. Essentially, the natural sciences seek to discover universal laws about the physical world while the social sciences seek to "understand" human behavior in context.

The search for universal truths ends and efforts to find "local" truth accelerates. To illustrate, studies that attempt to demonstrate that use of tutorial software is better than traditional teacher-centered instruction would end while efforts to create effective instructional packages (which may include tutorials as well as teacher presentations and many other approaches) for a specific context would increase.

Messy, in-context research becomes much more valuable and well-controlled research conducted "out of context" is devalued. Much of the model for programmed instruction was generalized from studies of pigeons. Even then, the pigeons were not even living the normal life of a pigeon. Instead, they were kept in small metal cages, isolated from one another, and starved to keep their motivation high. A shift to an interpretivist framework would encourage more work in the learning environment (of humans, other than college sophomores who are already studied more than pigeons). The shorter the step from research setting to the application setting the better. The higher technical quality of out of context research would not outweigh the leap required to generalize to classrooms. In this new milieu, interest in traditional quantitative research methods would wane because they are not suited to in-context research while a range of qualitative methods would become much more popular because they are well suited to scholarship in the real world.

Sources of truth expand well beyond the traditional research study. The dominant view has been that any valid effort to assess the success of instructional technology in schools must be based on traditional control group-experimental group research with the dependent variable almost always some form of objective test. A shift to an interpretivist framework would expand the sources of information drastically to include teacher reflections, journals, student debriefing, interviews, case studies, and much, much more. The current edition of *The Handbook of Qualitative Research* (Denzin & Lincoln, 2005) introduces many different methodologies and data gathering strategies. My own book, *Qualitative Research Methods in Education and Educational Technology* (Willis, 2009) covers a wide range of scholarly methods including many from the arts and humanities, and from philosophy.

Awareness and understanding of alternative traditions would become a virtue. In an age of information overload it is very easy to narrow our search for new knowledge down to those sources we have come to agree with. An interpretivist perspective calls for more attention to work in traditions different from the one we find most comfortable.

Some goals of research would become "nonsense." Wittgenstein's expansion of logical choices from True or False to True, False, or Nonsense allows us to ignore and move beyond some questions because they are nonsense. Consider this question: "Is computer-based learning more effective than traditional learning?" That question, which has been "answered" in the literature hundreds of times, is *nonsense* and we need not spend any more time on it because the choice is never between something called computer assisted or computer-based learning and "traditional learning," whatever that is. There is actually no such thing as CAI, at least not in the sense that there is a prototypical example of CAI that can be used in research to represent "all" CAI. However, that is what is needed if we are to have conclusions that can be generalized to all other forms of CAI. There are many differences between different instances of CAI even when they are based on the same theory such as behaviorism. There are many, many more differences when considering CAI based on different theories of learning.

Research becomes less an effort to support theory and more of an effort to create successful instruction. The theory-driven debates of American psychology over the past 100 years generated thousands of studies designed to support one theory over another. A shift to research, and professional practice, from an interpretivist perspective undermines the reason for theory-testing research. The reason is the search for universal laws of human behavior. If we give up that search and opt for a goal that focuses on understanding in context, new forms of research will continue to emerge that are more collaborative and focused on building a shared reality among participants as well as on the design, development, and use of effective instruction for a particular context (see Willis, 1993, 2007, 2009, for a discussion of this issue.)

Instead of requiring that all research begin with hypotheses and predictions, approaches that leave open the opportunity to discover things as the research progresses would be accepted. The main reason for requiring detailed research hypotheses before starting research is to avoid the dreaded subjectivism that could creep in if predictions are not made before data is gathered. This problem is eliminated in the intepretivist approach since all research is subjective. A more flexible approach to research would allow us to pursue new directions in the midst of a research project when the data suggests our original interests are not the most relevant. This also applies to instructional design as well. Traditional models of ID begin with the creation of instructional objectives and then proceed to the tasks of designing and developing materials to teach those objectives. But, what if the design team discovers halfway through the ID project that the objectives you began with are not the most relevant or most important? Most ID models discourage even thinking about such a possibility, but an ID model based on interpretive foundation would.

DISCUSSION

Much of this chapter has explored the implications of alternative positions on three core questions: the role of language, the definition of reality, and the question of what constitutes research. The major theoretical foundations used today in IT and ID, including behaviorism/empiricism and constructivism/interpretivism, take strong positions on all three of the questions. As you consider what sort of relationship you want to build with theoretical foundations, you may want to consider the most common options.

Relationships Based on Technical Ignorance

This is probably the most common approach to core issues and theoretical foundations. Many people have very successful careers in instructional technology and related fields without ever worrying about the definition of reality. They find questions like "When a tree falls in the forest does it make a noise if no one is there?" irrelevant and irritating (or Nonsense in Wittgenstein's terminology). There is much to be said for this position, but ignorance does not avoid the problem of making choices among the positions. It simply transfers the responsibility for making the choice to others—the people who taught you instructional design principles, for example, or the designers of the educational technologies that you use. Ignorance of the issues involved makes us technicians who use, fix, and repair but who find it more difficult to creatively design and develop.

Vocabulary Conversion

Another approach is to adopt the vocabulary of alternative approaches that have become popular but to maintain your current practices. To some extent, Rorty's use of objective and subjective in ways very different from the intent of empiricists illustrates this approach. However, the best example of this was related to me by Dr. Larry Hovey at Texas Tech University. A few years ago when individualized learning was a popular concept, he visited the classroom of a teacher who had mentioned to him how well work was progressing on creating individualized learning plans for each student in his classroom. When Larry visited the room, he noted that all the students in the classroom were completing the same assignment, and taking the same test upon completion. When he looked at the file cabinet containing the individualized lesson plans for each of the 27 students, he discovered 27 identical lesson plans. The teacher was using the new vocabulary of individualized learning but was still teaching as he had for years.

Promiscuity

Another option is to use different models and approaches in different situations. This approach, which would be very problematic for those in the empiricist tradition, is not as big a problem for interpretivists. That is because positions are less concrete, more flexible, and more open to interpretation and modification in the interpretivist tradition. Rand Spiro is somewhat promiscuous in his approach to selecting instructional strategies (Spiro, Feltovich, Jacobson, & Coulson, 1992). He recommends using direct instruction methods when teaching simple, well-structured content and cognitive flexibility hypertext when teaching complex, ill-structured content. Spiro reaches across the two traditions of direct-instruction (behavioral) and constructivist pedagogy, and makes a strong case for using constructivist approaches, based on interpretivist philosophies of science, for some types of content, and behavioral approaches, based on objectivist-empirical philosophies of science, for others. Perhaps, over the next decade a synthesis of these two traditions, at the philosophical and social science theory levels, will occur and Spiro will not be promiscuous anymore.

Dedicated Priesthood

Most practitioners probably fall into one of the three categories discussed already. However, many researchers are members of a dedicated priesthood who practice within a single theoretical and philosophical framework throughout their career. These scholars often advance their field but one wonders whether the narrowness of focus limits the contributions they make.

Integrative Polytheorism

For me the most promising position to take is integrative polytheorism. It is based on Rorty's democratic pragmatism and thus accepts that all of us are influenced, and to some extent limited, by our own experiences and expectations. We will practice within the framework of our culture, the theoretical frameworks we were taught in graduate school, and the models we currently understand. However, we are polytheorists because we remain open to, and interested in, alternative theories. [This is not always profitable. As a young graduate student brought up as a radical behavior analyst in the 1970s I was accustomed to the warm, supportive atmosphere of the annual meetings of applied behavior analysts. I decided to attend a meeting of humanist psychologists specializing in human relations training to get a feel for an alternative paradigm. The meeting was so acrimonious, the comments and interaction so hateful and combative, that I did not explore that area of psychology again for over a decade.] We are integrative because we accept that our approaches are open to revision (and revolution). As integrative polytheorists we may integrate approaches from other traditions into our own, or, when it seems appropriate, adopt an entirely different tradition and bring to it aspects of our former tradition that seem to add strength. Looking back over the last 20 years I can see that I began as a radical behaviorist in the tradition of applied behavior analysis and gradually transitioned to my current position. I rely heavily on interpretivist and constructivist theory but I remain open to work from other theoretical frameworks including critical theory and behavioral theories.

REFERENCES

Alessi, S., & Trollip, S. (1991) *Computer-based instruction: Methods and development.* Englewood Cliffs, NJ: Prentice Hall.

Angeles, P. (1992). *Philosophy: The Harper Collins dictionary.* New York: Harper Collins.

Antonenko, P., Toy, S., & Niederhauser, D. (2004). *Modular object-oriented dynamic learning environment: What open source has to offer.* Retrieved from http://www.contempinstruct.com/books/opensourcemoodle.pdf

Bigum, C., & Green, B. (Eds.). (1992). *Understanding the new information technologies in education.* Geelong, Australia: Center for Studies in Information Technologies and Education, Deakin University.

Chalmers, A. (1982). *What is this thing called science?* (2nd ed.). St. Lucia, Queensland, Australia: University of Queensland Press.

Chalmers, A. (1990). *Science and its fabrication.* Minneapolis: University of Minnesota Press.

Criswell, E. (1989). *The design of computer-based instruction.* New York: Macmillan.

Delnty, G., & Strydom, P. (Eds.). (2003). *Philosophies of social science.* Philadelphia: Open University Press.

Denzin, N., & Lincoln, Y. (Eds.). (1994). *Handbook of qualitative research.* Thousand Oaks, CA: SAGE.

Denzin, N., & Lincoln, Y. (Eds.). (2005). *Handbook of qualitative research* (3rd ed.). Thousand Oaks, CA: SAGE.

DeVaney, A. (1993). Reading educational computer programs. In R. Muffoletto & N. Knupfer, (Eds.). *Computers in education: Social, political & historical perspectives.* Cresskill, NJ: Hampton Press.

Dick, W., & Carey, L. (1985). *The systematic design of instruction.* Glenview, IL: Scott, Foresman.

Duffy, T., & Jonassen, D. (Eds.). (1992). *Constructivism and the technology of instruction.· A conversation.* Hillsdale, NJ: Erlbaum.

Evans, J. (Ed.). (2005). *Literacy moves on: Popular culture, new technologies, and critical literacy in the elementary classroom.* Portsmouth, NH: Heineman.

Feenberg, A. (2002). *Transforming technology: A critical theory revisited.* Oxford, England: Oxford University Press.

Feyerabend, P. (1975). *Against method: Outline of an anarchistic theory of knowledge.* London: New Left Books.

Fleming, M., & Levie, W. (Eds,). (1993). *Instructional message design: Principles from the behavioral and cognitive sciences.* Englewood Cliffs, NJ: Educational Technology Publications.

Forman, G., & Pufall, P. (Eds.). (1988). *Constructivism in the computer age.* Hillsdale, NJ: Erlbaum.

Goodman, G. (Ed.). (2008). *Educational psychology: An application of critical constructivism.* New York: Peter Lang.

Guba, E. (Ed.). (1990). *The paradigm dialog.* Newbury Park, CA: SAGE.

Guba, E., & Lincoln, Y. (1994). Competing paradigms in qualitative research. In N. Denzin & Y. Lincoln (Eds.), *Handbook of qualitative research* (pp. 105–117). Thousand Oaks, CA: SAGE.

Jonassen, D., Howland, J., Moore, J., & Marra, R. (2002). *Learning to solve problems with technology: A constructivist perspective.* Englewood Cliffs, NJ: Prentice-Hall.

Kemp, J., Morrison, G., & Ross, S. (1994). *Designing effective instruction.* New York: Merrill, Macmillan College.

Kuhn, T. (1970). *The structure of scientific revolutions.* Chicago: University of Chicago Press.

Lakatos, I. (1974). Falsification and the methodology of scientific research programs. In I. Lakatos & A. Musgrave (Eds.), *Criticism and the growth of knowledge,* (pp. 91–196). Cambridge, England: Cambridge University Press.

Lieber, J. (1991). *An invitation to cognitive science.* Oxford, England: Basil Blackwell.

Moll, L. (Ed.). (1992). *Vygotsky and education: Instructional implications and applications of sociohistorical psychology.* Cambridge, England: Cambridge University Press. '

Phillips, D., & Burbules, N. (2000). *Postpositivism and educational research.* Lanham, MD: Rowman & Littlefield.

Popper, K. (1972). *The logic of scientifc discovery.* London: Hutchison.

Rosenberg, A. (2008). *Philosophy of social science* (3rd ed.). Boulder, CO: Westview Press.

Schulte, J. (1992). *Wittgenstein: An introduction.* Albany, NY: SUNY Press.

Simon, H. (1991). *Models of my life.* Cambridge, MA: MIT Press.

Smith, J. (1989). *The nature of social and educational inquiry: Empiricism versus interpretation* (W. Brenner & J. Holley, Trans.). Norwood, NJ: Ablex.

Smith, J. (1993). *After the demise of empiricism: The problem of judging social and educational inquiry.* Norwood, NJ: Ablex.

Spiro, R., Feltovich, P., Jacobson, M., & Coulson, R. (1992). Cognitive flexibility, constructivism, and hypertext: Random access instruction for advanced knowledge acquisition in ill-structured domains. In T. Duffy & D. Jonassen (Eds.), *Constructivism and the technology of instruction: A conversation* (pp. 57–75). Hillsdale, NJ: Erlbaum.

Monk, R. (1990). *Ludwig Wittgenstein: The duty of genius.* New York: The Free Press.

Muffoletto, R., & Knupfer, N. (Eds.). (1993). *Computers in education: Social, political & historical perspectives.* Cresskill, NJ: Hampton Press.

Rorty, R. (1979). *Philosophy and the mirror of nature.* Princeton, NJ: Princeton University Press.

Rorty, R. (1982). *Consequences of pragmatism.* Minneapolis, MN: University of Minnesota Press.

Rorty, R. (1991). *Objectivity, relativism, and truth: Philosophical papers* (Vol. 1). New York: Cambridge University Press.

Rorty, R. (2000). *Philosophy and social hope.* New York: Penguin.

Wegeriff, R. (2007). *Dialogic education and technology.* New York: Springer.

Willis, J. (1993). Technology and teacher education: A research and development agenda. In H. Waxman & G. Bright (Eds.), *Approaches to research on teacher education and technology* (pp. 35–50). Charlottesville, VA: Association for the Advancement of Computing in Education.

Willis, J. (2007). *Foundations of qualitative research.* Thousand Oaks, CA: SAGE.

Willis, J. (2009). *Qualitative research methods in education and educational technology.* Charlotte, NC: Information Age Publishing.

Winograd, T., & Flores, F. (1987). *Understanding computers and cognition: A new foundation for design.* Norwood, NJ: Ablex.

Wittgenstein, L. (1953). *Philosophical investigations.* New York: Macmillan.

Wittgenstein, L. (1966). *The Blue and Brown Books* (2nd ed.). New York: Harper & Row.

SECTION II

The Family Resemblances of C-ID

Jerry Willis

There is a tendency in our society to try and break complex things down into simpler pieces or "components." For example, there are hundreds of lists—*the 10 characteristics of effective problem solvers, the 8 characteristics of genius, the 7 characteristics of a high-performance work team, the 10 (or 8 or 6) characteristics of successful leaders*—that purport to tell us the basic elements that make up something complex like leadership or literacy or ethics or whatever.

Breaking the complex down into simpler elements has been with us at least since Democritus, in the fifth century B.C.E., proposed that everything in the world is made of atoms. The word *atom*, by the way, comes from the Greek word for *indivisible,* which is just what the atomic theorists were trying to do—get down to the basic components of matter that cannot be further broken down into even more elemental components.

That idea has also been pursued in the field of instructional technology with regard to both theories of instruction and theories of instructional design. Robert Gagne's nine events of instruction come immediately to mind as well as B. F. Skinner's operant and classical conditioning. Even, ADDIE, the traditional generic model of ID, is often treated as a statement of the essential components or elements of ID: Analysis, Design, Development, Implementation, and Evaluation. In their paper proposing that ADDIE ID models be used to create distance education resources, Lee, Owens, and Benson (2002) said "the instructional design process is characterized by the five phases of ADDIE" (p. 406) and they further asserted that ID is also "characterized by the relationship of those phases to each other" (p. 406). The authors acknowledge that there are other types of ID models, including C-ID models, but they concluded that

> despite the challenges, the ADDIE model remains the most frequently used instructional design model to develop training in business and industry. The objectives focus of the ADDIE model and the performance improvement focus of training in the workplace go hand-in-hand. (p. 407)

They then take a position that theoretical debates (e.g., objectivist versus constructivist approaches to ID) are not important to "instructional designers with a performance focus." Instead of debating theoretical issues they simply

> focus on selecting the most appropriate strategy to achieve the stated performance objectives, regardless of whether that strategy is objectivist or constructivist in nature. They incorporate the tenets of constructivist instructional design while following the methodology of objectivist instructional design. (p. 407)

The positions taken by Lee, Owens, and Benson (2002) are relatively common in the ID literature and I will make a few extended comments on them because they represent one effort to bridge the gap between traditional ISD or ADDIE ID models that were created on a behaviorist foundation, and C-ID models that are based on constructivist and interpretivist foundations. I disagree with the authors at several levels but there are two points that seem central to our differences. The first has to do with the question of whether ID is best considered from Pedagogy or a Process perspective. The authors equate constructivist ID approaches with following the basic principles of constructivist pedagogy, and they cite Jonassen's (1996) guidelines for teaching constructively as "the tenets of constructivist instructional design" (p. 407). Jonassen's four guidelines (use authentic tasks in context; provide real-world, case-based learning opportunities; foster reflective practice, and promote collaborative/social learning) are about teaching and learning that involves using constructivist strategies. Lee, Owens, and Benson thus treat C-ID from a Pedagogical perspective. On the other hand, most of their article is about how to make the ADDIE ID model better. Virtually all of their proposals for achieving that goal are based on a view of ID as a Process. C-ID, in their view, can be integrated into a behavioristic, objectivist ID model like ADDIE because C-ID only involves consideration of a few more instructional strategies. I believe this is a fundamental error because C-ID is not primarily about instructional strategies. I often use behavioral learning strategies, including just-in-time direct instruction, tutorials, and much more, in instructional resources I design constructively. The difference between traditional ISD models and C-ID models is not primarily at the level of what instructional strategies are used. The differences are in the process of ID. Lee, Owen, and Benson implicitly acknowledge this when they spend most of their article discussing how to improve the ADDIE model by making changes in its process. For example, they propose that while doing a Front-End Analysis in the ADDIE model, a task like "Identify the background, learning characteristics and prerequisite skills of the audience" be changed to "What Web-based technology skills do the intended learners possess?" (p. 412). They thus focus on process issues, not pedagogy issues, in their enhancement recommendations. I agree with this focus; process is the critical aspect of an ID model. That is why the bulk of this book focuses on the process of ID.

A second fundamental disagreement I have with Lee, Owen, and Benson has to do with their assumption that you can ignore theory and go directly to the task of "selecting the most appropriate strategy to achieve the stated performance objectives." This statement, in itself, is heavily laden with theoretical assumptions. It assumes, for example, that there exists in the real world a *most appropriate instructional strategy* and that the designer is in the best position to choose the "most appropriate strategy." It also assumes that anything worth teaching or learning can be expressed completely in a "performance objective."

The authors are using a particular family of theories about learning and knowing that includes behaviorism, information processing theory and cognitive science when they make such a statement. I believe that saying you do not pay attention to theory is often one of the major indicators of a lack of awareness of the deep and powerful influence theories have on the way we think. I also believe that is the case here. We cannot make a decision about what is "the best instructional strategy" or whether we should base our design work on accomplishing "performance objectives," without being guided by theory. The question is not whether we use a theory or not, it is whether we are aware of using a theory and whether we realize that there are competing theories that, if used, would guide us to a different set of decisions.

GIVING UP ESSENTIALISM

The positivist search for laws of human behavior has focused the attention of American education (and the social sciences) on the search for absolute answers to questions that other theoretical foundations suggest do not exist. Bent Flyvbjerg's (2001) book is one recent expression of the view that the search for essential Truths about human behavior (and instructional design) is a search for the Holy Grail that will never be fulfilled.

> We should avoid social sciences that pretend to emulate natural science by producing cumulative and predictive theory. The natural science approach simply does not work in the social sciences. No predictive theories have been arrived at in social science, despite centuries of trying. This approach is a wasteful dead-end. (p. 38)

Flyvbjerg (2001) proposes an alternative foundation for social science research that is quite compatible with constructivist and intrpretivist theory. Adopting his proposals in education would require an alternative to what I would characterize as Essentialism—the search for things like "the One Best instructional strategy for." Because all knowledge is theory-bound and theories about human behavior cannot be confirmed conclusively as absolutely True and applicable across all contexts, decisions about what instructional approach to use will always be subjective. Our theories, our experiences, and the process we use to make the decision, will always influence our choices.

However, if we cannot achieve the Essentialist's goal of using verified laws of behavior to select the best pedagogy, how are we to make decisions about education? I think there are three theoretical concepts that can help us: Schön's reflective practice, Aristotle's phronesis, and Wittgenstein's family resemblances.

SCHÖN'S REFLECTIVE PRACTICE

Certainty is difficult to give up, even when it is illusory. That is one reason why positivist approaches to both educational research and practice are so appealing to all of us, and why it is so difficult to give them up. We all enjoy being certain about the decisions we make. One helpful framework for thinking about professional decisions, including those instructional designers make, is Donald Schön's (1987) concept of reflective practice. He contrasts his approach to a technical-rational approach:

> Technical rationality rests on an objectivist view of the relation of the knowing practitioner to the reality he [or she] knows. On this view, facts are what they are, and the truth of beliefs is strictly testable by reference to them. All meaningful disagreements are resolvable, at least in principle, by reference to the facts. And professional knowledge rests on a foundation of facts.

In the constructionist view, our perceptions, appreciations, and beliefs are rooted in worlds of our own making that we come to accept as reality. Communities of practitioners are continually involved in what Nelson Goodman ... called "worldmaking.... When practitioners respond to the indeterminate zones of practice by holding a reflective conversation with the materials of their situations, they remake a part of their practice world and thereby reveal the usually tacit process of worldmaking that underlie all their practice. (p. 36)

Schön (1987) argued that most professional decisions cannot be made from a technical-rational perspective. He believed each context has many unique characteristics that must be taken into consideration by the professional practitioner in order to make a good decision. His recommendation for how to do that was *reflective practice*. Thus, decision making is a thoughtful, reflective, and often social process rather than a technical process of matching a current problem to a tried-and-true solution to that problem. Each instance of a problem has unique characteristics that make the technical-rational approach problematic.

ARISTOTLE'S PHRONESIS

A second theoretical concept that is useful here is that of phronesis. Kessels and Korthagen (1996) make the point that the recent history of Western thought has been dominated by the idea that theoretical, abstract knowledge is superior to "concrete skills or the tacit knowledge of good performance" (p. 18). They link the superior position of abstract knowledge to the influence of Plato and his emphasis on epistemic knowledge. Plato's knowledge, *episteme*, is general, abstract, and universal. The "facts" of epistemic knowledge:

are of a general nature; they apply to many different situations and problems, not only to this particular one. Consequently, they are formulated in abstract terms. Of course, these propositions are claimed to be true; prefereably their truth is even provable.... Because they are true, they are also fixed, timeless, and objective... It is this knowledge that is considered of major importance, the specific situation and context being only an instance of the application of the knowledge. (p. 18)

The search for essentials, for basic components or skills, and for laws of human behavior, is the search for epistemic knowledge. We are so accustomed to viewing science, including the social sciences, as engaged in the search for this type of knowledge that it can be difficult to think of another foundation for professional practice and for research. That is where phronesis comes in. One type of knowledge advocated by Aristotle was phronesis. It is situated in a contxt and is dependent on the context. It is "practical wisdom" rather than abstract, universal wisdom. This type of knowledge is:

> not abstract and theoretical, but its very opposite: knowledge of concrete particulars…. In practical prudence, certitude arises from knowledge of particulars. All practical knowledge is context-related, allowing the contingent features of the case at hand to be, ultimately, authoritative over principle. (p. 19)

It is not much of an exaggeration to say that Plato's epistemic knowledge is the foundation for traditional instructional design while Aristotle's phronesis is the foundation for constructivist instructional design. Traditional ADDIE and ISD models assume the knowledge needed to make decisions about design is empirically derived, true, and generalizable across contexts. C-ID models assume the knowledge needed to make decisions about design is conditional, contextual, and contingent on the local situation.

Both reflective practice and the concept of phronetic knowledge have significant implications for how an ID model is both developed and disseminated. Together they suggest that there are no Essentials, no required components that *must* be included if an ID model is to be judged good and worthy of our attention. We cannot, for example, judge a designer to be deficient, lazy, or incompetent just because they elect to skip some of the components of the ADDIE model when leading a design team. There are no Essential components of design the must *always* be completed. If we accept the contextual nature of knowing, a logical conclusion is also that we should not present our ID models as a set of linear, fixed steps that must be followed exactly as the designer specifies. If ID models are examples of phronetic knowledge rather than epistemic knowledge, we must accept that the user will have to make many decisions *in the context of practice* about how to use a model.

WITTGENSTEIN'S FAMILY RESEMBLANCES

A third theoretical concept can help instructional designers make decisions about both what and how to use ID models in their practice. It is Ludwig Wittgenstein's concept of *family resemblances*. In dealing with the problem of how to define something, Wittgenstein (1978; Zitzen, 1999) rejected the traditional approach of trying to create a set of defining characteristics that precisely distinguish a concept or idea. Instead, he argued that concepts are best defined through family resemblances. All members of a particular family may share a set of family characteristics but no single member of a family will necessarily have all these characteristics. Yet, when the clan gathers for some family function and there are both blood relatives and in-laws present, it is often quite easy to spot the blood relatives. They have family resemblances, characteristics that tend to be more common among that clan.

I believe the concept of family resemblances applies nicely to the question, "What differentiates a C-ID model from other types of ID models?" There are no essential elements that must be present, no required components or steps that, if left out, automatically confirm that this is not a C-ID model, or that this particular ID project was not practiced according to C-ID principles. Aristotle's phronetic knowledge is not the same as epistemic knowledge; the context of practice will influence how things are done.

And, while there are no Essential principles of C-ID that must be included in every project, there are family resemblances. In an earlier paper (Willis, 1995) I suggested a set of family resemblances for C-ID. Below is a modified set of family characteristics I believe characterize the emerging family of C-ID models today:

1. The C-ID process is *reflective*
2. The C-ID process is *recursive* and nonlinear.
3. The C-ID process is a *participatory*, social, collaborative process.
4. The C-ID process is *emergent* rather than precisely preplanned.

Each of the chapters in this section of the book address one or more of the C-ID family resemblances. The application of Donald Schön's reflective practice theory is addressed in chapter 12, for example, while chapter 10 discusses the use of participatory design approaches.

While the chapters mentioned thus far focus on one of the family resemblances, they typically cover other aspects of C-ID as well. However, in addition to these relatively focused chapters, there are three more that address broader topics that are relevant to several family resemblances. For example, a group of Italian scholars, led by Luca Botturi at the University of Lugano, wrote chapter 8 on fast prototyping. Fast or rapid prototyping is a way of designing that involves end users in the design process and facilitates the reflective, recursive and emergent nature of C-ID as well as provides a powerful way for those who will use the resulting product to participate in the ID process.

Virginia Richardson, from the University of Arizona-Tucson, wrote chapter 9 on the general topic of doing research on practice. Richardson distinguishes between two types of research on practice: formal research and practical inquiry. Formal research is, essentially, what I have called positivist research. Practical inquiry, on the other hand, is much closer to C-ID. Richardson's ideas about how to conduct practical inquiry and how the results can be both applied in the classroom where the research was conducted and shared with other professionals who practice outside that classroom are well worth considering.

Finally, chapter 11, by a group of Canadian scholars led by Katy Campbell, is a thoughtful and sophisticated analysis of the ethical and personal issues of instructional design. Campbell and her coauthors argue that "designers are not journeymen workers directed by management, but act in purposeful, value based ways with ethical knowledge, in social relationships and contexts that have consequences in and for action." What these authors have to say is important to designers regardless of whether they practice within a C-ID framework or some other model.

REFERENCES

Flyvbjerg, B. (2001). *Making social science matter: Why social inquiry fails and how it can succeed again* (S. Sampson, Trans.). Cambridge, England: Cambridge University Press.

Jonassen, D. (1996). Thinking technology. *Educational Technology, 34*(4), 34–37.

Kessels, J., & Korthagen, F. (1996). The relationship between theory and practice: Back to the classics. *Educational Researcher, 25*(3), 17–22.

Lee, W., Owens, D., & Benson, A. (2002, November). Design conceriations for Web-based learning systems. *Advances in Developing Human Resources, 4*(4), 405–423.

Schön, D. (1987). *Educating the reflective practitioner.* San Francsico: Jossey-Bass.

Willis, J. (1995). A recursive, reflective instructional design model based on constructivist-interpretivist theory. *Educational Technology, 35*(6), 5–23.

Wittgenstein, L. (1978). *Philosophical investigations* (G. E. M. Anscombe, Trans.). Oxford, England: Oxford University Press.

Zitzen, M. (1999). *On the efficiency of prototype theoretical semantics.* Retrieved January 27, 2006, from http://ang3-11.phil-fak.uni-dusseldorf.de/~ang3/LANA/Zitzen.html

CHAPTER 7

EMERGENT DESIGN AND LEARNING ENVIRONMENTS

Building on Indigenous Knowledge

D. Cavallo

The empirical basis of this paper is a 2-year project to bring new learning environments and methodologies to rural Thailand. Pilot projects were mounted outside of the education system, with the specific purpose of breaking "educational mind-sets" that have been identified as blocks to educational reform. A salient example of such a mind-set is the assumption that the population and teachers of rural areas lack the cognitive foundations for modern technological education. The work required a flexible approach to the design of digitally based educational interventions. Analysis of design issues led to a theoretical framework, Emergent Design, for investigating how choice of design methodology contributes to the success or failure of education reforms. A practice of "applied epistemological anthropology," which consists of probing for skills and knowledge resident in a community and using these as bridges to new content,

Previously published in *IBM Systems Journal, 39*(3/4).

was developed. Analysis of learning behaviors led to the identification of an "engine culture" in rural Thailand as an unrecognized source of "latent learning potential." This discovery has begun to spawn a theoretical inquiry with significant promise for assessment of the learning potential of developing countries.

The central thrust of this paper is the presentation of a new strategy, which I call Emergent Design. The paper describes an approach used for educational intervention; the claim is a more general one, however, in that the strategy is appropriate in settings for technologically enabled paradigmatic change. I claim that the more traditional approaches to systems design, implementation, and deployment have not produced desired results in situations where the goals and needs are for systematic change. When the desired changes cannot be reliably foreseen, and particularly when the target domain is computationally too complex for automation and thus relies on the understanding and development of the people involved, then top-down, preplanned approaches have intrinsic shortcomings and an emergent approach is required.

Educational environments definitely possess these characteristics. However, in the emerging business and cultural environment, many other domains do as well. I have utilized this approach previously in the design and implementation of enterprise architectures and process re-engineering. The most notable example (Cavallo, 2000) is a health care delivery environment where the Emergent Design of the architecture and applications of the systems for health care delivery, administration, and patient use enabled a broad change in medical practice. The approach to the design of the educational intervention I describe here resembles that of architecture, not only in the diversity of the sources of knowledge it uses but in another aspect as well—the practice of letting the design emerge from an interaction with the client. The outcome is determined by the interplay between the understanding and goals of the client, the expertise, experience, and aesthetics of the architect, and the environmental and situational constraints of the design space. Unlike architecture, where the outcome is complete with the artifact, the design of educational interventions is strengthened when it is applied iteratively. The basis for action and outcome is through the construction of understanding by the participants.

The technological ramifications are immense. We often build inappropriate technology because the domain changes too quickly, or the designers' understandings and aesthetics vary too much from the users' understandings, needs, and goals. At other times projects fail because, even though the technology might be appropriate, the deployment is flawed. Design cycles that cannot adapt to rapidly changing conditions miss emergent phenomena that either need correction because they are undesirable, or need capitalization if desirable. A resultant long-term problem is lack of belief in the true possibilities for technology because it did not live up to expectations. This is certainly the case in education, although business uses also share this outlook, as evidenced by complaints about lack of productivity gains through technology.

Perhaps more importantly, traditional approaches to *learning* of *and* through technology have not mobilized the indigenous knowledge *and* expertise among many people. The growing "digital divide"—concerns about the potential of a widening gap between rich *and* poor in the new, knowledge-based global economy due to a lack of modern, technological skills among people in lower social-economic strata, *and* a growing concern about the potential of educational systems to ameliorate this situation—all point to a serious problem becoming seemingly permanently intractable.

This chapter describes an approach to technology design and use that provides hope for a different, more positive outcome. The same technology that can be a primary factor in widening the divide, may be the best hope for eliminating the divide. The Emergent Design approach enabled the discovery and utilization of latent, engineering expertise and creativity among people in rural Thailand. Rather than being bereft of social capital necessary to succeed in the new economy, these traditionally poor, rural people are conceivably better situated for success so long as the technology and methodology used is expressive, appropriable, and constructionist.

While the claims here are broad, I choose to focus on one concrete example, that of an effort in educational reform. Educational institutions, although relatively young, have proven extremely resistant to change Tyack & Cuban, 1995). Moreover, schools for the most part have not used new computational technology in innovative ways. This, despite a lot of hype for the possibilities of technology in education, has caused many to doubt the potential in the technology. The problem, though, is not with the technology per se, but rather with the design, deployment, and uses of the technology.

TECHNOLOGY AND THE REFORM OF EDUCATIONAL ENVIRONMENTS

Educational reform efforts, over a long period of time, have offered many different blueprints. Yet none has had the substantial effect for which it was designed. Why is this the case?

We need to look at the way in which education reforms are usually carried out. Some set of individuals decide there is a problem needing addressing (such as low math and science scores) or a change deserving implementation (such as the introduction of a new item like ethics to the curriculum). A group convenes. They call in the various experts, stakeholders, practitioners, and other usual suspects. They design a blueprint for their reform. The blueprint contains a curriculum, materials, texts, assessment, teacher training, and so on.

This chapter presents the view that these blueprints have failed simply because they are blueprints. Many analysts researching this situation, most recently David Tyack and Larry Cuban (1995) have shown how the process fails. Whatever blueprint is proposed, it is inevitably going to be transformed in the course of appropriation, ending more in conformance with what the designers originally hoped to reform. The institution tends to reform the reform, perhaps retaining the rhetoric but rendering it toothless. Tyack and Cuban brilliantly term the overriding mindset the "grammar of school." Like a grammar, they describe a deeply held organizing system that allows only certain expressions (or actions) as legitimate and renders some expressions nonsensical if they deviate from the underlying system.

Tyack and Cuban (1995) made clear that whether reforms are big or small, from the "right" or from the "left," national or local in scope, they do not work. Some might deserve to fail because of the nature of their content. But while content may or may not be a limiting factor, they fail because of the form in which they were designed.

What is needed is an alternative approach that is not a blueprint. This naturally raises the question of whether having no blueprint means the abrogation of all design and planning so that "anything goes." In the same way that a jazz group can improvise within the structure of a piece while remaining coordinated and within the theoretical principles of the genre, so too can an emergent design remain consistent within a core set of principles.[1]

This chapter describes a form of intervention in learning that is very different from the model of reform studied by Tyack and Cuban (1995). It offers hope for addressing the great educational needs created by the digital age by drawing on two of its important innovations: (1) digital technology and (2) the approach to management of organization and of organizational change that has come in the wake of the technology.

More precisely, this work draws on the combination of these two innovations. A distinction must be made because, as I show, the temptation to use either of them alone has led to failure. It is the combination that offers an optimistic vision for the future of learning—the combination of these two products of the digital age along with a theoretical framework based on the work of pre-digital-age thinkers who knew what to do but did not have the means to do it. Among these the most central is Paulo

Freire (1972), but also represented are John Dewey (1938) and, although he did not focus on education per se, Jean Piaget (1977).

Limitations of a Single Focus

A focus solely on technology leads to technocentrism, that is, a view that it is the technology and not what we do with it that has impact (Papert, 1990). Such a focus also leads to a narrowness of vision. In other words, we simply place the technology into the existing structure and thus are not able to see the possibilities that extend beyond the existing organization. Merely adding technology reinforces an experimental paradigm out of place. This paradigm tries to modify one element at a time, holding the others constant. When using such an approach when introducing technology, what one holds constant—rather than maintaining experimental purity—merely serves to neuter the potential for educational change catalyzed by the technology. Thus, an erroneous view of the technological and learning potential results.

In his book *The Productive Edge*, Richard Lester (1998) describes mistakes made within conventional mindsets about business, productivity, and change that resonate with conventional mind-sets of education and school reform. Lester describes how many companies, in an attempt to improve productivity, quality, or some other often highly quantifiable attribute, would attempt to apply a seemingly scientific method by researching a new methodology or so-called best practice; they then attempted to test whether adding this method to their own operation would generate positive results, holding all other things constant. In a vast majority of cases such applications of new methods failed to produce positive results. This not only called into question whether the new methodologies truly had value, but also led to an experimental fatigue from being repeatedly forced to adapt to the change program of the month.

What Lester (1998) demonstrated was that there are not typically such things as decontextualized best practices that can be grafted onto existing organizations and thereby produce results. Rather, each company has its own complex culture, full of subtleties, and successful companies are the ones that can innovate, cultivate, adapt, and use methods that can thrive in their particular environment. The successful approaches fit more with the Emergent Design concept advocated here than with the more traditional top-down, change-one-variable-at-a-time approaches thought to be more in the line of scientific management. Digital technology enabled the customization of process to culture rather than forcing culture to be responsive to management dictates and "the one best method" (Taylor, 1998). The critical point is that adoption and implementation of new methodologies needs to be based in, and grow from, the existing culture, and typically fails when it is merely imposed from above without such cultural considerations. Interestingly, incremental approaches to educational reform closely resemble the less successful methods that Lester describes.

The reform of educational management, usually in the form of administrative decentralization, does not break the stranglehold of the grammar and ends up with reversion to type. By themselves, ideas such as decentralization of control and decision-making, or intradistrict competition, do not generate new content and methods. On the contrary, they merely push the same practices down the hierarchy without fundamentally changing practice. Thus, the only substantial change is in administration, not innovation in the learning environment.

The Need for New Principles

Saying one needs to base new methodologies and the change process to resonate with, and build upon, existing culture does not mean that deep change is not intended, nor that any type of change is desired. In the case of learning environments, the primary principles we brought were: constructionism, technological fluency, immersive environments, long-term projects, applied epistemological anthropology, critical inquiry, and Emergent Design.

Constructionism builds (Harel, 1991; Papert, 1991) upon principles in constructivism. While constructivism holds that the learner constructs new knowledge based on the existing knowledge he or she has, constructionism builds on this idea by maintaining that this process happens particularly well when the learner is in the process of constructing something. For example, in our work with LEGO^2-LOGO (Cavallo, 1996a, 1996b; Kafai & Resnick, 1996) we witnessed many children, including those who had previously done extremely poorly in school, understand complex ideas in mechanics, physics, and mathematics through constructing LEGO robots to accomplish various tasks.

The idea of building technological fluency draws on the image of being fluent in a language (Cavallo, 1996a, 1996b; Papert & Resnick, 1995). When one is fluent in a natural language one can think, express, communicate, imagine, and create with that language. In the same way, we like to develop fluency through the construction of, and with technology as a means of, personal and group expression. We try to develop fluency with technology in order to help people become more eloquent and effective in their expression. Just as fluency changes the focus to a more holistic use of natural language, this also changes the focus of learning with technology.

Just as the idea of fluency is adopted from language, so too is the concept of immersive environments. Being immersed in the culture and environment facilitates learning a foreign language. So too does working with others in a culture where the knowledge of technology and construction is deeply embedded facilitate the development of technological fluency.

Building artifacts of interest to learners aids the construction and the development of fluency. In order to delve deeply enough to unearth the underlying concepts and principles, we enable students to work on projects over a long period of time. Rather than rushing through a broad curriculum in a shallow manner, we prefer to encourage diving deeply into the projects. This takes time. This also differentiates our practice from other project-oriented approaches in education, where the project is preplanned by the curriculum designers and not emergent from the interests of the learners, and where it lasts only a short period of time in order to fit the traditional classroom situation.

Applied epistemological anthropology is a term I have given to the practice of unearthing the meaning learners attribute. This applies on both a cultural and individual basis. In order to facilitate the construction of new knowledge on the existing knowledge of the learners, one must first help discover the existing frameworks as best one can. This practice itself is facilitated through the construction of objects of interest to the learner, where the learner has as much freedom of expression as possible. When the freedom of expression exists, then the learner has the space in which to express himself or herself in a manner faithful to the learner's thoughts. This is a key element in the design of technologies for learning. Through the construction, and mediated by discussion, the underlying thoughts become more evident. This enables the teacher or facilitator to better design and implement learning interactions. This leads to the necessity of a more emergent approach.

Critical inquiry is the process of engaging in a conversation with one's world in order to understand and act upon it (Freire, 1972). Through critical inquiry we collaboratively determine upon which projects to work. Also through critical inquiry we try to understand the phenomena of study in sufficient detail so as to construct artifacts modeling the phenomena or designed to ameliorate the situation, as well as to understand *and* debug the artifacts of construction.

Emergent Design is what manages the overall process. Due to the emphasis on approaching learning by building on the existing knowledge of the learners through their expressive construction of projects of their own choosing, this process by definition has strong emergent tendencies. However, design is also emphasized as the others in the community who work with learners—be they teachers, parents, or other community members—also play an active role in assisting to assist and guide the learner in the process. The idea of design extended to a "grassroots" level enters because, just as Emergent Design is practiced to facilitate organizational change, so too is Emergent Design practiced in the interaction between teachers and learners.

There is no claim that the methodology described in this chapter is the only way to achieve the desired results. However, it does represent a sort of "proof of existence" of a way to achieve the desired results. What this does serve as is an existence proof of a way. That this way at least initially demonstrates interesting, unexpected, and quite positive results among populations that previously did not exhibit such results in traditional settings should serve to question the existing school grammar as well as to facilitate other such experiments. I close this section with a story that illustrates the prevailing educational mind-set as well as effective informal methods that can be leveraged.

DOS Commands and Flower Gardens

On my first visit to Thailand my hosts took me to a nonformal education (NFE) site in a Buddhist temple. I saw a computer class held at the NFE temple school. A child was being taught DOS commands. The logic behind such an introduction to computers, following the typical school curriculum grammar of using sequential building blocks of knowledge, is that it provides the requisite basis for later, more difficult learning. However, the useful learning never comes! And in the meantime, the formalistic nature of the beginning work confuses and frustrates the novice.

The student's teacher assigned four commands for him to learn and practice. The first was dir, to get a listing of files in his directory. The second was copy, to copy a file from one location to another. The third was format, to format his A drive (fortunately, it was not the C drive). I do not remember the fourth but it was made irrelevant by the reformatting of his disk.

This confounding situation led the student to stop me with a plaintive question, "What is the problem here? It worked before but now it no longer works. I am following my teacher's instructions, but this is not working properly." On the first iteration of practicing his commands by rote, everything was fine. Subsequently, however, none of the commands was giving the specified results. His directory was now empty. He could not copy his file. I explained to him that the result of using the format command is that it reformats the entire disk, meaning it wipes clean what was on it. Thus, there were no more files in his directory to list or to copy.

Despite several attempts at various ways of explaining it, including recreating the example on the computer and showing how dir, copy, and format work with a newly created set of files, I am not sure he understood my explanations. One reason for this is that my explanations meant that what his teacher had said, done, and assigned no longer made sense, which would be quite disorienting. Another possibility is that no matter what the explanations and examples, learning commands this way is too decontextualized to make sense. One is merely learning by rote what someone else says is important without any conception of why or how it might be used.

The split between conventional "School thinking"[3] and cultural learning was shown vividly in the contrast between the computer class at the temple and how the monks themselves teach flower gardening. Beautiful flowers are grown and displayed at all the Buddhist temples in Thailand. They are impressive, colorful, and fragrant. After my visit and the experience with DOS teaching run amok, I inquired about how people learned to cultivate such gorgeous gardens. A monk explained that when initiate monks enter the temple, they work alongside more experienced ones and learn by demonstration, by asking questions, in the best sense of learning by doing. I mischievously asked whether any classroom instruction was involved. The monk looked at me askance, but politely answered no, they felt there was no need. I tried to explain that this was the approach we also preferred for learning computational ideas. That is, that new learners work on projects of their own, are in an environment with others working on similar, but perhaps more complex, projects, and can observe and ask others questions—in essence they are immersed in a culture of computing just as the monks are immersed in their culture.

The depth of resistance to these ideas was illustrated by the way the teacher who was translating my remarks into Thai misrepresented the explanation, creating an initial misunderstanding between the monk and me. After listening to my translator, the monk politely responded that they would never do what I suggested. Considering that I had just suggested that we create environments for learning computational ideas in the same manner that the monks learn gardening, I could not understand how he disagreed. So I inquired again about what was said. The teacher told me that she told the monk I had suggested that they teach gardening in a classroom just as we teach computers in a classroom. When I re-explained what I had really intended, the teacher could not believe I meant it. Rather than immediately retranslating, she passionately protested. Surely classrooms were the modern and most effective means of teaching. How could I, from a modern western university, suggest that the monk's method could be better? It took quite a while to get her to tell the monk what I thought. In retrospect, this was a powerful learning moment for my translator, although thoroughly and necessarily unplanned.

Scenes From Project Lighthouse

The context of this chapter is Project Lighthouse, a bold intervention to initiate radical change in the educational processes in Thailand. As its name suggests, Project Lighthouse is not a blueprint for education or education reform. Rather, it attempts to highlight actual possibilities for powerful *learning environments* in Thailand, particularly in settings where traditional education has not succeeded. A primary goal is to break mind-sets about what education must be by providing concrete examples. The following are samples of activities from Project Lighthouse over a 17-month period. The scenes provide a concrete basis for the discussion that follows.

Bangkok, March 1997

In the first scene, Seymour Papert and I, from the MIT Media Lab, were meeting with leaders of the Suksapattana Foundation.[4] We were designing a proposed intervention intended to provoke a radical reform of the educational system in Thailand. The meeting came about because a group of industry leaders and government officials had come to believe that, unless they achieved a total transformation of their educational sys-tem, Thailand would not merely stagnate economically, but also that they would lose all the gains of the previous decade. More critically, the leaders worried that there was a growing and more intractable divide between rich and poor that would destroy the fabric of Thai society. They further believed that in the absence of an educated, thoughtful, literate populace, it would be impossible to support their nascent democracy and prevent a return to autocratic, and corrupt, military rule.

The Thai leaders believed that the existing school was not a hospitable medium for developing alternative forms of learning. Moreover, they felt that to change it directly would cost too much and take too long. They believed the existing schools to be too rigid, too reliant on rote instruction, and staffed by too many teachers who were barely educated themselves.

They had set bold and ambitious goals for their educational system. They had developed a new national education plan as an essential part of their national development plan. This education plan, combined with a special commission from the Office of the Prime Minister devoted to edu-cation reform (Office of the National Education Commission, 1999), spec-ified the new goals. The goals were thoughtful and admirable. They included:

- Becoming learner-centered
- Developing critical-thinking ability

- Fostering innovation *and* creativity
- Developing collaborative spirit and skills
- Learning how to learn
- Providing familiarity, ability, and comfort in working with technology
- Developing "happy" *learning, that is, a joy* for learning

However, none of the plans specified how to achieve such a system. They did not discuss how to operate in this new paradigm or how to make the transformation. Thus, while the goals were lofty, the implementation of both the new system and the method of reforming the current one, were mired in the existing, undesirable paradigm.

The goal behind our endeavor, Project Lighthouse, was to break mind-sets by creating technologically rich learning environments that would demonstrate the "out-of-the-box," yet practical, possibilities for children in Thailand.

However, it was not clear what to do. Moreover, there was little agreement on, or acceptance of, our proposal. Some believed that we should focus on gaining the acceptance of the national curriculum developers as the current system moved only through the curriculum. Others believed we from MIT should train the trainers who would then train the teachers who would then work with the students. Others felt we should place computer labs in more schools *and* train teachers to work there. Finally, there was near unanimous agreement that the existing teacher corps was incapable of working in a new, learner-centered, project-based, technologically rich environment. Virtually everyone told us that the teachers were barely educated themselves and might not be able to learn to use the technology, let alone teach with it.

We proposed creating four pilot projects where we could quickly demonstrate significant results in some of the most critical areas of need.[1] These were:

- Alternative learning environments within nonformal education
- Rural village learning centers
- Teacher development
- Alternative learning environments for at-risk youth such as street children in urban areas and girls at risk for or exiting prostitution

There were two major objections to our proposal. First, commentators argued that our proposal did not fit the prototype 5-year plan, which spells out all activities over that time period. How could people know what to do if we did not provide such a plan? People wondered if perhaps either we were not serious or did not know what we were doing. Second, people told us that the quality of the teacher corps was so low that they would be incapable of carrying out an ambitious endeavor such as ours.

However, we argued that it would be counter-productive, if not impossible, to develop any specific plan. It was not merely that we were not familiar enough with Thailand to know what would be the right things to do. More profoundly, what was needed was a philosophy of design based on recognizing that no one could know beforehand what would resonate, how people would appropriate new learning technologies and methodologies, what learners would choose as projects, how villagers would react to the intervention, and so on.

At the meetings we tried to show that there is a fundamental contradiction between having learning environments that function through connecting to, building upon, developing, and deepening the interests of the learners, and planning everything centrally in a top-down manner where all activities are predetermined for all learners and all locations. What is needed is a philosophy of design for educational innovation as different from traditional ideas of reform as the content of the new innovation would be from traditional educational content. The theoretical framework that evolved from this and similar experiences of Project Lighthouse is Emergent Design.

The phrase Emergent Design puts a spotlight on the need (which has not been recognized by education policymakers) to study the conceptual space where the purposeful stance implied by the word "design" mates with the openness implied by the word "emergent." This mating underlies modern approaches to organizational practice.

The emphasis on emergence as the guiding principle does not imply that this is an anything-goes environment reacting to the whims of the participant teachers and learners. As described above, we brought a very disciplined set of principles, methodologies, tools, activities, models, and exemplars for learning environments. However, to deliver a pre-set curriculum with pre-chosen problems, explanations, and sequence of events would be not only antithetical to the underlying learning philosophy, but also it would be incapable of taking advantage of the very benefits that the technology affords.

Nong Baot Village, BuriRam Province, Northeastern Thailand, January 1998

The second scene from Project Lighthouse took place in Nong Baot in the northeast of Thailand, the poorest region of the country. It is approximately 100 kilometers from the Cambodian border. *The New York Times* (Mydans, 1997) described it as having "two harsh seasons, flood and drought." The economy is based on agriculture but, due to the harsh weather, little can be grown. Nong Baot survives by cultivating one rice crop per season. There are some small vegetable plots used primarily for subsistence, because there is not enough water to grow enough crops to sell. Lately, some groups of villagers have tried to cultivate fish farms by creating small reservoirs during the rainy season. This, too, provides food for them for only a brief time, because the water is gone within a few months.

Nong Baot is an area that suffers from logistical problems that have stifled the potential for economic development (Sachs, Gallup, & Mellinger, 1998). It is tropical and does not have ready access by water to the rest of the world. These factors inhibit the development of industry. The soil is poor and there are no mineral deposits. Thus, it has remained an area of minimal means and wealth.

Education in this area likewise has been minimal. There is little incentive to remain in School. Many people leave school as soon as they are legally eligible, claiming that School has no relevance to their lives. Children need to work in the fields or in other occupations to help their families. Few people go on to attend a university.

Within this scenario, I conducted an introductory Logo immersion workshop to develop technological fluency. Unlike most projects that try to bring technology to remote or impoverished areas, my goal was to have the attendees quickly build projects and create programs.

The workshop had a mix of participants: villagers, teachers, and a few local economic development workers from the Population and Development Agency. I began the workshop by showing what a computer is, how you turn one on and off, and how you operate one, because this was the first time the villagers had ever personally seen a computer, except for viewing one on television.

In the evening the MIT participants held discussions with the villagers to get to know them and their situation. I asked why the villagers said they wanted us to place computers and Internet connections in their village. They told us that water is very scarce in this region. Worse, there is either too much of it during the 2-month rainy season or there is none of it during the rest of the year.

It was in the discussion in the evenings after this workshop that the village leader expressed the need of the people to gain more control over their lives and the belief that certain uses of the technology could help them. The people described many of their problems as economic, caused by the harsh climate where there was either too much or too little water. They wanted access to expert knowledge, but most importantly they wanted to be in control of gaining the access to and making the decisions about what to do with the knowledge. They felt the local authorities did not involve them in the thought process and decision-making whenever the villagers asked for assistance. This left the villagers feeling dependent and without the hope for their own progress. To make matters worse, due to the appearance of new problems with the cattle and the water, the villagers believed the advice and proposed remedies they were given to be harmful rather than helpful.

Introducing the First Phase

The villagers wanted to end this cycle of dependency and lack of control by gaining access to information and gaining control of the situation via the technology. Even though I had to introduce the workshop by demonstrating what a computer is, including how to turn one off and on, through the symbolic value of the computer they viewed competency with the technology as a plausible path to this control. Although there could not have been any real experience with how computational technology could provide this path, they had heard enough about computers and the Internet to believe it had potential for them.[6] The computer was a symbol of modern technology and a connection to the modern world.

In short, the villagers were able to experience what we did in the spirit of "cultural leverage." As a result, the participants were soon building their own projects, first in Microworlds Logo, then adding robotics with LEGO-Logo. What at first was a foreign and potentially intimidating technology, now became a source of fun and pride in product. The villagers worked in multigenerational groups, from young children to the elders in their seventies and eighties. The teenagers and children did more of the programming, being more open to new technologies. The adults contributed their wisdom, maturity, and experience. They made all of the decisions jointly. They were doing programming and engineering, working on projects of their own design.

Moving to the Second Phase

When I returned in August, the situation was quite different. In a brainstorming session about potential projects, we quickly converged on the critical need for access to water for household and agricultural use. We discussed ideas broadly at first, looking for areas they felt were major problems or, from a more optimistic point of view, areas they thought could provide major benefit if we could find means to create solutions within these areas.

Naturally, there were many trade-offs in the dimensions of each project. The MIT participants wanted to address major problems, but some problems perhaps were extremely difficult to solve. We wanted to achieve some quick successes to help change mindsets about possibilities, produce real results, and develop belief in what we were attempting. However, easy and quick successes are most likely rare or trivial. We believed in the potential of the technology to help think about and design potential beneficial projects, but the villagers were technical neophytes.

We did not want to just design solutions ourselves, because this would neither develop the villagers' own technological fluency and capabilities nor empower them in the long term. So we needed to choose initial projects that were approachable by people with their scant experience, yet were real enough to actually provide tangible benefit, while simultaneously providing a rich learning experience. Through my experience both in developing technical solutions to real-world problems, as well as working with adults and children learning to engineer and program, we worked to develop a group consensus on the initial set of projects. Knowing that we were committed for the long term relieved tension from feeling a need to accomplish everything immediately. The initial projects were to design a dam to create a reservoir for farming; investigating alternative strategies for rice cultivation; redesigning the irrigation system; developing new means to collect, store, purify, and distribute rain water; and creating new vegetable plots. In this discussion I focus primarily on the dam project.

A First, Important Project

We began work calculating the potential and the reality of building a dam. In each of the past 2 years the villagers had tried to construct a dam to create a reservoir. It was hoped that the dam would retain water at the end of the rainy season that could be used for agriculture in the dry season. In each of the past 2 years the project had failed, since the reservoir did not contain the water. Now both the villagers and the rural teachers worked to develop the new project together. I took a supporting, mentoring role rather than a direct role in the project myself, believing that the only sustainable benefit would be for them to develop the package of skills themselves.

They had not previously calculated the potential benefit from the dam. When we engaged in brainstorming about this topic with them, together we calculated that the villagers would more than double their yearly income if they could harvest a second vegetable crop. We walked through the flood plain and took some digital photographs. We measured the distances between relevant objects in the terrain using the odometer on a motorcycle. We uploaded the photographs into Microworlds Logo and the groups began making visual representations of the area.

To my surprise this was a totally new experience, not merely for the villagers but also for the teachers. While the fact that the villagers could not do this on their own might not be surprising, the teachers could not do so either. They had certainly taken school courses and passed school exams on this type of knowledge, yet in practice they could not make a map. Together, the teachers and villagers created accurate computer representations of the areas, preserving distances, maintaining relationships and ratios as they created various views at different scales and calculated the relevant distances between important objects.

Then, a remarkable thing happened! Immediately upon creating the maps, we discovered a mistake repeated in each of the previous 2 years. The villagers had been building the dam in the wrong place! The original location benefited from natural terrain to create the reservoir; however, it was about two kilometers from the village water pump used for irrigation. Once the villagers constructed their own map of the area, they realized they could not create a reservoir large enough to cover the distance to the pump. Even if the dam had functioned properly, it would not have provided the expected benefit, because it was prohibitively expensive to relocate the pump and the irrigation hoses.

Discovering an Exceptional Student

As the design project continued, we observed how the efforts of one of the participants was exceptional. He told me that he had not had any success in school and left as soon as it was legal. He primarily helped his family with the farming. We had only introduced computers to the village within the current year. He spent this time working on programming--not by taking classes, but by programming his own projects.

What was so striking was that he had quickly become quite an adept software hacker.[7] Atypical of many of our experiences with more educated people, he, as well as others in other parts of Thailand, dived in and figured out how to build the projects he wanted. If something did not work, he was not daunted. Rather, he debugged the system and worked until it was satisfactory.

We discovered that he spent considerable time working with engines. By learning how to build and repair engines and by working on the farm with few resources, he had developed a bricolage spirit. That is, he would make what they needed with what little he had. If something did not work, he fixed it. If he did not have the right tool or material, he improvised. He took this spirit and applied it to computational technology.

As this skill and experience became apparent, he and others took me to visit their farms. At the farms, everyone who could used a small Kubota diesel engine to power a wide variety of local technological contraptions. They used the little motors to power rice mills, well-water pumps, irrigation pumps, one-person tractors, field vehicles, and even lightweight trucks. The barns contained little pulley systems for lifting the motor from one device to another. The logic of each machine was open and obvious. The innovation and creativity were remarkable. The utility was tremendous. The people had taken objects for other, often quite specific, purposes and combined them in a general-purpose melange particular to their needs, resources, and budgets. The experience and expertise of those who worked with these engines and devices was quite impressive.

Thai Combustion-Engine Culture

Virtually all commentators on Thai education and on the Project Lighthouse proposal believed the quality of rural teachers was extremely poor and they would be unable to work successfully in the proposed technologically rich, learner-centered environment. These same commentators bemoaned the problems and capabilities of the overwhelming majority of rural students as well. Lack of faith in the intelligence and capability of economically disadvantaged children is an unfortunately widespread belief that is all too difficult to dislodge.

Contrary to the perceptions that rural and impoverished students are not capable learners, rural teachers are not competent technologists, and Thai culture is not amenable to innovation, collaboration, deep learning, and technical expertise, we discovered that there are deep intellectual roots and significant innovation practiced and learned over at least many decades, and presumably much longer. Indeed, although not written about in academic circles, there is a strong tradition of so-called "peasant technology"[8] in Thailand, particularly using and adapting the internal-combustion engine to satisfy local concerns and constraints. Our final scene in this section focuses on this "engine culture."

Perhaps the best example of this innovation is the creation of the long-tailed boat. There are many areas throughout the country where waterways are the principal means of travel. Significantly, this is also the case on the rivers and canals of Bangkok. In the past, as people desired to transport more and heavier goods, human-powered boats became problematic. In the north, one innovator decided to experiment with placing motors onto the boats. After several attempts with various types of inboard and outboard motors, he settled upon using an automobile engine with a long driveshaft so that the propeller was far from the boat.

Typical outboard motors did not work well, because they churned too much water into the long canoes that everyone used. The many reeds in the rivers also jammed the propellers too often, negating their benefit. The driveshaft, or long-tail, not only solved the churning problem, but also served as a rudder for steering and enabled the pilot to lift the shaft from the water to avoid entanglement with the reeds. The use of an auto engine leveraged existing knowledge about repairs and benefited from not requiring parts manufactured outside of Thailand, which would be difficult and expensive to obtain. People quickly adopted this technological innovation throughout the country (Phongsupasamit & Sakai, 1989).

Tuk-tuks are another similarly inspired innovation. Small motorcycle engines are placed onto the pedicabs, again to alleviate human stress and increase speed. Other rural innovators have also adapted engines to create low-cost, one-person tractors, irrigation pumps (including one ingenious invention to pump over roadways, since the native soil had the tendency to crumble into irrigation tunnels), and devices to help operate wells in drought-stricken areas.

For the most part, not only did these innovations not occur in universities, research labs, or corporate departments, such circles barely took notice of them. Rather, they were a grassroots effort, based in the interests, needs, and practices of Thai culture. People created and adapted new technologies to alleviate their burdens and to create new opportunities.

These innovations could not have achieved such widespread use if a culture of practice and knowledge had not also developed to spread and support them. In order to use engines widely, a group of people capable of maintaining them had to exist. This group did not do well in school and did not receive its training in school. Rather, almost exclusively they learned to diagnose and repair engines in informal learning cultures. Making this diagnosis and repair more difficult is the fact that among this social stratum in Thailand, there are not a lot of materials, parts, diagnostic equipment, or written manuals. These mechanics have to become *bricoleurs* (Levi-Strauss, 1966; Papert, 1980) that is, they must adapt materials at hand to satisfy their goals, even if it is not the accepted way to accomplish the goal nor the proper materials for the task.

DISCUSSION

What makes this story compelling is that these mechanics, while respected for their mechanical abilities, were not regarded as academically capable. Conventional wisdom stated that people in this group may be good with their hands but they were not good with their heads. Moreover, the belief in the dichotomy that different people with different skills are required in order to be good with their heads remains.

However, in the context of Project Lighthouse, the capability of these motorcycle and engine mechanics was immediately evident. Not only did they learn the new computational technology quickly, they were also quite adept at adapting it and applying it to solve local problems. This was the case with designing dams, improving irrigation, and devising alternative methods of cultivation of rice and other crops in BuriRam.

Still, moving from one technology, engines, to another, computing devices, while impressive, would not necessarily be remarkable except that in order to accomplish the tasks with computational technology they had to competently handle some sophisticated mathematics, biology, engineering, physics, and computer science. What is remarkable then is that:

- They accomplished projects requiring competence in these recognized bodies of knowledge.
- They accomplished this in extremely short timeframes.
- They leveraged their mechanical expertise and "hacking" spirit to build a computational technological fluency.
- They then utilized the technological fluency to gain competence in bodies of knowledge previously inaccessible to them.

To make maps they had to measure distances and perform calculations over these distances. To provide zoom-in and zoom-out views, they had to maintain proportions and adjust accordingly. To design the reservoir they had to again measure, calculate areas and volumes, and determine water usage for various crops over time while accounting for evaporation and drainage. To think about placement of the dam they had to think about how water flows over terrain. To design the irrigation system they had to think about networks and shortest paths. To determine which project to do or which decision to make within a project, they had to calculate costs and benefits, factoring in more subjective factors as well, and create compelling arguments to convince others. To test various rice cultivation methods and to create decision-support systems to assist them in the cultivation and care, they had to delve deeply into the supporting science. To create new LEGO robotic-controlled apparatuses to assist in farming and environmental sensing (or just for play), they had to go deeply into the underlying engineering, control, mechanics, and physics. What unified these various endeavors was the formal language for description.

Moreover, the dam design is but one example from one site. Other sites also had similar results. For example, in the north where water was not a problem, people worked on issues of soil erosion, developing and testing alternatives to slash-and-burn farming, experiments in nutrition, and cultivation of new crops. They also worked on social issues such as substance abuse, public health monitoring and awareness, and creating community on-line magazines. The point is not that everyone should design a dam, but rather that at each site the learners could work-within the same methodology and same set of tools on projects of interest and import to them. That each site developed uniquely is an important result of this work.

The significant accomplishment in this work is demonstrating a significant gain in accomplishment among a population that had not previously exhibited such competence in educational institutions. This work demonstrates how to build on and enhance local knowledge. Within the design of this learning environment, the learners:

- Work from local knowledge and interests
- Bridge to other knowledge domains
- Liberate their local knowledge from its specific situated embodiment

While others have demonstrated the ability of people to develop technology and use science without the benefit of schooling (Levi-Strauss, 1966), the key point here is that the constructionist use of computational technology leveraged this ability and helped people apply their knowledge to new and varied situations in an extremely short period of time. The knowledge did not remain limited to the particular technology such as combustion engines, but rather they could use the malleable computer technology as a tool for understanding other domains. Moreover, bodies of knowledge such as mathematics and physics were opened to them in new and more accessible ways.

The role of the computer in this process is to draw on a set of skills that can be transferred to something different. Combustion engines provided a means for developing technical and diagnostic expertise; applications remained mechanical, however. Through computational tools, learners design and construct and thereby make the forms of knowledge they have more general. Developing technological fluency enables them to break out of the specific context and represent their knowledge in forms they can draw on in many contexts. Neither traditional education nor nonconstructionist use of technology enabled the recognition and leverage of this indigenous expertise.

Success was due to the existence of several critical elements in the design and affordances of the technology. The technology is a malleable, expressive tool for construction. We do not merely use the computer as a means of delivery of information or as a means of communication, although both of those uses are beneficial. Rather, the users program, in languages localized to their own language, building their own idioms, on projects of importance to them. By working on a variety of projects over time, they develop a technological fluency. The combination of relative freedom of expression and self-selected projects of interest facilitated the mobilization and leverage of indigenous knowledge.

The Design of Technologically Rich Learning Environments and the Reform of Education

Discovering the engineering expertise and hacking spirit among so many Thai people who had previously not succeeded in school is a major benefit from the Emergent Design approach utilized in this project. Not only had Thai educators not built on this talent and intelligence, they did not even recognize it. The typical school reforms, despite their intention to promote creativity, problem solving, technological capability, and so on, also are generally incapable of discovering and leveraging such local knowledge. This is due to their top-down, preplanned, standardized, curricular approach.

There is no way to know beforehand for every site what will resonate and what local concerns and local knowledge exist. What one can assume is that there always is something. Using the Emergent Design framework, combined with principles of learning environments and open, programmable, technological tools, this "something" can be built upon and leveraged.

The work suggests a conclusion with a very broad sweep: The latent learning potential of the world population has been grossly underestimated as a result of prevailing mind-sets that limit the design of interventions to improve the evolution of the global learning environment.

NOTES

1. In the case of this work, the principles of the learning environment include constructionism, technological fluency, computer immersion, long-term projects, learner-centered activities, and connected projects. Later in this section I will provide a brief description of each concept. While it is beyond the scope of this chapter to delve into detail for each, the sense of the work will emerge through the description here.
2. Trademark or registered trademark of the LEGO Group.
3. I am adopting the convention of capitalizing the word "School" when referring to School as an institution containing the prevailing mind-set around organization, process, learning, and teaching.

4. The Suksapattana Foundation was created by Thai MIT alumni in honor of the fiftieth anniversary of the King of Thailand's ascension to the throne. They procured funding and coordinated a number of socially beneficial projects in honor of His Majesty the King of Thailand.
5. This chapter will only describe some of the activity within one village *learning* center. For more detail, please see Cavallo (2000).
6. While I do not have strong evidence for this belief, I base this statement on a number of conversations with villagers and others who worked with them.
7. I mean this in the original, positive sense of "hacking" where the term signifies informal and creative engineering expertise, and not someone practicing malicious destruction.
8. This is the term in use in Thailand and so I adopt it, but prefer the idea of "indigenous technology" as it is respectful and not pejorative.

REFERENCES

Cavallo, D. (1996a). *New Initiatives in youth development: Technology works enterprises. International Conference on the Learning Sciences.* Lawrence Mahwah, NJ: Erlbaum.

Cavallo, D. (1996b). *Leveraging learning through technological fluency.* Master's thesis. Cambridge, MA: MIT Media Laboratory.

Cavallo, D. (2000). *Technological fluency and the art of motorcycle maintenance: emergent design of learning environments.* PhD thesis, MIT Media Laboratory, Cambridge, MA.

Dewey, J. (1938). *Experience and education.* New York: Collier Books.

Freire, P. (1972). *Pedagogy of the oppressed.* New York: Herder & Herder.

Gruber, H. E., & Voneche, J. J. (Ed.). (1977). *The Essential Piaget.* New York: Basic Books.

Harel, I. (1991). *Children designers: Interdisciplinary constructions for learning and knowing mathematics in a computer-rich school.* Norwood, NJ: Ablex.

Lester, R. K. (1998). *The productive edge: How U.S. industries are pointing the way to a new era of economic growth.* New York: W. W. Norton.

Levi-Strauss, C. (1966). *The savage mind.* Chicago: University of Chicago Press.

Office of the National Education Commission. (1999). *Education in Thailand 1998.* Bangkok, Thailand: Office of the Prime Minister, Kingdom of Thailand.

Papert, S. (1980). *Mindstorms: Children, computers, and powerful ideas.* New York: Basic Books.

Papert, S. (1990). *Computer criticism vs. technocentric thinking: E&L Memo #1.* Cambridge, MA: MIT Media Laboratory, Epistemology and Learning Group.

Papert, S. (1991). Introduction. In S. Papert & I. Harel (Ed.), *Constructionism.* Norwood, NJ: Ablex.

Papert S., & Resnick, M. (1995). *Technological fluency and the representation of knowledge. Proposal to the National Science Foundation.* Cambridge, MA: MIT Media Laboratory.

Phongsupasamit, S., & Sakai, J., (1989). Studies on engineering design theories of hand-tractor ploughs. *Proceedings of the Eleventh International Congress on Agricultural Engineering, Dublin,* 1617–1626.

Resnick, M., Bruckman, A., & Martin, F. (1996). Pianos, not stereos: Creating computational construction kits. *Interactions 3,* 6.

Sachs, J., Gallup, J. L., & Mellinger, A. (1998, July). *Geography and economic development.* Paper presented at the Annual World Bank Conference on Development Economics, World Bank Group.

Tyack. D., & Cuban, L. (1995). *Tinkering toward Utopia: A century of public school reform.* Cambridge, MA: Harvard University Press.

Mydans, S. (1997, January 21). How country air took the seat out of one stop. The *New York Times,* Section A, p. 4. Retrieved from query.nytimes.com/gst /fullpage.html?res=9A05E4DC1638F933A25752C0A961958260

CHAPTER 8

FAST PROTOTYPING AS A COMMUNICATION CATALYST FOR E-LEARNING DESIGN

**Luca Botturi, Lorenzo Cantoni,
Benedetto Lepori, and Stefano Tardini**

This chapter proposes a renewed perspective on a known project management model, fast prototyping, which was adapted for the specific issues of e-learning development. Based on extensive experience with large e-learning projects, we argue that this model has a positive impact on e-learning project team communication, and that it provides a good basis for effective management of the design and development process, with specific stress on human-factor management. The chapter stems from the experience gained at the eLab, e-learning laboratory (www.elearninglab.org), a lab run jointly by the Università della Svizzera Italiana (USI—University of Lugano) and the Scuola Universitaria Professionale della Svizzera Italiana (SUPSI—University of Applied Sciences of Southern Switzerland) in Switzerland. It contains three case studies of different applications of the fast prototyping model, and has a strongly practical focus.

Previously published in Bullen, M., & Janes, D. (Eds.). (2006). *Making the transition to e-learning: Strategies and issues* (pp. 266–283). Hershey, PA: Idea Group.

INTRODUCTION: SOME ISSUES IN LARGE E-LEARNING PROJECTS

The transition to e-learning in higher education institutions, at course, program or institutional level, always requires a radical change in the organization. This means that instructors, teaching assistants and subject-matter experts are faced with a new situation, in which a lot of the assumptions on which they previously relied are brought into discussion. Moreover, they need to work in teams with other professionals—graphic designers, Web programmers, instructional designers, and so forth,—who might not share their professional language and understanding of the topic and of teaching and learning as such (Botturi, 2006). In many cases the team members are novices in the field of e-learning and do not have sound design practices or established routines for their tasks, and the team cannot rely on shared "common ground" for mutual understanding (Clark, 1996).

From the point of view of teaching staff, we should consider at least two main layers: (a) that of knowledge/skills; and (b) that of the attitudes required to implement effective and efficient e-learning experiences. In the first layer, the main issues are concerned with a radical change in the teaching development context, moving from a craftsmanship model—the teacher looking after the whole teaching process, from conception to delivery, from materials development to evaluation—to an industrial model, where many different people, with different professional backgrounds are to collaborate in order to design and implement the e-learning experience (Bates & Poole, 2003). In the second layer, an instance of the well-known process of diffusion of innovation is found: people fear innovation, and resist it unless positive conditions occur (Rogers, 1995).

The design model, which embodies the overall approach to e-learning, plays a key role in tackling these issues. This chapter addresses them in the context of large e-learning projects, where a fast prototyping model has been adopted, stressing two areas of intervention in the two layers.

1. The first area is collaboration in working groups, where people with different backgrounds and expectations are to collaborate in order to develop e-learning applications. In fact, the design, development and delivery of an e-learning course or program is a team activity that requires a high level of coordination and cooperation, and integration in the organization's culture (Engwall, 2003). The people who take part in the process should feel at ease if they are to express real commitment to the project and establish trust in each other. This is particularly true for teachers and instructors who play the key role in an online course, as they are mainly responsible for content production and course delivery.

2. In the second layer, fast prototyping provides e-learning projects with the attribute of *trialability*, so important in fostering the adoption of innovations: trainers not accustomed to the e-learning field are offered a concrete experience of what courseware could be; this, in turn, helps them leave aside prejudices and negative attitudes. The following section will provide some background about the management of e-learning projects and the institutional context of the Swiss Virtual Campus (SVC), from which our case studies are drawn. We will then introduce some reference to the design models from Instructional Design (ID) research, and then move on to present the eLab fast prototyping model, which will be described and discussed through three case studies.

Background: Institutional Context

This chapter mainly focuses on the introduction of information and communication technologies in traditional campus-based universities; namely, we will deal with the projects promoted by the Swiss Virtual Campus (SVC, www.virtualcampus.ch) program to introduce e-learning in Swiss higher education institutions (Lepori & Succi, 2003). The SVC program understands *e-learning* as defined by the Commission of the European Community: "the use of new multimedia technologies and the Internet to improve the quality of learning by facilitating access to resources and services as well as remote exchanges and collaboration" (CEC, 2001). This definition includes all e-learning models that could be situated on the continuum between fully face-to-face teaching and fully distance education through the Internet (Bates, 1999).

SVC projects bring together a network of higher education institutions for the development of shared e-learning resources. Project team members usually speak different languages and have a different background and education; moreover, for most of them, it is their first experience in e-learning. These situations are characterized by the lack of established routines and of common ground, so that developing a shared understanding and setting clear goals is often an issue.

There is a growing body of literature concerning the adoption of e-learning in European universities showing a consistent pattern (e.g., Collis & Van der Wende, 2002; Lepori & Succi, 2004; Van der Wende & Van der Ven, 2003): in most cases e-learning is introduced in a very decentralized way and as an instrument to improve existing face-to-face activities rather than to radically transform them (Collis & Van der Wende, 2002); moreover, only in some cases does the introduction of technologies lead to the creation of new educational offerings and of specialized subunits: e-learning is generally embedded into the existing curricula and departments (Lepori, Cantoni, & Succi, 2003).

There are some features here that are not easily compatible with conventional ID models and practice, especially in e-learning (Lepori & Perret, 2004):

1. e-learning is rarely implemented as stand-alone, online courses, but more often as units within existing face-to-face activities; this requires considerable integration of course production and delivery.

2. e-learning is embedded in a context where competencies and attitudes towards technology are very diverse, ranging from early adopters to a significant share of innovation-averse people (Rogers, 1995; Surry & Farquhar, 1997); thus, we cannot assume from the beginning that all people involved in a project have sufficient competencies in educational technologies, nor that they share the same vision concerning their adoption and usefulness. Communication and sharing views is thus a central issue.

3. The academic culture traditionally attributes a central role to the professor, not only in deciding the main guidelines for course content, but also in managing and fine-tuning it during the delivery; a work division between the production of contents (by experts of the subject), their technical implementation and their delivery (possibly with tutoring) is not compatible with this culture. It is thus necessary to involve professors in all development phases, but this makes project management more difficult, since academic hierarchies interfere with it.

4. University education is far from being homogeneous in aspects like the level of standardization of contents, the type of delivery, the level of students, etc.

Thus, each e-learning application has to be tuned to its specific context. Projects in the first phase of the SVC program, launched in 1999, were seriously beset by these issues. The SVC financed the development of online courses aimed at university students and produced by large consortia of Swiss universities: the underlying logic was to gather the contributions of different professors on the same subjects to produce high-quality courses to be used throughout Switzerland, thus achieving economies of scale. An accompanying study showed that this model—largely inspired by the production of on- line courses in distance universities—was in most cases at odds with higher education and academic culture (Lepori & Perret, 2004). As a result, development was delayed, most projects did not complete all the units foreseen, and a lot of energy was spent in experimenting and in discussing technical issues. The average cost per project was very high (for a single university course the costs in many cases exceeded U.S.$700.000). Also, project management proved to be difficult due to the size of the projects and to academic conflicts, while project coordinators were mostly relegated to an executive role. We could say that the failure of the model proposed by the SVC led most projects to go back to more traditional academic models, well-suited for research, but not for e-learning course development.

During the preparation of the second phase of the SVC (CUS, 2002), the eLab, the e-learning support centre of the Università della Svizzera italiana (USI—University of Lugano) and of the Scuola Universitaria Professionale della Svizzera Italiana (SUPSI—University of Applied Sciences of Southern Switzerland) developed a critical reflection on possible development models for e-learning courses in traditional universities. Management science has proved that the best management model for a project depends to a large extent on two elements: (a) the kind of application to be developed and (b) the specific institutional context, considering not only organizational issues and resources, but also the organizational culture and the relationship with institutional strategies (Engwall, 2003). The SVC experience thus far and an extensive body of empirical research (Lepori & Rezzonico, 2003; Lepori & Succi, 2003) showed that most classic ID models rely on assumptions which are, to a large extent, incompatible with the mainstream academic culture in traditional campus-based universities, and in many cases the success of e-learning projects was hindered by these incompatibilities.

Our effort therefore concentrated on developing a different approach: the goal was to provide simple guidelines that could fit into the existing cultural frameworks and enhance communication in our teams. This model was included in the e-learning management manual (Lepori, Cantoni, & Rezzonico, 2005), which was distributed to all new SVC projects started in Summer 2004.

In order to set the context for the presentation of the model, the next section will introduce some current ID models and clarify some of their assumptions in relation to the context of SVC projects and of the introduction of e-learning in traditional higher education institutions.

ID Models and Their Assumptions

The tradition of ID has collected a huge number of models that guide the design and development processes of instructional units (Andrews & Godson, 1995). Each model emphasizes a peculiar aspect of the process, striving to achieve prescriptive value without overlooking the eclectic (and often hectic) reality of practice. Classic ID models, starting from ADDIE, up to ASSURE (Heinrich, Molenda, & Russel, 1993) and the Dick, Carey, and Carey model (2001), take a linear perspective: they describe the ID process as a structured and orderly step-by-step activity, characterized by a progressive advancement through Analysis, Design, Development, Implementation and Evaluation; the process also includes a cycle of revision for each edition or delivery of the training (Figure 8.1).

Such models, which have behaviorist roots and were mainly developed in the military context, still represent the foundations of ID as a discipline, and have provided inspiration for many projects. They offer clear guidance, emphasize the intrinsic logic of design, and rely on two main assumptions:

1. The assumption of quality information: the designer can work on complete information (from the Analysis phase) and the designer can rely on the fact that the instructional context is stable (i.e., there are no unforeseen events)

2. The assumption of expertise: the designer can master the process and will not make errors, and all the team members and

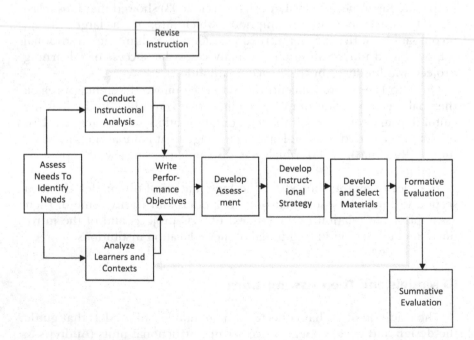

Figure 8.1.   Sample linear model (adapted from Dick, Carey, & Carey, 2001).

stakeholders will give their contributions as required, at the right moment and in a clear and unambiguous manner

In the history of ID, a specific emphasis in the education of instructional designers was derived from the second assumption: it was more process-oriented and tool-oriented than communication-oriented. Experience such as that in the analysis of SVC projects has shown that these assumptions do not always hold in the academic setting: often stakeholders, professors and instructors cannot express precise requirements, and it can happen that the analysis overlooks some relevant details; also the actors in the ID process may make errors. These are exactly the pitfalls that we identified for e-learning design, a setting in which technologies bring more complexity and uncertainty.

More recent works in ID have proposed a heuristic approach, less prescriptive and more practice-oriented (Figure 8.2). Morrison, Ross, and Kemp (2003) proposed a model that includes all the *steps* proposed by Dick, Carey, and Carey (2001) as *elements* in a progressive discovery model:

> The elements are not connected with lines or arrows. Connections could indicate a sequence, linear order. The intent is to convey flexibility, yet some order in the way the nine elements may be used. Also some instances may not require treating all nine elements. (p. 8)

The designer will decide which ones are relevant and which do not require particular consideration. This provides play for adapting to new technological situations in the e-learning domain.

The assumption behind this model is that the designer has strong meta-cognitive skills:

> she or he can shape and re-shape the process according to the situation. From a relational point of view, the designer also needs strong leadership skills, as he or she has to steer the design and development process with a good deal of improvisation, without relying on the solid guidance provided by linear models.

The R2D2 model (Willis, 1995) takes a similar perspective, borrowing a strong emphasis on communication and negotiation from constructivism, and placing itself at the opposite ideal end of linear models. R2D2 has four overarching principles:

1. Recursion: the steps/elements are revisited at different times, and decisions can be made anew, shaping a spiral-like flow.

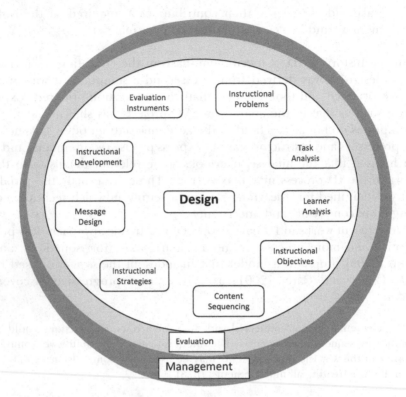

Figure 8.2.   Sample heuristic model (adapted from Morrison, Ross, & Kemp, 2003).

2. Reflection, which is contrasted with the linear design rationality of linear models: according to Willis (Colón, Taylor, & Willis, 2000) "(r)eflective design places less faith in preset rules and instead emphasizes the need ... to thoughtfully seek and consider feedback and ideas from many sources"

3. Non-linearity, that is, R2D2 does not present a set of steps, but rather one of focal points, close to the idea of elements in Morrison, Ross, and Kemp (2003, see above).

4. Participatory design: the whole idea behind this model is that the ID process is not only the designer's job, but rather team work, in which different people collaborate.

Communication and negotiation acquire a primary role here.

The drawback of this model is that much is left to interaction, and very little guidance is provided for complex or problematic situations. Namely, when few recognized common practices exist, the discussion may expand without converging. On the other hand, R2D2 and other constructivist models are focused on the fact that instruction lives in a specific context, and its conception, design and development should be strongly rooted in it. The community dimension is here taken as the focal point, and the model aims at providing a controlled space for discussion, maximizing sharing and mutual understanding in the design team, helping it develop a common background and hence to become—at least to a certain extent —a community of practice (Cantoni & Piccini, 2004; Wenger, 1998; Wenger, McDermott, & Snyder, 2002).

The eLab fast prototyping model tries to merge the three perspectives (linear, heuristic and constructivist) by providing a method organized into brief steps for the development of a "physical" focus of discussion—namely, a prototype. Its major aim is to have a development model soft enough to adapt to each project, but at the same time sufficiently structured to keep development time and costs reasonable. This was necessary also because the budget of the second series of SVC projects was significantly reduced.

FAST PROTOTYPING: THE ELAB MODEL

The Model

The eLab chose to tackle these issues in e-learning projects in higher education with a well-shaped and sound prototype-based design and development model. The originality of the approach lies in considering fast prototyping as a communication catalyst: the main advantage of a fast prototyping model is to enhance discussion in the team in a focused way, by concentrating on facts and results, and not on theories or prejudices about learning technologies. Enhanced and focused communication fosters the development of mutual understanding among the different professionals involved in the project, and the creation of trust—two important conditions for a successful development. The goals for which the eLab model was developed are:

1. To make the design and development process flexible with respect to ideas emerging from the progressive understanding of the project among team members, by providing moments in which new inputs can be taken into account.
2. To make the design and development process adaptable to new needs emerging from tests and results, given that the use scenario is varied (multiple institutions), partly undefined (e.g., changes in

curricula because of higher education reforms), and not available in detail at the outset of the project.

3. To allow teachers, instructors and subject-matter experts to focus on the teaching and learning activities, and not on the technologies themselves, fostering *trialability*.

4. To enhance communication with external partners.

The adapted fast prototyping model for e-learning is structured in two cycles: (a) the inner or *product cycle*, and (b) the outer or *process cycle* (Figure 6.3).

The design and development process starts with the identification of high-level learning goals and of a specific strategy, for example, teaching level B1 English with a game-based strategy; or teaching the basics about color perception with a case-based approach. This is a team effort, often accomplished in writing the project proposal.

These elements are embedded in a scenario, a narrative and semi-formal description of the instruction, which sets some parameters, namely: target students, communication flow and support, the organization of the

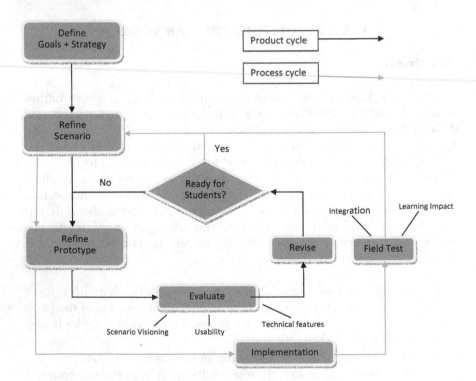

Figure 8.3.   eLab adapted fast prototyping model.

schedule in terms of time allocation and as a blend of face-to-face and distance learning activities, the use of multimedia and interactive technologies. The scenario is therefore an informal definition of the instructional and technical requirements for the project. It is paramount that the scenario is agreed upon by all team members, as it serves as leverage for the evaluation and revision of the prototype. The development of a shared scenario, guided by the instructional designer, is in itself an important activity for the project: team members who are led to discuss in concrete terms and to try to see the final product with the students' eyes.

The Product Cycle

The scenario is the starting point for the product cycle which starts with prototype development and is aimed at developing a product that fits the scenario. By *prototype* we mean structured courseware, with real content, already implemented as if it were to be used in a real setting. A prototype often includes only a part of the content, or leaves out some features, but the main point is that it is actually usable in the related scenario.

The project team then internally evaluates the prototype in two ways:

1. The eLab staff evaluates it with standard procedures that assess its technical features and usability, and produces a list of improvements that are proposed to the team.

2. Other non-technical team members try out the prototype's fit to the scenario description in a focus group in which they envision its use in the scenario they developed.

This double revision process provides full-spectrum feedback, and makes project members move one step further in the development of a shared understanding: while developing the scenario they merely described a wish-situation; now, the prototype has them evaluate single features (e.g., navigation structures, exercise feedback, etc.) and make decisions. Moreover this discussion helps the designers to get an insight into the non-technical partners' understanding of the training.

After the evaluation, the prototype is consequently revised, and a decision is made as to whether it is ready for real testing. If it is not, another product cycle is performed, starting from a refinement of the scenario according to the new possibilities explored during evaluation; if it is, the process moves onto the process cycle. When this happens greatly depends on the single project, as discussed in the case studies below.

The Process Cycle

The process cycle is basically a field test. Its first step is the refinement of the scenario (a virtual description) into the description of an actual use setting: one single institution, a specific group of students, which technical facilities, etc. The prototype is accordingly revised and adapted, and is then implemented and integrated into the course. The testing is constantly monitored, and the final evaluation of the process cycle happens in three steps: (a) first, with a standard questionnaire, delivered to the students, which measures Kirkpatrick Levels 1–3 (satisfaction, learning, transfer; cf. Kirkpatrick, 1998); (b) secondly, by analyzing the performance of students in the course exam or assessment; and (c) with a focus group that collects feedback from the instructors.

The evaluation provides new input for the project team, which can decide to make revisions and perform another test, to conclude the implementation and produce the final courseware, or even to switch back for another product cycle, if the real situation has proved very different from the scenario.

The following case studies show the impact of this model in three SVC projects supported by the eLab. Case studies will be analyzed with respect to the type of e-learning application development, the subject matter, the institutional partners and team members, budget and expected results.

EAD—ECOLOGY IN ARCHITECTURAL DESIGN

The goal of this project is to develop a blended learning course on the integration of ecological issues into architectural design, both at the level of buildings and of human landscape. The project leader is the Accademia di Architettura of the USI (Academy of Architecture). The project started in July 2004 and immediately adopted a fast prototyping model: at the beginning of September 2004 the first prototype module (*Building –Climate*) had already been developed by the project leaders, moving from the product to the process cycle; the module was then tested with more than 100 USI bachelor students in the winter semester 2004 (October 2004–February, 2005). At the same time as the test phase, the module was evaluated and discussed by all other project partners.

In the case of EAD, the rapid development of the prototype module aimed to rapidly create a concrete basis for communications about the course, thus avoiding long and useless discussions focused only on abstract ideas about e-learning. By being shown an example of how the modules could be designed and could appear, all the people involved in the project, even those who were not expert in e-learning, could get a concrete idea of the course. In fact interesting discussions soon arose among project partners, in particular about the issues of information design and of graphics and layout. The emergence of these discussions also shows one of the possible drawbacks of catalyzing communications through fast prototyping: the risk of focusing on specific details and losing touch with higher-priority issues, thus creating a situation of being unable to "see the wood for the trees" (Cantoni & Piccini, 2004). In the EAD project team, for instance, the issue of graphics and layout catalyzed most of the discussion, partly because of the scientific background of the team members. This fact can be analyzed from two opposite perspectives: on one hand, it can be seen as a drawback, in that, as already mentioned, focusing the discussion on details does not allow the overall picture to be seen, thus hindering discussion and decisions about more important issues; on the other, it can be turned into an advantage as well, in that the most important decisions can be made by the experts without long discussions. Of course, the responsibility for leading the discussion onto relevant issues and taking advantage also of discussions about details is up to the project manager.

However, on the basis of these discussions and of the results of the test phase, the prototype module of the EAD course was then refined, and other modules developed according to a template that was approved by all partners. In June 2005, exactly halfway through the project schedule, eight modules out of twelve had been developed and were ready to be delivered to students for a second test phase. Thus, half of the whole project time could be spent on implementing the last modules and testing and refining the whole course.

Colore

The goal of this project is to develop a set of content, resources and exercises both on the fundamentals of color (physics, perception, processes) and on color applications in different domains of the arts and visual communication. The project leader is the Dipartimento Ambiente, Costruzione e Design of SUPSI (Department of Environment, Construction and Design). The prototype of the first module (*History of color*) was developed in the first two months by the project leader and presented during a project meeting in November 2004.

The prototype immediately acted as a fuse in a powder keg: facing a concrete object, the project partners made their thoughts clear and hidden misunderstandings emerged at once: would the online resources be tailored to a specific partner's needs, or would they be more general-purpose? Would they foster offline activities, such as lab experience, or would the project invest in creating highly interactive online materials? The prototype triggered useful discussions, not only about the design and the graphical layout of the course modules, but also about more general issues regarding the project, such as the division of work, the assignment of tasks, the future use of the course, and so on.

Immediately after the first prototype, a second prototype module was developed (*Physiology of color*) and the general structure of the learning environment was designed. It is worth noticing here that in this case the rapid development of a prototype did not help the production of learning materials as such, but played a very important role in revealing some critical issues about the project itself, which had remained hidden during the drafting of the project proposal. Also in this case, focusing on general issues concerning the whole project might be seen as a waste of time, because they risk slowing down the project's progress; however, if these issues had not been faced immediately, a longer delay would have occurred, with very negative consequences on the development of the project. The project manager's task in this case was to have the discussion converge on key decisions, without letting it flare up into an argument.

ARGUMENTUM—E-COURSE OF ARGUMENTATION THEORY FOR THE HUMAN AND SOCIAL

Sciences

The goal of this project is to create a set of customized autonomous blended-learning courses about argumentation theory in different social contexts and for different educational purposes and targets. The Faculty of Communication Sciences of the USI leads the project. In this case, the prototype module was the introductory module; this module has a rather particular status within the whole course, since it presents a general introduction to argumentation theory, and was not conceived in order to be integrated into specific courses, but was instead to be freely accessible to everybody on the Internet. The introductory course was developed (both in Italian and in English) by the project leader during the first 4 months of the project. It was presented to all the other partners during a project meeting in February 2005, and tested with about 60 USI master's students in the first half of the summer semester 2005 (March–April, 2005). In parallel with the test phase, three other courses were developed by the project leader to be used and tested in the second half of the semester in three different USI masters. The presentation of the prototype allowed the project team to reach a rapid agreement about the graphical appearance of the course, so that the three other modules could be developed quickly.

The main function of the prototype modules was to help the project partners understand the possibilities, the opportunities and the limitations of the Learning Management System where the course runs. As a matter of fact, while the project leader had previous experiences with e-learning projects, the partners had not. Unlike the previously presented projects, the fast development of the prototype modules did not aim primarily at getting to a shared information structure of single modules, nor at fostering discussion, but rather at leveling the knowledge of the project team members about the technologies employed by showing them their main features and possibilities for use. In this way, the project partners could see, for instance, how maps could be used as tools for the metaphorical representation of contents, for accessing them and for orientation during the navigation into the course; what kind of learning material could be used for what purpose (e.g., PDF files for case studies, HTML pages for general contents, video files for interviews with experts, etc.); how discussion activities could be implemented in the course, and so on. As a consequence, leveling the knowledge of the project team by showing them some examples led to a shared concept about the general structure of the course and of the learning materials.

Critical Discussion and Conclusions

Fast prototyping has been around quite a while, especially in human-computer interaction and computer science, as a way to foster user-centered design. Moving from the issues which have emerged in large multilingual and multi-institutional e-learning projects in the SVC program, our approach has considered the same approach as a communication catalyst. Fast prototyping can enhance e-learning development by improving both team communication and team commitment; it supports the development of a shared understanding of what is being discussed and designed and gives team members the opportunity to try out the e-learning experience in the first person and to be involved from the very beginning. As for any development model, fast prototyping is not a panacea that ensures effectiveness and efficiency. Fast prototyping shows its advantages where (a) the project is quite big; (b) team members are not accustomed to working together; and/or (c) many of them have little experience in e-learning course development. Moreover, experience so far has highlighted a few conditions that seem to be required in order to make fast prototyping a sensible choice (or even a necessary one).

1. Fast prototyping costs. What is developed risks being rejected and "demolished," even if in critical and fruitful demolition. In order to be cost effective, a sound ratio between prototype scale and the final product is needed; when this is not feasible, examples taken from other experiences may be used.

2. Fast prototyping is particularly helpful in order to provide a shared understanding of what the final e-learning course is likely to be; it offers the development team a common background where many misunderstandings can be avoided. Being multi-disciplinary, e-learning teams bring together people with very different backgrounds, who need to share a simple, effective and efficient way of collaborating, each of them providing her or his own contribution, while acknowledging the expertise of others. It is important to note that committing to a human-centered approach implies that also the choice of fast prototyping itself has to be negotiated and shared among team members.

3. E-learning is a new world: it happens quite often that people working in course development do not have extensive experience. Fast prototyping provides them with a common language and an initial experience of e-learning. In fact, while point (b) above underlines the usefulness of fast prototyping to reduce team heterogeneity in general, (c) stresses its being a tool that enhances e-learning competencies inside the team.

These conditions are necessary but not enough to provide a sound fast prototyping experience. Two pitfalls in particular are to be mentioned here, both concerned with the prototyping speed. The first pitfall is the "quick and dirty" effect, that is, a very rapid, but low quality development may negatively affect further developments, hindering understanding, collaboration and commitment. The second one is just at the opposite pole in the "speed" scale: the non-fast prototyping case. Here the prototyping phase is extended so much that it only delivers a late contribution, which often has to be accepted as time resources do not allow substantial revisions. Continuous and endless prototype revisions turn into the biggest obstacle in the actual e-learning course development. Successful e-learning projects are always team efforts (Botturi, in press), and depend absolutely on the quality of team collaboration. The SVC experience has shown that classic ID models are often at odds with academic tradition when introducing e-learning technologies into higher education institutions, generating conflicts and misunderstandings. If properly managed and applied to a context that can benefit from it, a fast prototyping approach can provide an opportunity to enhance communication by providing a concrete focal point—the prototype—for discussion and design.

This model can leverage on the human factor in order to achieve better designs and finally better e- learning applications.

REFERENCES

Andrews, D. H., & Goodson, L. A. (1995). A comparative analysis of models of instructional design. In G. Anglin (Ed.), *Instructional technology. Past, present, and future* (pp. 161–182). Englewood, CO: Libraries Unlimited.

Bates T. W. (1999), *Managing technological change: Strategies for college and university leaders*. San Francisco: Jossey-Bass.

Bates, T. W., & Poole, G. (2003). *Effective teaching with technologies in higher education*. San Francisco: Jossey-Bass.

Botturi, L. (2006). E2ML. A visual language for the design of instruction. *Educational Technologies Research & Development, 54*(3), 265-293.

Cantoni, L., & Piccini, Ch. (2004). *Il sito del vicino è sempre più verde. La comunicazione fracommittenti eprogettisti di siti internet*, Milano, Italy: FrancoAngeli.

CEC. (2001). *The eLearning action plan: Designing tomorrow's education*, COM(2001)172, Brussels, 28.3.2001. Retrieved June 24, 2005, from http://europa.eu.int/comm/education/policies/ntech/ntechnologies_en.html

Clark, H. H. (1996). *Using language*. Cambridge, MA: Cambridge University Press.

Colón, B., Taylor, K. A., & Willis, J. (2000). Constructivist instructional design: Creating a multimedia package for teaching critical qualitative research. *The Qualitative Report, 5*(1–2). Retrieved June 8, 2005, http://www.nova.edu/ssss/QR/QR5-1/colon.html

Collis, B., & Van der Wende, M. (2002). *Models of technology and change in higher education*. CHEPS report, Toegepaste Onderwijskunde.

CUS. (2002), *Campus virtuel suisse–Programme de consolidation visant à renouveler l'enseignement et l'étude (2004–2007)*. Berne. Retrieved June 24, 2005, from http://www.cus.ch/Fr/F_Projekte/F_Projekte_Campus/S_projets_campus_2004.html

Dick, W., Carey, W., & Carey, L. (2001). *The systematic design of instruction* (6th ed.). New York: HarperCollins College.

Engwall, M. (2003). No project is an island: Linking projects to history and context. *Research Policy, 32,* 789–808.

Heinrich, R., Molenda, M., & Russell, J. (1993). *Instructional media and new technologies of instruction* (4th ed.). New York: Macmillan.

Kirkpatrick, D. L. (1998). *Evaluating training programs: The four levels*. San Francisco: Berrett-Koehler.

Lepori, B., Cantoni, L., & Rezzonico, S. (Eds.) (2005). *Edum eLearning Manual*. Lugano, Switzerland: University of Lugano. Retrieved from www.edum.ch

Lepori, B., Cantoni, L., & Succi, C. (2003). The introduction of e-Learning in European universities: models and strategies. In M. Kerres & Voss B. (Eds.), *Digitaler campus. Vom Medienprojekt zum Nachhaltigen Medieneinsatz in der Hochschule*. Münster, Germany: Waxmann.

Lepori, B., & Perret, J.-F. (2004), Les dynamiques institutionnelles et les choix des responsables de projets du Campus Virtuel Suisse: Une conciliation difficile. *Revue Suisse de Sciences de l'Education, 2/2004,* 205–228.

Lepori, B., & Rezzonico, S. (2003), Models of eLearning. The case of the Swiss virtual campus. *Proceedings of the International Conference on New Learning Environments 2003,* Lucerne, Switzerland.

Lepori, B., & Succi, C. (2003). *e-Learning in higher education. Prospects for Swiss universities.* 2nd EDUM report, Lugano. Retrieved June 24, 2005, from www.edum.ch

Lepori, B., & Succi, C. (2004). eLearning and the Governance of Higher Education in Continental Europe. *Proceedings of ELEARN 2004,* Washington, DC.

Morrison, G. R., Ross, S. M., & Kemp, J. E. (2003). *Designing effective instruction* (4th ed.). New York: Wiley & Sons.

Rogers, E. M. (1995). *Diffusion of innovations* (4th ed.). New York: The Free Press.

Surry, D. W., & Farquhar, J. D. (1997). Diffusion theory and instructional technology. *Journal of Instructional Science and Technology, 2*(1). Retrieved June 24, 2005, from http://www.usq.edu.au/electpub/e-jist/docs/old/vol2no1/article2.htm

Van der Wende, M., & Van der Ven, M. (2003). *The use of ICT in higher education. A mirror of Europe,* Utrecht, Holland: LEMMA.

Wenger, E. (1998). *Communities of practice: Learning, meaning, and identity.* New York: Cambridge University Press.

Wenger, E., McDermott, R., & Snyder, W. (2002). *Cultivating communities of practice.* Boston: Harvard Business School Press.

Willis, J. (1995). A recursive,reflective instructional design model based on constructivist-interpretivist theory. *Educational Technology, 35*(6), 5–23.

PROJECT WEB SITES

Argumentum: www.argumentum.ch
Colore: www.coloreonline.ch
EAD: www.ead-project.ch

CHAPTER 9

CONDUCTING RESEARCH ON PRACTICE

Virginia Richardson

This article explores two forms of research on practice: formal research and practical inquiry. Formal research is undertaken by researchers and practitioners to contribute to an established and general knowledge base. Practical inquiry is undertaken by practitioners to improve their practice. It is suggested that practical inquiry is more likely than formal research to lead to immediate classroom change; that these two forms of research are fundamentally different, and that both are useful to practice, but in different ways.

The purpose of this chapter is to explore issues related to the nature of research on practice, its uses by researchers and practitioners, and its benefits to educational practice. Since most of the research on educational practice has been conducted with elementary and secondary school teachers, the primary focus will be on teaching practice.[1] I will first describe how conceptions of teaching have shifted from a view of teachers as the recipients and consumers of research to the current view of the teacher as producer or mediator of knowledge. The newer conception of teaching has affected research on teaching in terms of what is examined, how the

Previously published in *Educational Research* (1994), *23*(5), 5–10.

research is conducted, and who conducts the research. This leads into a discussion of two forms of research on practice: practical inquiry undertaken by researchers or practitioners designed to contribute to an established and general knowledge base. New understandings of the teacher change process show us why practical inquiry is more likely than formal research to lead to immediate classroom change. I conclude by suggesting that these two forms of research are fundamentally different, that practical inquiry may be foundational to formal research, and that both forms of research are useful to practice, but in different ways.

RECENT RESEARCH ON EDUCATIONAL PRACTICE

Research on the practice of teaching is undergoing change; change that reflects considerations of power and voice, the nature of knowledge, and research methods. In the 1970s, researchers on teaching sought to uncover generic teaching behaviors that were related to student learning as measured on standardized tests (for summaries of this work, see Brophy & Good, 1986; Gage, 1985). While there were many concerns about the assumptions inherent in the process-product research paradigm (see, e.g., Doyle, 1977; Fenstermacher, 1979), the strongest criticism of this research followed an awareness of how it was being used (and abused) by policymakers, staff developers, teacher educators, and school and school district administrators (McCloskey, Provenzo, Cohn, & Kottkakmp, 1991; McNeil, 1988; Richardson-Koehler, 1988). The scholarly community soon realized that research on teaching should not be conducted in the absence of considerations of two questions: Who owns the knowledge on teaching practice, and who benefits from the research (see Apple, 1993)? Elementary and secondary classroom teachers were notably missing in answers to both questions.

Research on the practice of teaching has recently shifted from a focus on effective behaviors toward the hermeneutic purpose of understanding how teachers make sense of teaching and learning. Two forms of this research are discussed here: one conducted by individuals who are labeled researchers by themselves and others; and the other by teachers. The questions of power and control have now shifted somewhat to considerations of who creates, constructs, or reconstructs knowledge about teaching practice, whether such knowledge can move beyond the local to the more general, and the degree to which local or general knowledge may be used in the improvement of practice.

Studies conducted by researchers on teaching examine the nature of teachers' knowledge, beliefs, perceptions, and other such constructs in relation to learning to teach, teachers' classroom actions, and changes in practice. The research methods involve the use of participant observation within the context of classrooms and schools—the settings that constitute teachers' realities. These studies are often conducted in collaboration with practitioners, a practice that is thought to enhance the validity of the hermeneutic studies of teachers conducted by outside researchers (Connelly & Clandinin, 1990; Elliott, 1988; Goldenberg & Gallimore, 1991).

At the same time, there has been a strong movement toward teacher research that gives voice to practitioners, allows them to communicate their wealth of knowledge to other practitioners, and helps them improve their practice (see Cochran-Smith & Lyyle, 1990; Hollingsworth & Sockett, 1994). One could perhaps suggest that it is a teacher who knows best what it means to be a teacher. It is often argued that teacher research is fundamentally different from research conducted by outside researchers (e.g., Cochran-Smith & Lytle, 1993), if only because research concluded by teachers may be more useful to teachers for the improvement of practice. In fact, there are some discussions of teacher research that border on suggesting that teacher research, at least in the form of stories, is the only valid form of teaching research (Carter, 1993). This controversy is explored in a series of four articles and responses concerning the use of formal research on teachers' thinking in the improvement of teacher practice (Clark, 1988: Clark & Lampert, 1986; Floden & Klinzing, 1990; Lampert & Clark, 1990).

A VIEW OF TEACHING AND TEACHER CHANGE

The areas of interest in current formal research on teaching practice are teachers' thinking, knowledge, and beliefs, as well as classroom actions that are considered to be momentary solutions to enduring dilemmas (e.g., Carter, 1990; Kagan, 1990; Pajares, 1992). This research has led to the conclusion that teachers are driven by a type of thinking that differs from linear prescriptions and propositions derived from formal research and theory. Researchers in the new tradition, for example, have found that experienced teachers—even expert teachers—do not plan in the linear manner prescribed in many teacher education programs. In summarizing the descriptive literature on teacher planning, Borko and Niles (1987) suggest that teachers seldom start with objectives. Instead, they focus primarily on content and activities, and objectives are embedded within the activities.

Teachers' knowledge and beliefs are viewed as practical (Elbaz. 1983), personal (Clandinin, 1986; Lampert, 1985), situated (Leinhardt, 1988), craft like (Grimmett & Mackjnnon. 1992), embodied (Johnson, 1987), and relational (Hollingsworth, Dybdahl, & Minarik, 1993). It has also been suggested that such knowledge is structured in images (Calderhead, 1988), narratives or stories (Connelly & Clandinin, 1990; Gudmundsdottir, 1991), and cases or events (Doyle, 1990). This knowledge is seen as often tacit and always experiential (Kagan, 1990). Teachers' subject matter knowledge has been explored in relation to the ways in which it is combined with knowledge about student learning and the specific context to drive its curricular representation in the classroom (e.g., Grossman, 1990; Munby & Russell, 1992; Shulman, 1987; Wilson & Wineburg, 1988).

The conception of teaching underlying these projects rejects the dominant norm among many educators and policymakers that the teacher is a recipient and consumer of research and practice. Rather; the teacher is seen as one who mediates ideas, and constructs meaning and knowledge, and acts upon them. The ideas may come from many different sources such as staff development, other teachers, research and practice articles, and reflection on experience. New understandings are developed on the basis of these ideas as they interact with existing understanding.

The view of change in practice has also been enhanced by recent research. The conception of change in teaching practice that has dominated the educational Literature until quite recently focuses on teaching activities, practices, and curricula that are suggested or mandated by those who are external to the setting in which the teaching is taking place: administrators, school district officials, policymakers, and/or staff developers. This image of change has led to the conclusions that change hurts (Fullan, 1991) and that teachers are recalcitrant and don't change easily (e.g., Duffy & Roehler, 1986). However, as pointed out by Morimoto (1973):

> When change is advocated or demanded by another person, we feel threatened, defensive, and perhaps rushed. We are then without the freedom and the time to understand and to affirm the new learning as something desirable, and as something of our own choosing. Pressure to change, without an opportunity for exploration and choice, seldom results in experiences of joy and excitement in learning. (p. 955)

The newer conception of teacher change implies a different process—one that is not mandated by others but is undertaken voluntarily. Research that has led to this conception concentrates on teachers' practical reasoning and everyday work life, and suggests that teachers actually change all the time. These changes, while often minor adjustments in program, can also be quite dramatic (Richardson, 1994). The norms of the workplace and the systemics of the context also affect teachers' considerations of change in practice (Placier & Hamilton, 1994). When a teacher tries new activities, he or she assesses them on the basis of whether they "work," that is, whether they fit within the teacher's set of beliefs about teaching and learning, engage the students, and allow the teacher the degree of classroom control he or she feels necessary. If the activity does not work, it is quickly dropped or radically altered.

The decision as to whether a new activity works, however, is often not conscious, is highly personal, and may be based on experiences and understandings that are not relevant to the particular setting in which instruction is taking place. Teachers make decisions on the basis of a personal sense of what works, but without examining the beliefs underlying a sense of "working," teachers may perpetuate practices based on questionable assumptions and beliefs. In response to this new understanding of change, a normative conception of teacher as inquirer has evolved that provides a vision of a teacher who questions his or her assumptions and is consciously thoughtful about goals, practices, students, and contexts.

The research that led lo these conceptions of teaching and teacher change provides an understanding of how practitioners make sense of their settings and actions; however, it does not respond to the type and immediacy of knowledge needs that a teacher confronts in everyday classroom work. While the hermeneutic research on practice often attempts to bring two knowledge needs together—that is, the daily and immediate knowledge needs and the more general conceptual questions related to the nature of teaching practice—it may be difficult. if not impossible, to create a formal knowledge base that responds to immediate classroom needs. Formal research provides us with new and useful ways of thinking about teaching, and may eventually enter into teachers' practical reasoning and affect their practices (Fenstermacher & Richardson, 1993). What, however, will help teachers with immediate knowledge needs that do not necessarily call for generalized propositions?

This and other questions have provided rationale for the renewed interest in teacher researcher notions. These are research processes that are either taken over by practitioners or actively and equally involve them. I will now turn to a discussion of teacher research and will suggest that a form of teacher research, practical inquiry, has the potential to respond to teachers' immediate knowledge needs.

Teacher Research

Teacher research or teacher researcher as it is sometimes referred to, is a somewhat confusing concept. Cochran, Smith, and Lytle's (1990, 1993) works provide thorough descriptions of the notion and also highlight the tensions and contradictions in the field. There are several motivations captured in the teacher researcher notion, each representing some combination of political and improvement-to-practice concerns. One is that teacher inquiry will help to improve teaching; that teachers are as good as, if not better than, researchers in producing more valid and relevant research for their own classrooms. On the other hand, many researchers feel, as did Judd (quoted in Grinder, 1981) that teacher research could reduce the antagonism between researchers and practitioners and motivate teachers to accept and use all forms of research in their teaching. Still other scholars (e.g., Schön, 1991) feel that teacher research, or at least the equal involvement of teachers in research, would increase the validity of hermeneutic research. This last rationale responds to what Giddens (1976) terms the double hermeneutic in social science research and is related, here, to research on practice. The first hermeneutic is the conception of reality constructed by the teacher, and the second is the reconstruction of that meaning into new frames of reference by the researcher. Active collaboration, it is felt, leads to a shared or mutual reconstruction that is agreed upon by both practitioner and researcher as is exemplified in the work of Connelly and Clandinin (1990) and Cochran-Smith and Lytle (1990).

These multiple motivations have led to a number of meanings of the teacher researcher concept. Several meanings may be present in the same article as an author both argues, politically for teachers' voice and describes various forms of teacher research. One approach to teacher research, for example, is the notion that *teaching is research*. This argument suggests that the work of teaching is like the work of the researcher. The teacher experiments with a treatment or an activity, collects data, and makes decisions on the basis of data and judgments as to whether the activity "works" (Neilsen, 1990). A second approach relates to various conceptions of *teacher as reflective practitioner* (Dewey, 1933; Louden, 1991; Russell & Munby, 1992; Schön, 1983; Van Manen, 1977). Schön's (1983) notion of reflection-in-action and its potential to improve practice has often been referred to as teacher research. A third conception of teacher research relates to *action research*. In this form of teacher research, teachers as a group may become more systematic in thinking about their work, collect and analyze data related to perceived problems in their classrooms and schools, and, thereby, understand and improve their practice (Carr & Kemmis, 1986; Elliott, 1976-77; Noffke, 1992). A fourth notion of teacher researcher is qualitatively different from the preceding three. It is *teacher as formal educational researcher*. There is an expectation that this form of research will contribute no the knowledge base for use by others within the community through publication, workshops, or educational material. While few would suggest that this is the only form of teacher research, many do advocate the publication of teacher research (e.g., Cochran-Smith & Lytle, 1993; Duckworth, 1986).

The first three notions of teacher research described above may be thought of as "practical inquiry" whereas the fourth category is "formal research." While formal research is quite well understood in terms of accepted methodologies and considerations of quality, practical inquiry is not. An important project for the coming years for those involved in research on teaching will be to better understand how teachers make inferences in practical reasoning, the nature of practical inquiry as a form of research, and how practitioners can be helped to improve their practical inquiry. The next section will continue this exploration by providing an initial, though tentative, analysis of the differences between practical inquiry and formal research.

PRACTICAL INQUIRY AND FORMAL RESEARCH

Practical inquiry is conducted by practitioners to help them understand their contexts, practices, and, in the case of teachers, their students. The outcome of the inquiry may be a change in practice, or it may be an enhanced understanding. This type of research is not conducted for purposes of generalization, expanding the larger community's knowledge base, or publication. The term inquiry has been defined by Clift, Veal, Johnson, and Holland (1990) as "a deliberate attempt to collect data systematically that can offer insight into professional practice" (p. 54).

There is no formal research methodology associated with practical inquiry, although recent acceptance of qualitative research has accompanied an increased interest in and advocacy of practical inquiry The telling of narrative and dialogical conversations about practice, and writing of journals have been advocated for this type of inquiry. Connelly and Clandinin (1988), for example, suggest these forms of inquiry in a book for teachers on how to inquire into their own personal practical knowledge. Another example of practical inquiry is the maintenance of extensive records on observations of student progress by teachers who use literature to teach reading (Short, 1992). Reflection on the moral basis for action also constitutes an element of practical inquiry. Practical inquiry, then, is not conducted for purposes of developing general laws related to educational practice, and is not meant to provide the answer to a problem. Instead, the results are suggestive of new ways of looking at the context and problem and/or possibilities for changes in practice.

Formal research is what we usually think of in educational research. It is designed to contribute ton general knowledge about and understanding of educational processes, players, outcomes, and contexts and the relationship between or among them. This type of research is generally written about in research and research methodology chapters and books and is broken down by methodological types: experimental, correlational, survey, case study, qualitative, and evaluation.

The major distinction between these two forms of research is that practical inquiry is conducted in one's everyday work life for purposes of improvement, and formal research is meant to contribute to a larger community's knowledge base. As suggested by Nespor and Barylske (1991), those who conduct formal research "participate in a bigger network, one that is constituted by the circulation of more stable, movable, and combinable representations than that of other persons" (p. 809). There is a community understanding of appropriate rhetorical forms of representation, as well as adequate research designs and methodology in formal research, whereas the individual or group of practitioners involved in practical inquiry need only respond to a personal sense: of validity and further questioning. Both forms, however, may be conducted by the practitioner and at times, practical Inquiry may be turned into formal research.

I would speculate, at this point, that formal research methodology, because of its need to narrow and focus for purposes of generalizability, is too confining for practical inquiry. Further, any formalization of the process of practical inquiry turns it into formal research, thus potentially causing it to lose its value in informing day-today classroom actions. However, there are intellectual virtues, such as regard for evidence, that are important in both forms of research. 1 developed these understanding during my involvement in a project in which I was both a teacher practitioner (in this case, staff developer), and the co-principal investigator of a formal research project designed to study the process.[2] My experience with the knowledge needs in these two roles provides some instantiation of the differences between practical inquiry and formal research.

AN EXAMPLE OF PRACTICAL INQUIRY AND FORMAL RESEARCH

Patricia Anders and I, along with several colleagues, examined the beliefs of a number of upper elementary teachers of reading comprehension, the relationship of these beliefs to their classroom practices (Richardson, Anders, Tidwell, & Lloyd, 1991), and the nature of change in teaching (Richardson, in press). As an element of this study we implemented a staff development program that focused on the teachers' beliefs and practical reasoning, and current research on reading comprehension. Anders and I were the primary staff developers and, as teacher- practitioners in this role, were faced with a need to conduct practical inquiry.

The staff development process was new for us and the teachers. We did not walk into this process with a set of neat prescriptions for practice, but worked from the teachers' own understandings and rationale for their practices, a process described as practical arguments by Fenstermacher

(1986). We were faced with expectations for developing sound approaches to the staff development process as well as producing research on the process that would be useful to others in the research on teaching and teacher education communities.

As staff developers, we engaged in practical inquiry. We struggled with the nature and form of the new process. At the same time, we attempted to collect and analyze data that would inform others. We were conducting practical inquiry and formal research simultaneously and became aware of the differences. A great deal of data were collected and used in both research processes. These data included videotapes of the staff development sessions; extensive belief interviews of teachers, staff developers, and principals; and videotapes and written observations of the teachers in their classrooms.

Engaging in practical inquiry involved us in viewing videotapes of staff development sessions following a session. We were concerned with individual teacher responses and interactions: Did we interpret responses validly? How could we have missed what teacher A was trying to say? There is some underlying anger here; how should we deal with it in the next session? Are we talking too much? Too little? In this inquiry we were not looking for propositional, law-like statements. We wanted to understand *that* context and *these* participants so that we could meet our goals. These knowledge needs relate to those described by Habermas (1973) in his second category of "knowledge constitutive interests": interests that are practical and serve the purpose of meaningful communication and dialogue.

At the same time, and at a very different level of analysis, we were engaged in formal research. We were looking for themes in the processes and responses, clues for generalizations that we would begin to look for in videotapes of other sessions, for ways of examining concepts that had been identified in the literature, and new concepts that seemed to emerge from our interaction with the data. We were concerned, in this type of investigation, with methodological procedures accepted by our research community such as triangulation and reliability, generalizability and validity.

Practical inquiry provided us with immediate information that would help with the next class, with an individual teacher, and in understanding and describing the process to ourselves and the participants. The process of formal research did not provide immediate feedback to us as staff developers. It did, however, help us later on to understand the process in terms of both what it meant to "work" and "not work," as well as how we might conceptually approach the process next time around.

The results of practical inquiry then, informed our practice in quite different ways than did the formal research. It also, however, informed our formal research by helping to provide a sense of the important and worthwhile questions for formal research and for the goals of our program. One could suggest, then, that practical inquiry may be foundational to formal research that will be truly useful in improving practice.

IMPROVING PRACTICAL INQUIRY

In this article, I have suggested that a focus on teaching and hermeneutical inquiry has brought many researchers closer to practice and that this research has helped to describe what it means to think like a teacher and what is entailed in teacher change. However, many teachers still consider research on teaching to be irrelevant to their day-to-day practice, am attitude that makes it difficult for such research to help to improve practice. This may be because formal research cannot provide teachers with knowledge for their immediate needs within their unique contexts.

The formal research program, however, is helping us understand the ways in which teachers develop knowledge that they use in solving the immediate needs of the classroom. This understanding may lead to the development and acceptance of a different form of research, one that maps on to practitioners' methods of acquiring and constructing practical knowledge and their goals in improving classroom practice.

While we have developed procedures for generating and maintaining methodological and ethical standards for the conduct of formal research, the standards and means for developing and improving practical inquiry are not well understood. House, Mathison, and McTaggart (1989) explored a similar problem in evaluation studies. They suggested that we have three distinct situations with respect to the development of inferences:

> (a) the researcher draws inferences from an evaluation study and expects the practitioner to apply it; (b) the practitioner draws inferences from an evaluation study but modifies those inferences based on his or her particular domain of application, (c) the practitioner draws inferences based on his or her own experience and applies it in context. (p. 15)

They then suggest that the third type of inference has largely been ignored, but is the must crucial in terms of the improvement of practice: "we think that the third situation is more important than the fist two as far as the *conduct and improvement of professional practice* are concerned and that the validity concerns for practitioner inferences have been very much ignored" (p. 15). They strongly advocate research in this area.

I would agree. We know little about how to work with teachers in helping them improve their practical inquiry, although such work has begun (see, e.g., Cliff, Houston, & Pugach. 1990; Connelly & Clandinin, 1988; Fenstermacher & Richardson, 1993; Grimmett & Erickson, 1988). This work focuses on helping teachers reflect on their own beliefs, personal practical knowledge, and practices. We also are beginning to develop a foundational knowledge related to practical knowledge that will help in this endeavor (for a summary; see Fenstermncher, 1994).

Berliner (1992) counsels that educational psychologists should not view themselves as psychologists who are interested in education. Such an approach, he feels would miss "a chance to profit from the knowledge of practitioners" (p. 145). I would go one step further and suggest that the practitioners are not just those in elementary and secondary schools, but they are all of us. Many of us, after all, are educational practitioners in addition to being researchers. We may teach in higher education or con- duct staff development programs or workshops. Understandings of our own practice and of how we make inferences related to our practice can help us develop ways to improve the inferencing process through practi- cal inquiry.

The practical inquiry that Anders and I engaged in as staff developers, for example, responded to immediate needs, was clinical, and called on relational knowing—that is, having knowledge about specific participants (including ourselves) in our staff development process, This inquiry was enhanced by a strong and extensive database such as videotapes of the staff development process. Although these data were initially collected for purposes of formal inquiry; they were invaluable in our practical inquiry. We also found that our practical inquiry was strengthened by the presence of a dialogue partner. The manner that we developed in approaching the staff development process also helped us focus on practical inquiry. Anders and I were aware that we, as staff developers, were learning as much it not more in this process than the teacher participants. Thus, we allowed ourselves to admit failure, to experiment with different processes, and talk about our practical inquiry with the teacher participants.

In addition to understanding our own knowledge needs and inquiry processes as practitioners, we can also conduct formal research on practi- cal inquiry, as suggested by House, Mathison, and McTaggart (1989). This look into our own practice through practical inquiry and formal research, as well as into the practical inquiry of other practitioners, will provide information on ways of improving the process. These methods may then be appropriated by practitioners in other fields and/or grade levels to improve their practice through practical inquiry.

ACKNOWLEDGMENT

This chapter is a revision of a paper presented at the Invited Symposium New Approaches to the Intersection of Education and Psychology:

Alternative Paradigms, Emphases, and Organizational Schema at the American Psychological Association, August 1992, Washington, DC.

NOTES

1. However, other educational practitioners who work in teacher learner relationship with students, such as staff developers and teacher educators, have benefited, and will continue to benefit from research on teaching practice.
2. This 3-year project was funded, in part, by the Office of Educational Research and Improvement, U.S. Department of Education (Richardson & Anders, 1990).

REFERENCES

Apple, M. (1993). *Official knowledge: Democratic education in a conservative age*. New York: Routledge.

Berliner, D. (I992). Telling stories of educational psychology. *Educational Psychologist, 27*(2). 143–161.

Borko, H., & Niles, J. (1987). Descriptions of teacher planning: Ideas for teachers and research. In V. Richardson-Koehler (Ed.), *Handbook of research for educators* (pp. 167–187). New York: Longman.

Brophy, J. & Good. T. (1986). Teacher behavior and student achievement. In M. Wittrock (Ed.), *Handbook of research on teaching* (3rd ed., pp. 328–375). New York: Macmillan.

Calderhead, J. (1988). The development of knowledge structures in learning to teach. In J. Calderhead (Ed.), *Teachers' professional learning* (pp. 51–64). Basingstoke, England: Taylor & Francis.

Carr, W., & Kemmis, S. (1986). *Becoming critical: Education, knowledge and action research*. Philadelphia: Falmer.

Carter, K. (1990). Teachers' knowledge and learning to teach. In R. Houston (Ed.), *Handbook of research on teacher education* (pp. 291–310). New York: Macmillan.

Carter, K. (1993). The place of story in the study of teaching and teacher education. *Educational Researcher, 22*(1), 5–18.

Clandinin, J. (1986). *Classroom practice: Teacher images in action*. London: Falmer.

Clark, C. (1988). Asking the right questions about teacher preparation: Contributions of research on teacher thinking. *Educational Researcher, 17*(2), 5–12.

Clark, C., & Lampert, M. (1986). The study of teacher thinking: Implications for teacher education. *Journal of Teacher Education, 37*(5), 27–31.

Clift, R., Houston, W. R., & Pugach, M. C. (Eds.). (1990). *Encouraging reflective practice in education*. New York: Teachers College Press.

Clift, R. Veal, M., Johnson, M., & Holland, P. (1990). Restructuring teacher education through collaborative action research. *Journal of Teacher Education, 42,* 52–62.

Cochran-Smith, M., & Lytle, S. (1990). Research on teaching and teacher research: The issues that divide. *Educational Researcher, 19*(2), 2–11.

Cochran-Smith, M., & Lytle, S. (1993). *Inside outside: Teacher research and knowledge.* New York: Teachers College Press.

Connelly, M., & Clandinin, J. (1988). *Teachers as curriculum planners: Narratives of experience.* New York: Teachers College Press.

Connelly, M., & Clandinin, J. (1990). *Stories of experience and narrative inquiry.* Educational Researcher, *19*(5), 2–14.

Dewey, J. (1933). *How we think.* Boston: Heath.

Doyle, W. (1977). Paradigms for research on teacher effectiveness. *Review of Research in Education, 5,* 3–16.

Doyle, W. (1990). Themes in teacher education research. In W. Houston (Ed.), *Handbook of research on teacher education* (pp. 3–24). New York: Macmillan.

Duckworth, E. (1986). Teaching as research. *Harvard Educational Review, 56,* 481–495.

Duffy, G., & Roehler, L. (1986). Constraints on teacher change. *Journal of Teacher Education, 35,* 55–58.

Elbaz, F. L. (1983). *Teacher thinking: A study of practical knowledge.* London: Croom Helm.

Elliott, J. (1976–77). Developing hypotheses about classrooms from teachers' practical constructs: An account of the teaching project. *Interchange, 7*(2), 2–22.

Elliott, J. (1988). Educational research and outsider-insider relations. *Qualitative Studies in Education, 1*(2), 155–166.

Fenstermacher, G. D. (1979). A philosophical consideration of recent research on teacher effectiveness. *Review of Research in Education, 6,* 157–185.

Fenstermacher, G. D. (1986). Philosophy of research on teaching: Three aspects. In M. Wittrock (Ed.), *Handbook of research on teaching* (3rd ed, pp. 37–49). New York: Macmillan.

Fenstermacher, G. D. (1994). The knower and the known: The nature of knowledge in research on teaching. In L. Darling-Hammond (Ed.), *Review of Research in Education* (pp. 1–54). Washington, DC: American Educational Research Association.

Fenstermacher, G. D., & Richardson, V. (1993). The elicitation and reconstruction of practical arguments in teaching. *Journal of Curriculum Studies, 25*(2), 101–114.

Floden, R., & Klinzing, H. (1990). What can research on teacher thinking contribute to teacher preparation? A second opinion. *Educational Researcher, 19*(4), 15–20.

Fullan, M. (1991). *The new meaning of educational change.* New York: Teachers College Press.

Gage, N. (1985). *Hard gains in the soft sciences: The case of pedagogy.* Bloomington, IN: Phi Delta Kappa.

Giddens, A. (1976). *New rules of socio-logical method.* London: Hutchinson.

Goldenberg, C., & Gallimore, R. (1991). Local knowledge, research knowledge, and educational change. A case study of early Spanish reading improvement. *Educational Researcher, 20*(8), 2–14.

Grimmett, P., & Erickson, G. (Eds.). (1989). *Reflection in teacher education.* New York: Teachers College Press.

Grimmett, P., & Mackinnon, A. (1992). Craft knowledge and the education of teachers. In G. Grant (Ed.), *Review of Research in Education* (pp. 385–456). Washington, DC: American Educaitonal Research Association.

Grinder, R. (1981). The "new" science of education: Educational psychology in search of a mission. In F. H. Farley & N. J. Gordon (Eds.), *Psychology and education: The state of the union.* Berkeley, CA: McCutchan.

Grossman, P. (1990). *The making of a teacher: Teacher knowledge and teacher education.* New York: Teachers College Press.

Habermas, J. (1973). *Knowledge and human interests* (J. Shapiro, Trans.). Boston: Beacon.

Hollingsworth, S., Dybdahl, M., & Minarik, L. (1993). By chart and chance and passion: The importance of relational knowing in learning to teach. *Curriculum Inquiry, 23*(1), 5–35.

Hollingsworth, S., & Sockett, H. (Eds.) (1994). *Teacher research and educational reform* (NSSE Yearbook). Chicago: University of Chicago Press.

House, E., Mathison, S., & McTaggart, R. (1989). Validity and teacher inference. *Educational Researcher, 18*(7), 11–15, 26.

Johnson, M. (1987). The body in the mind: *The bodily basis of meaning, imagination, and mind.* Chicago: University of Chicago Press.

Kagan, D. (1990). Ways of evaluating teacher cognition: Inferences concerning the Goldilocks principle. *Review of Educational Research, 60,* 419–469.

Ladwig, J. G. (1991). Is collaborative research exploitative? *Educational Theory, 41*(2), 111–120.

Lampert, M. (1985). How do teachers manage to teach? Perspectives on problems in practice. *Harvard Educational Review, 55,* 178–184.

Lampert, M., & Clark, C. (1990). Expert knowledge and expert thinking in teaching. A response to Floden and Klinzing. *Educational Researcher, 19*(5), 21–23.

Leinhardt, G. (1988). Situated knowledge and expertise in teaching. In J. Calderhead (Ed.), *Teachers' professional learning* (pp. 146–168). Basingstoke, England: Falmer.

Louden, W. (1991). *Understanding teaching: Continuity and change in teachers' knowledge.* New York: Teachers College Press.

McCloskey, G. N., Provenzo, E. F., Cohn, M. M., & Kottkamp, R. B. (1991). Disincentives to teaching. Teacher reactions to legislated learning. *Educational Policy, 5,* 251–265.

McNeil, L. M. (1988). *Contradictions of control: School structure and school knowledge.* New York: Routledge & Kegan Paul.

Morimoto, K., (with Gregory, J., & Butler, P.). (1973). Notes on the context for learning. *Harvard Educational Review, 43,* 245–257.

Munby, H., & Russell, T. (1992). Transforming chemistry into chemistry teaching: The complexities of adopting new frames for experience. In T. Russell & H.

Mumby (Eds.), *Teachers and teaching: From classroom to reflection* (pp. 90–123). London: Falmer.

Neilsen, L. (1990). Research comes home. *The Reading Teacher, 44,* 248–250.

Nespor, J., & Barylske, J. (1991). Narrative discourse and teacher knowledge. *American Educational Research Journal, 28,* 805–823.

Noffke, S. E. (1992). Action research and the work of teachers. In R. Clift & C. M. Evertson (Eds.), *Focal points: Qualitative inquiries into teaching and teacher education* (Teacher Education Monograph No. 12, pp. 1–21). Washington, DC: ERIC Clearinghouse on Teacher Education.

Pajares, M. (1992). Teachers' beliefs and educational research: Cleaning up a messy construct. *Review of Educational Research, 62,* 307–332.

Placier, P., & Hamilton, M. L. (1994). Prophecies unfulfilled: The complex relationship between school context and staff development. In V. Richardson (Ed.), *Teacher change and the staff development process: A case in reading instruction* (pp. 121–128). New York: Teachers College Press.

Richardson, V. (Ed.). (1994). *Teacher changes and the staff development process: A case in reading instruction.* New York: Teachers College Press.

Richardson, V., & Anders, P. (1990). *Final report of the reading instruction study.* Tucson: University of Arizona, College of Education. (ERIC Document Reproduction Service No. ED 324 655)

Richardson, V., Anders, P., Tidwell, D., & Lloyd, C. (1991). The relationship between teachers' beliefs and practices in reading comprehension instruction. *American Educational Research Journal, 28,* 559–586.

Richardson-Koehler, V. (1988). What works doesn and doesn't. *Journal of Curriculum Studies, 20*(1), 71–79.

Russell, T., & Munby, H. (Eds.). (1992). *Teachers and teaching: From classroom to reflection.* London: Falmer.

Schön, D. (1983). *The reflective practitioner.* New York: Basic Books.

Schön, D. (1991). Concluding comments. In D. Schön (Ed.), *The reflective turn* (pp. 343–360). New York: Teachers College Press.

Short, K. (1992). "Living the process": Creating a learning community among educators. *Teaching Education, 4*(2), 35–42.

Shulman, L. S. (1987). Knowledge and teaching: Foundations of the new reform. *Harvard Educational Review, 57,* 1–22.

Van Manen, M. (1977). Linking ways of knowing with ways of being practical. *Curriculum Inquiry, 6,* 205–228.

Wilson, S., & Wineburg, S. (1988). Peering at history through different lenses: The role of disciplinary perspectives in teaching history. *Teachers College Record, 89,* 525–539.

CHAPTER 10

THE USE OF PARTICIPATORY DESIGN IN THE IMPLEMENTATION OF INTERNET-BASED COLLABORATIVE LEARNING ACTIVITIES IN K-12 CLASSROOMS

Marcos Silva and Alain Breuleux

Participatory design is a concept originating out of the Scandinavian nations (Schuler & Namioka, 1993) that has as its objective the inclusion of the user in the design and implementation of any new technology. It is a user driven design in that it "places the needs and abilities of the worker at center stage along with the other needs of the firm" (Emspak, 1993, p. 21). Participatory design grew out of the realization that traditional systems design was unable to effectively introduce new technologies in the workplace and factory floor.

Previously published in *Interpersonal Computing and Technology: An Electronic Journal for the 21st Century* (1994), 2(3), 99–128.

Constructivist Instructional Design (C-ID): Foundations, Models, and Examples, pp. 223–241

Or, as stated by Greenbaum (1993) when discussing the introduction of participatory design, "over the last 30 years the pages of management and system journals have been peppered with articles bemoaning the fact that so many systems don't work or fail to do things that both managers and users expect them to" (p. 30).

The introduction of new technologies into classrooms has faced similar problems. Indeed, promises made decades ago about computer technology remain for the most part unfulfilled: one of the reasons being lack of teacher participation in its introduction (Plomp & Akker, 1988; Schultz & Higginbotham-Wheat, 1991). Moreover, although computer technology is present in schools, it is under-utilized. This has not had the effect, however, of banishing technology from the classroom. With the growing interconnection between K–12 local area networks and university or regional wide area networks, exhortations for new uses of old technologies along with the introduction of new technologies in classrooms are likely to occur. Of these new technologies, use of local and wide area networks by students is currently one of the most popular.

Furthermore, traditional learning techniques are being scrutinized for effectiveness. In the last 20 years, the popularity of new learning techniques emphasizing critical thinking and collaboration has increased. Most of these new techniques fall under the rubric of cooperative learning methodologies.

There are three issues that merit discussion: the potential in participatory design as the means to introduce and implement new technologies while promoting the use of collaborative learning; the impact that collaborative learning methodologies may have in the use of these new technologies, or inversely, the role that new technologies may have in shaping or promoting collaborative learning; and the need to examine the process when introducing and implementing computer networking technologies in the classroom.

Four distinct, but inextricable developments are responsible for bringing these issues to the fore: Collaborative learning techniques, a greater sensitivity to the needs of the user, the advent of K–12 local or wide area networking, and the evolution of the Internet. Because each event has been examined in isolation, the process arising out of their interaction has often been ignored. While these developments may be complementary, the merging and introduction process the classroom remains unknown.

REASONS FOR THE USE OF PARTICIPATORY DESIGN

There are five reasons to investigate the use of participatory design when implementing new technologies in the classroom. First, the introduction of any new technology into classrooms is difficult, especially in light of previous statements made about them in the past. Again, a parallel can be made with industry where technologies are introduced without worker participation. Research on the use of participatory design in industry suggests that "local participants increased their competence on new technology and became more willing to take initiatives around it" (Clement & Van den Besselaar, 1993, p. 34). Perhaps teacher and even student involvement may foster a better understanding of the needs of the user with an optimal integration of the technology with everyday tasks.

Second, since many new projects may depend on collaborative learning activities, the decision of the teacher to combine resulting classroom tasks with use of the Internet requires an approach that maximizes their participation and cooperation. Arguably, only then will they understand and, therefore, endorse the technology. Because participatory design relies on full cooperation between users and systems analysts, it offers a ready made theoretical blueprint for initiating the activity and process.

After all, a fundamental tenet of participatory design is the belief that user participation gives workers the power to influence matters that directly concern them in their work (Clement & Van den Besselaar, 1993, p. 36).

Third, a participatory design approach has the potential to create a setting where opportunities for the researcher to share in and understand the concerns and perspectives of the participants become possible. Participatory design methodology has an affinity with research methodologies that place emphasis on interaction between researcher and participants. This is similar to its use in industry where a the role of the system analyst-management consultant is transformed into a user-facilitator, the role of the educational researcher is changed from that of an expert to that of an equal participant who happens to have expertise (Carmel, Whitaker, & George, 1993, p. 46).

Fourth, participatory design is attuned to current trends in education where attention to the learner and teacher, as opposed to the expert, instructional methodology, or technology, is primary. Inclusion of student and teacher needs, through their active participation in the design and objectives of the project, arguably harmonizes the need to introduce new technologies with new research approaches. Or, as stated by Schultz and Higginbotham-Wheat (1991), "There should be frequent feedback from teachers as implementation takes place. Teachers can identify problems before they become disasters" (p. 212).

And last, use by children of local area networks logically linked to wide area networks or the Internet has been somewhat controversial; the media has focused on isolated cases where children accessed pornographic, violent, or dangerous information. Naturally, parents, school administrators, and teachers, who have little knowledge of the network, may feel concern. Their participation in the project will allow them to understand the safeguards placed on the project and secure their cooperation.

REASONS FOR THE PROMOTION OF K–12 NETWORKING

There are several reasons why educators and researchers are promoting use of networks and computer mediated communication as a medium for education in K–12-higher education classrooms. The unreserved justifications to integrate these new resources in the classroom are surprising given the newness of K–12 local and wide area networks. Although network facilities are commonplace in higher education environments, they remain rare in K–12 schools. Nevertheless, it is possible to discern a trend to design projects that incorporate the use of networks for teaching and learning with more traditional educational tasks.

K–12 Computer Networks and Collaborative Learning

One of most common justifications given for the establishment of educational networking projects is the belief that use of computer networks fosters collaborative learning. In other words, computer networks are ideal vehicles for collaborative learning tasks and activities (Bump, 1990; Davits, 1988; Din, 1991; Levin & Cohen, 1985; Owen, 1991; Owen, 1993; Resnick, 1992; Riel, 1989; Riel, 1990, 1990b; Riel, 1992a; Robinson, 1993; Sloan & Koohang, 1991; Tinker, 1993). Moreover, because of the flexibility and potential of the networks, collaboration may be effected among students in the same classroom or among students dispersed among remote classrooms (Resnick, 1992), the former being the more common approach.

Indeed, initial results of projects emphasizing computer- network-based collaborative learning has led Bump to assert that "the most intense collaboration occurs when computers are electronically linked to each other to form networks" (1990, p. 49). Riel lends support to the above in her claim that the true potential of computer networks lies in their ability to create new forms of group interactions that are essentially of a collaborative nature (1990b, p. 449).

As connections to the Internet by K–12 schools become more commonplace, educators will have new opportunities to integrate collaborative learning techniques with new curricular activities, projects, and instructional methodologies. Sellers (1994), for example, in her guide to educational networking, emphasizes the shift from teacher-as-expert model to one of shared responsibility for learning arising from the use of computer- mediated communication. A similar viewpoint is made by Hunter who argues that with the advent of the National Research and Education Network (NREN), educators will have a resource where they will be able to direct and establish network projects, software, and structures to support and foster collaborative learning (United States Office of Science and Technology Policy, 1992, Director, p. 26).

It is interesting to speculate whether the above perspective could have had an indirect influence on the final text of American legislation calling for the establishment of research and education gigabyte networks. For instance, one of the primary purposes of the U.S. High-Performance Computing Act of 1991 is to "invest in basic research and education, and promote the inclusion of high-performance computing into education institutions at all levels" (United States Congress, 1991, Sec. 3(H)). The said inclusion of high-speed computing, however, must integrate collaborative projects among members of the research and education community (United States Office of Science and Technology Policy, Director, 1992, p. 1). Again, collaboration is seen as being essential to the purpose and success of the project. Admittedly, collaboration among educators and researchers does not necessarily imply the structured learning methodology found in collaborative learning theories. Nevertheless, collaboration in the context used does suggest recognition that the Internet is a suitable medium to undertake collaborative tasks. This in turn intimates that it may be prudent to structure collaboration in such a manner so as to best exploit learning activities and the sharing of knowledge, especially in K–12 environments.

The influence of collaborative learning is stated more directly in the National Information Infrastructure Act of 1993 (H.R. 1757), originally proposed by Rep. R. Boucher (D-VA). In the H.R. 1757, it is possible to find a call for educators and researchers to develop and evaluate educational software specifically designed for collaborative use over the Internet (1993, Sec. 307, (A),4). Still, although the intent to promote collaborative learning approaches remains implicit as opposed to explicit, the Bill does suggest that the Internet offers a virtual collaborative environment.

However, in the National Information Infrastructure: Agenda for Action, the Clinton administration's attempt to define its vision of the electronic superhighway, it is possible to find an explicit call for the express use of collaborative learning methodologies. A section in the report specifies "Students and teachers can use the NII to promote collaborative learning between students, teachers, and experts" (United States, White House, Information Infrastructure Task Force, 1992). It is possible to discern, therefore, an acknowledgment that a collaborative environment necessitates a well structured approach to ensure optimal use of its resources. And arguably, because of the nature of the network, collaborative learning techniques may be the approach that most optimally ensures maximum use of the potential in gigabyte networks.

It is interesting to note that calls for greater collaboration do not, as a rule, attempt to restrict communication among specific groups. On the contrary, current and future networking projects are being designed and implemented under the assumption that networks should foster greater communication among groups having different skills, professions, and status. The number of joint post-secondary and K–12 networking projects are indicative of this trend (Clement, 1992b, 1992c; Rude-Parkins & Hancock, 1990).

In Canada, a project similar to the NREN called the Canadian Network for the Advancement of Research, Industry, and Education (CANARIE) has been established (CANARIE Associates, 1992; Canadian Network for the advancement of Research, Industry, and Education Business Plan Working Group, 1992, p. 7; Silva & Cartwright, 1992). And not surprisingly, those responsible for the implementation of CANARIE have made collaborative research, development of new partnerships, and support for education, notably higher education, a key component of the project. In fact, the CANARIE Business Plan Working Group claims that the possible linkages among schools, research centres, and universities is one of the principal benefits of the forthcoming Canadian electronic highway (Canadian Network for the advancement of Research, Industry, and Education Business Plan Working Group, 1992, p. 7). Even more, the Group asserts the said linkage is essential in guaranteeing the ability of users to cooperate in joint research while remaining physically remote from each other (p. 7). And, increased collaboration is viewed as indispensable if Canada is to remain competitive in the modern international marketplace (CANARIE Associates, 1992).

K–12 Computer Networks and Situated Learning

Another reason for the enthusiasm found for educational networking projects is the belief that students using computer networks are able to contextualize and cognitively situate learning tasks (Lave & Wenger, 1989; Levin, Riel, Miyake, & Cohen, 1987; Mabrito, 1992; Riel, 1985; Tinker, 1993). In other words, social interaction and physical activity are viewed as being an integral part of the learning process. Or, the essence of learning is the result of sharing purposeful, patterned tasks (Roschelle, 1992). In collaborative writing projects, for example, contextualization was possible because of the effect of having an immediate audience responding to the text (Din, 1991; Riel, 1985); writing was no longer a solitary activity devoid of the social interaction present in most non-classroom activities. Also, the effect of collaboratively writing for a remote audience contextualized the work; the effort had meaning and significance resulting in superior work and deeper learning (Cohen & Riel, 1989).

An additional benefit of telecommunication projects is that through use of networking protocols, students are able to form partnerships with experts in a domain. These partnerships can be so structured to resemble what Brown, Collins, and Duguid (1989) call cognitive apprenticeships. The aim of the apprenticeship is to "embed learning in an activity and make deliberate use of the social and physical context" so that the learning is "more in line with the understanding of learning and cognition that is emerging from research" (p. 32).

In networked environments, these apprenticeships are called teleapprenticeships (Levin, Riel, Miyake, & Cohen, 1987; Teles, 1993). The Writer in Electronic Residence project (Owen, 1993), where a professional writer works directly with the students through telecommunications, is illustrative of this approach. Teleapprenticeships, therefore, are mediated by access to peers and professionals in networked environments (Teles, 1993). This is why Clement claims that the value of wide area networks lies in their potential to support collaborative projects linking educators and students that provide "meaningful learning experiences connected to the curriculum" (1992a, p. 18).

In essence, computer networks create virtual classrooms and laboratories where spatial and geographic concerns become secondary (Harasim, 1993; Silva & Cartwright, 1993). Of importance here is that these virtual meeting places can offer the student and teacher the context necessary to imbue the information with meaning. Once learning is contextualized and situated, therefore, knowledge is meaningful and, as a result, can be processed at a deeper cognitive level giving rise to greater understanding.

Note, however, that the drive to implement collaborative network-based projects also stems from the recognition that this approach more closely resembles work procedures in modern industry (Hunter, 1992, p. 25). Mabrito argues along parallel lines in his contention that computer networks have the potential to simulate the workplace of the future (1992, p. 317). This point of view, called "new work" by Mabrito, is explicit in the Clinton administration's National Information Infrastructure project and in Canada's CANARIE.

K–12 Computer Networks and Cognitive Growth

Given the above, it is inexplicable that more research has not been conducted concerning the development of higher-order thinking skills through the use of collaborative computer network projects via the Internet. Research on collaboration in non-networked computer tasks, however, indicate greater cognitive processing and growth. Admittedly, while it may be difficult to generalize the results from these small studies to gigabyte networking environments, they may offer some indication for future studies allowing researchers to structure their experiments accordingly.

Nastasi and Clements (1992), in their study of social processes as mediators of treatment effects on higher-order thinking, concluded that certain computer tasks—working with LOGO in small groups—may foster cognitive growth by promoting certain forms of social interactions, namely cognitively-based resolution of cognitive conflicts. Given the communicative potential of supernetworks, the possibility of inducing certain types of cognitive conflicts during specific tasks becomes very real. Higher-level reasoning and problem solving by students in similar tasks were also found by Johnson, Johnson, and Stanne (1986). Hooper (1992) lends further support to the above with his claim that intra-group reflection during computer-based instruction enhances future collaboration. And collaboration promotes interaction that in turn engenders deeper cognitive processing.

Educators tracking the impact of computer networks have also justified such projects on the grounds that children have exhibited greater emotive and social growth as a result of their opportunity to collaborate via networks. In her study of a collaborative networking project funded by AT& T, Riel (1990b, 1992a) found that children displayed greater self-esteem. In another project that linked two economically and racially different Detroit high schools, Ladestro (1991) claims that students experienced a breaking down of stereotypes and greater empathy for students of different backgrounds and socioeconomic groups.

Although these studies are mostly descriptive, they nevertheless support findings concerning self-esteem and empathy towards other groups reported by researchers concerned with the effects of collaborative learning techniques on social and emotional cognitive behavior. It is not unusual, therefore, to discover projects that have components that are designed specifically to foster prosocial development during collaborative learning tasks (Solomon, Watson, Schaps, Battistich, & Solomon, 1990).

The above outright optimism is somewhat tempered by research that examined mathematically-based groupwork with computers (Hoyles, Healy, & Pozzi, 1992). Although they agree that pupil-managed groups can effect positive outcomes during collaborative computer tasks, the authors warn that "groups must also be viewed as social systems, which, if they are to produce an agreed outcome, require a minimum level of mutual regard" (p. 256). That is, in groups where there are negative interpersonal relationships, the autonomous learning engendered by groupwork can encourage the "atomization of the group, a centration on computer products, a curtailment of negotiation and unhealthy competition" (p. 256). This finding requires further validation, especially since most collaborative learning approaches use heterogeneous or random groupings when selecting participants for classroom tasks (Davidson & Worsham, 1992b, p. xiii) where it becomes difficult to control the selection into groups of participants who have negative interpersonal relationships.

K–12 Computer Networks and Isolation of Teachers

The isolation of educators from fellow teachers and other researchers is an additional reason for the current level of support for networking projects. Gigabyte networks are seen as the tools capable of allowing educators to communicate, share, and access valuable knowledge. Those responsible the implementation of the Texas Education Network (TENET), for example, have argued that one of the major benefits of the network is the potential for greater collaboration between K–12 educators and post-secondary educators and researchers (Consortium for School Networking, 1992; Stout, 1992, p. A-130). In their survey of the use of networks in technologically privileged schools, Honey and Henriquez reported that educators listed less isolation as one of the benefits ensuing out telecommunication usage (1993, p. 16). Or, simply stated, greater opportunities for professional support and growth become possible; more equitable access to and dissemination of resources for staff development can be guaranteed.

Riel (1990b), in her study of the AT&T Learning Network, Electronic Learning Circles, argued along parallel lines. For instance, teacher participants showed a greater willingness to admit their ignorance on a particular subject and use the network to request information. Also, electronically linked teachers appeared more amenable to sharing and cooperating in the design of new instructional techniques and classroom organization. In this manner, support for educational restructuring is made available. Finally, like students, teachers demonstrated greater self-esteem as a result of their participation in the project.

K–12 Computer Networks and Academia

A further reason put forward in support of K–12 networking projects is that postsecondary institutions, which often provide networking support and access, will benefit from possible collaborations. The perception by postsecondary researchers that K-12 networking is "an enabling resource for research, scholarship, and (at least in local settings) education" (Clement, 1991, p. 15) is frequently found in academe.

Another benefit to postsecondary institutions is the realization that higher education, and a knowledgeable workforce, depend on well trained incoming students (Allum, 1991; Clement, 1991). Furthermore, increased participation in K–12 networking with postsecondary institutions may influence legislators to support national educational goals, which benefit all levels of education. The last benefit to postsecondary institutions lies in the possibility of strengthening their image and relationship to the community and private industry (Allum, 1991). This may be an invaluable resource at a time of increased efforts at fundraising activities by university administrators.

K–12 Computer Networks and Resource Sharing

More practical reasons are also forwarded as justification for increased investments in K–12 networks. In times of budgetary constraints, it is unrealistic to assume that schools are able to acquire all materials necessary to meet the demands of the curricula, or, more importantly, to meet the demands arising from new curricula. With computer networks, the possibility exists for greater resource sharing. Local and administrative databases, textual information, and school materials can be loaded on a central or remote server, and so eliminate costly duplication of materials. In addition, intellectual resources held by a school district can be disseminated and shared easily with other districts and intellectual resources found in research centers and universities become more easily accessible.

A more functional reason why teachers support collaborative learning networking projects rests in the cost of hardware and software. At present, most school districts do not have the means to offer individual students their own workstation. Most computer classroom activities, because of costs, demand that students share equipment. Collaborative learning environments are seen as a feasible approach that can maximize learning when students must, by necessity, work in groups.

Finally, with computer networks, educators and students have access to a vast warehouse of information. Databases, domain experts, full text reports, electronic books and journals, graphic images and sound, and software are some of the resources that are accessible and retrievable. With the advent of universal resource locators such as gophers, World Wide Web (WWW), Cello, and Mosaic, novice users and children can locate and retrieve electronic resources. Indeed, given the exponential growth of the Internet, exclusion from these resources may hinder educators from offering their students the best possible learning environment.

GROWTH OF THE INTERNET

The Internet, a world wide interconnected computer network of networks, is growing at a phenomenal rate (Hart, Reed, & Bar, 1992; Krol 1992; Lynch & Rose, 1993). Some estimates put its growth at 14% a month, its user base at 25 million users, and the number of computer hosts at 1,500,000. Indeed, almost every major research center and library in the world has an Internet connection giving the student a gateway to distributed resources and information.

Also indicative of the Internet's phenomenal growth is the increasing heterogeneity of its user population. Whereas but a few years ago the Internet was used exclusively by the research, government, and academic communities, today its user population includes school children, business persons, and the public (Hunter, 1992). For example, it is estimated that over 600,000 school children in the United States used the Internet to supplement their curricular activities during the 1991–1992 school year (Itzkan, 1992, p. 1).

It is the newness of the Internet that makes its growth appear astonishing. McGill University gained connectivity sometime in late 1988. Its general student population had full access only in 1991. The literature on the Internet also follows this pattern. Prior to 1989, articles on the Internet were mostly technical and of interest to a limited audience. Today, a cursory search for monographs and periodicals will retrieve a wealth of information directed at a far less technically oriented audience.

Using the Internet, students are able to access remote computers and search databases and online catalogues, transfer binary or text files, and exchange information with peers or experts. New software commonly called resource discovery tools (RDTs) or universal resource locators (URLs) allow students to access hypermedia information databases, the Library of Congress Vatican Exhibit and the University of California, Berkeley, Virtual Paleontology Museum being cases in point.

Issues of Concern

There are two simultaneous unfolding parallel events that should be of concern to educators: The constant growth of the Internet and the increasing awareness by teachers of the potential for learning, especially in collaborative teams, through the use of gigabyte networks. Consequently, hundreds of Internet based projects have been established (Batson, 1988; Eisenberg & Ely, 1993; Julyan, 1989; Kurshan, 1990; Murray, 1993; Quebec-Alberta Telecomputing Project, 1993; Riel, 1985; Sackman, 1993; Solomon, 1992; Tinker, 1993). Moreover, most of the research details the processes in the classroom after teachers and students are linked to a LAN or the Internet. That is, with the exception of the work by Willis (1991), there is a lack of research concerned with the process of implementing a technology as dynamic and as evolving as the Internet into the classroom.

By "process" we mean the method of introducing Internet based instructional activities to parents, teachers and students through a well structured set of actions and changes, and those steps which are necessary to successfully integrate this technology into classrooms so as to guarantee its most optimal use. This definition does not include planning for hardware and software, problems concerning telephone lines and gateways to the Internet, and logical links between networks. Rather, the definition of process emphasizes the methodology and design utilized to introduce and merge gigabyte telecommunications with regular K–12 classroom curricular activities, in particular, activities that employ collaborative learning tasks, whether locally or virtually.

Indeed, the disinterest shown by researchers toward the above problem becomes inexplicable given that some estimates place the number of U.S. children with some form of local regional, or Internet network activities at 5,000,000 (Harasim, 1993, p. 21). Naturally, this figure includes local area networks not logically connected to the Internet. Nevertheless, given the growing drive to interconnect K–12 administrative and educational with state or regional networks, the need for studies on the process of introducing Internet access to classroom is arguably necessary and potentially significant.

In our discussion of the above, we noted several assumptions. First and foremost is the assumption that the Internet is a potential collaborative education medium that enhances and promotes collaborative learning. That is to say, the Internet not only promotes collaborative learning, but is inherently collaborative. Resource sharing, communication, dissemination of information, and exchange of ideas are some of the services available via the Internet that may be forwarded in support of the above assumption. So powerful is the belief in the inherent collaborative potential of the Internet, that policymakers have consistently justified investment in national networks on the basis that it will foster greater collaboration among different sectors of society, namely industry, education and academia (CANARIE Associates, 1992; United States Office of Science and Technology Policy, Director, 1992; United States Congress, 1991).

Second, we believe that participatory design has applications outside of industrial settings, namely the K–12 sector. Participatory design, because of its cooperative approach to decision making, also offers researchers and educators the means to meet demands made by users of new technologies. Since successful implementation of new technologies may depend on user acceptance, a cooperative approach to planning is arguably worth attention, study, and research.

The third assumption is the belief in the efficacy of collaborative learning techniques. The present work assumes that research on collaborative learning has established a solid enough foundation (Davidson & Worsham, 1992a; Sharan, 1990; Slavin et al., 1985; Slavin, 1980, 1983, 1990) making it possible to argue for new technologies that enhance and promote the approach as opposed to demonstrating its validity.

Fourth, it is assumed that collaboration among different age and education levels is desirable. Again, we forward the assumption that carefully structured collaborative learning projects that foster cooperation among university, high school, and elementary age students is an effective learning approach.

The last assumption is the belief that the Internet will be an integral part of classroom activities in the very near future. Investments in K–12 logical networks will continue to increase giving teachers and students greater access to information and expertise. It is also assumed that this phenomenon is a positive trend and should be advanced and encouraged, especially since, given present trends, the advent of K-12 networks appears inevitable.

Limitations with the Participatory Method

Most of the research on participatory design is recent and has been undertaken by practitioners of this approach. As a result, there is a serious lack of studies on its weaknesses and flaws. Furthermore, most research is restricted to industrial settings, so it is not known if participatory design has widespread applicability. Finally, North American and European, especially Scandinavian, approaches appear to be splitting into somewhat similar but separate schools, making claims about the success of the method more problematic.

In addition, participatory design closely resembles the methodology employed in qualitative research. The qualitative research tradition can be described as naturalistic, ethnographic, or humanistic. A participatory approach is likewise part of this tradition (Kirk & Miller, 1986, p. 9). Moreover, similar to participatory design, qualitative research emphasizes the need to conduct research in natural settings so as to ensure a thorough understanding of the needs and perspectives of the participants. This intertwining of approaches resembles what Whyte calls participatory action research where "research and action are closely linked" (1991, p. 8). Not surprisingly, most well known participatory design projects have adopted an action research approach (Clement & Van den Besselaar,

1993, p. 29). As a result, participatory design has faced similar criticism leveled against qualitative and humanistic research.

Another criticism made against participatory design is imprecise definition of the concept of participation. For example, Elden and Levin, in their discussion of participatory action research stress that "the degree and nature of participation in all phases of participatory action research is a critical factor" (1991, p. 133). Indeed, they assert that not all participation is necessarily empowering, especially within a non-democratic organization. They argue that participation must be full participation for it to be truly empowering. However, they also state that empowering participation does not mean that every person in an organization is a full participant. Rather, participation is dependent on representation by union members, managers, and top management.

Finally, attempts to define the users of the system has likewise posed problems under a participatory method. Carmel, Whitaker, and George accept the idea that "an unambiguous definition of user is impossible" (1993, p. 40) and claim that the main difference between the many different participatory methodologies is the degree to which users are able to participate in the project. A single definition, therefore, is impossible. As a result, different participatory approaches, dependent on specific settings and conditions, offer many varying definitions.

CONCLUSION

As gigabyte K–12 networks continue to grow, educators will be faced with increasing demands for integration of Internet resources and services with traditional classroom activities. With the advent of the NREN and CANARIE, it is safe to assume that demands will increase significantly in the near future. Indeed, the current number of school children linked to local area networks and the Internet presages the direction of the above trend.

Given the above, it is almost inexplicable that researchers appear disinterested in the process of introducing and establishing Internet based K-12 projects. Granted, there is research supporting the use of network activities in the classroom. Nevertheless, there is a serious need for research concerned with the process of introducing, training, and establishing such projects. This need should not be underestimated given that teachers frequently object to their lack of participation concerning the introduction of new technologies.

In conclusion, the collaborative nature of the Internet calls for a participatory approach where users of the system have a say in the design and implementation of new procedures and technology. In this manner, many of the pitfalls, obstacles, and misunderstandings normally found when introducing new technologies may be avoided. Indeed, participatory design may offer the means to fully exploit the potential in new technologies and networking.

REFERENCES

Allum, K. F. (1991). Partners in innovation: School-college collaborations. *EDU-COM Review, 26*(3/4), 29–33.

Batson, T. (1988, February). The ENFI Project: A networked classroom approach to writing instruction. Academic Computing, 32–33, 55–56.

Brown, J. S., Collins, A., & Duguid, P. (1989). Situated cognition and the culture of learning. *Educational Researcher, 18*(1), 32–41.

Bump, J. (1990). Radical changes in class discussion using networked computers. *Computers and the Humanities, 24, 49–65.*

Canadian Network for the Advancement of Research, Industry, and Education Business Plan Working Group. (1992). *Report of the Business Plan Working Group.* (n.p.).

CANARIE Associates. (1992). CANARIE business plan. Ottawa: CANARIE, Retrieved from ftp://unbmvs1.csd.unb.ca/pub/net/CANARIE

Carmel, E., Whitaker, R. D., & George, J. F. (1993). PD and joint application design: A transatlantic comparison. *Communications of the ACM, 36*(4), 40–48.

Clement, A., & Van den Besselaar, P. (1993). A retrospective look at PD projects. *Communications of the ACM, 36*(4), 29–37.

Clement, J. (1991). K–12 networking benefits for higher education. *EDUCOM Review, 26*(2), 14–16.

Clement, J. (1992a). Constructing the K–12 collaboratory on the NREN. *EDU-COM Review, 27*(3), 18–20.

Clement, J. (1992b). Network-based collaborations: How universities can support K–12 reform efforts. *EDUCOM Review, 27*(1), 9–12.

Clement, J. (1992c). Surveying K–12 and postsecondary school networking partnerships. *EDUCOM Review, 27*(4), 44–46.

Cohen, M., & Riel, M. (1989). The effect of distant audiences on students' writing. *American Educational Research Journal, 26, 143–159.*

Consortium for School Networking. (1992). The National Research and Education Network and K-12 Education. *In Proceedings of the NREN Workshop, Monterey, California, September 16-18, 1992* (pp. A-113-A-126). [Washington, DC]: Interuniversity Communications Council.

Davidson, N., & Worsham, T. (Eds.). (1992a). Enhancing thinking through cooperative learning. New York: Teachers College Press.

Davidson, N., & Worsham, T. (1992b). HOTSICLE—Higher order thinking skills in cooperative learning environments. In N. Davidson & T. Worsham (Eds.), *Enhancing thinking through cooperative learning* (pp. xi–xx). New York: Teachers College, Columbia University.

Davits, D. (1988). Computer-supported co-operative learning systems: Interactive group technologies and open learning. *Programmed Learning and Educational Technology, 25,* 205–215.

Din, A. H. (1991). Computer-supported collaborative writing: The workplace and the writing classroom. *Journal of Business and Technical Communication, 5,* 123–150.

Eisenberg, M. B., & Ely, D. P. (1993). Plugging into the 'net. *ERIC Review, 2*(3), 2–10.

Elden, M., & Levin, M. (1991). Cogenerative learning: Bringing participation into action research. In W. F. Whyte (Ed.), *Participatory action research* (pp. 127–142). Newbury Park, CA: SAGE.

Emspak, F. (1993). Workers, unions, and new technology. In D. Schuler and A. Namioka (Eds.), *Participatory design: Principles and practices* (pp. 13–26). Hillsdale, NJ: Erlbaum.

Greenbaum, J. (1993). A design of one's own: Towards participatory design in the United States. In D. Schuler & A. Namioka (Eds.), *Participatory design: Principles and practices* (pp. 27–37). Hillsdale, NJ: Erlbaum.

Harasim, L. M. (1993). Networlds: Networks as social space. In L. M. Harasim (Ed.), *Global networks: Computers and international communication* (pp. 15–34). Cambridge, MA: MIT Press.

Hart, J. A., Reed, R. R., & Bar, F. (1992). The building of the Internet: Implications for the future of broadband networks. *Telecommunications Policy, 16,* 666–689.

Hoyles, C., Healy, L., & Pozzi, S. (1992). Interdependence and autonomy: Aspects of groupwork with computers. *Learning and Instruction, 2,* 239–257.

Honey, M., & Henriquez, A. (1993). *Telecommunications and K-12 educators: Findings from a national survey.* New York: Center for Technology in Education, Bank Street College of Education.

Hooper, S. (1992). Cooperative learning and computer-based instruction. *Educational Technology Research and Development, 40,* 21–38.

Hunter, B. (1992). Linking for learning: Computer-and-communications network support for nationwide innovation in education. *Journal of Science Education and Technology, 1,* 23–34.

Itzkan, S. J. (1992). How big is the global classroom? *Matrix News, 10*(2), 1, 7–8.

Johnson, R. T., Johnson, D. W., & Stanne, M. B. (1986). Comparison of computer-assisted cooperative, competitive, and individualistic learning. *American Educational Research Journal, 23,* 382–392.

Julyan, C. L. (1989). National Geographic Kids Network: Real Science in the elementary classroom. *Classroom Computer Learning, 10*(2), 30-33, 35–36,38, 40–41.

Kirk, J., & Miller, M. L. (1986). *Reliability and validity in qualitative research.* Beverly Hills, CA: SAGE.

Krol, E. (1992). *The whole Internet: User's guide & catalog.* Sebastopol, CA: O'Reilly & Associates.

Kurshan, B. (1990). Educational telecommunications connections for the classroom—Part 1. *Computing Teacher, 17*(6), 30–35.

Ladestro, D. (1991). A tale of two cities. *Teacher Magazine, 2,* 47–51.

Lave, J., & Wenger, E. (1989). *Situated learning: Legitimate peripheral participation.* Palo Alto, CA: Institute for Research on Learning.

Levin, J. A., & Cohen, M. (1985). The world as an international laboratory: Electronic networks for science instruction and problem solving. *Journal of Computers in Mathematics and Science Teaching, 4,* 33–34.

Levin, J. A., Riel, M., Miyake, N., & Cohen, M. (1987). Education on the electronic frontier: Teleapprentices in globally distributed educational contexts. *Contemporary Educational Psychology, 12,* 254–260.

Lynch, D. C., & Rose, M. T. (Eds.). (1993). *Internet system handbook.* Reading, MA: Addison-Wesley.

Mabrito, M. (1992). Real-time computer network collaboration: Case studies of business writing students. *Journal of Business and Technical Communication, 6,* 316–336.

Murray, J. (1993). K12 Network: Global education through telecommunications. *Communications of the ACM, 36*(8), 36–41.

Nastasi, B. K., & Clements, D. H. (1992). Social-cognitive behaviors and higher-order thinking in educational computer environments. *Learning and Instruction, 2,* 215–238.

Owen, T. (1991). Online learning links are language learning links: Writer-in-residence program at Simon Fraser University. *Output, 12*(1), 22–26.

Owen, T. (1993). *Wired writing: The writers in Electronic residence program. In R. Mason (Ed.), Computer conferencing: The last word.* Victoria, BC: Beach Holme.

Plomp, T., & Akker, J. J. (1988). *Computer integration in the curriculum: Promises and problems.* Paper presented at the Annual Meeting of the American Educational Research Association, New Orleans, LA.

Quebec-Alberta Telecomputing Project. (1993). Quebec-Alberta Telecomputing Project: Final evaluation report, 1992–93.

Resnick, M. (1992). Collaboration in simulated worlds: Learning through and about collaboration. *SIGCUE Outlook, 21*(3), 36–38.

Riel, M. (1985). The computer chronicles newswire: A functional learning environment for acquiring literacy skills. *Journal of Educational Computing Research, 1,* 317–337.

Riel, M. (1989). The impact of computers in classrooms. *Journal of Research on Computing in Education, 22,* 180–190.

Riel, M. (1990a). Building electronic communities: success and failure in computer networking. *Instructional Science, 19,* 145–169.

Riel, M. (1990b). Cooperative learning across classrooms in electronic Learning Circles. *Instructional Science, 19,* 445–466.

Riel, M. (1992a). Making connections from urban schools. *Education and Urban Society, 24,* 477–488.

Robinson, B. (1993). Telling tales: The use of electronic conferencing for collaborative story writing. In R. Mason (Ed.), *Computer conferencing: The last word.* Victoria, BC: Beach Holme.

Roschelle, J. (1992). What should collaborative technology be? A perspective from Dewey and situated learning. *SIGCUE Outlook, 21*(3), 39–42.

Rude-Parkins, C., & Hancock, M. (1990). Collaborative partnership for technology adoption: A working model in Louisville. *TechTrends, 35*(1), 3–5.

Sackman, G. (1993). Global Schoolhouse. Computer message sent on NET-HAP-PENINGS@IS.INTERNIC.NET, October 25, 1993, 13:07.

Schuler, D., & Namioka, A. (1993). *Participatory design: Principles and practices.* Hillsdale, NJ: Erlbaum.

Schultz, C. W., & Higginbotham-Wheat, N. (1991). Practitioners' perspectives of computers in the classroom. In T. M. Shlechter (Ed.), *Problems and promises of computer-based training* (pp. 199–214). Norwood, NJ: Ablex.

Sellers, J. (1994). Answers to commonly asked "primary and secondary school Internet user" questions. Retrieved from ftp://ds.internic.net/rfc/rfc1578.txt

Sharan, S. (Ed.). (1990). *Cooperative learning: Theory and research.* New York: Praeger.

Silva, M., & Cartwright, G. F. (1992). The Canadian Network for the Advancement of Research, Industry, and Education (CANARIE). *The Public-Access Computer Systems Review, 3,* 4–14.

Silva, M., & Cartwright, G. F. (1993). The Internet as a medium for education and educational research. *Education Libraries, 17*(2), 7–12.

Slavin, R. E. (1980). Cooperative learning. *Review of Educational Research, 50,* 315–342.

Slavin, R. E. (1983). When does cooperative learning increase student achievement? *Psychological Bulletin, 94,* 429–445.

Slavin, R. E. (1990). *Cooperative learning: Theory, research, and practice.* Englewood Cliffs, NJ: Prentice-Hall.

Slavin, R. E., Sharan, S., Kagan, S., Hertz-Lazarowitz, R., Webb, C., & Schmuck, R. (1985). *Learning to cooperate, cooperating to learn.* New York: Plenum Press.

Sloan, F. A., & Koohang, A. A. (1991). The local area network and the cooperative learning principle. *Computers in Schools, 8,* 207–208.

Solomon, D., Watson, M., Schaps, E., Battistich, V., & Solomon, J. (1990). Cooperative learning as part of a comprehensive classroom program designed to promote prosocial development. In S. Sharan (Ed.), *Cooperative learning: Theory and research* (pp. 232–260). New York: Praeger.

Solomon, G. (1992, March). The most complete guide ever to telecommunications. *Electronic Learning,* 18–28.

Stout, C. (1992). TENET: Texas Education Network. In *Proceedings of the NREN Workshop, Monterey, California, September 16-18, 1992* (pp. A-127-A-134).

Teles, L. (1993). Cognitive apprenticeship on global networks. In L.M. Harasim (Ed.), *Global networks: Computers and international communication* (pp. 271–281). Cambridge, MA: MIT Press.

Tinker, R. F. (1993). *Educational networking: Meeting educators needs.* Paper presented at the INET'93, the annual Conference of the Internet Society, San Francisco, CA, (pp. ECB-1-ECB-11).

United States Congress, House. National Information Infrastructure Act of 1993. 103rd Congress, 1st session, H.R. 1757. Retrieved from ftp://ftp.cpsr.org/cpsr/nii/hr1757_july_1993.txt

United States Congress, Senate. High-Performance Computing Act of 1991. 102nd Congress, 1st session, S. 272. Retrieved from ftp://ftp.nic.merit.edu/nren/hpca.1991/nrenbill.txt

United States Office of Science and Technology Policy, Director. (1992). The National Research and Education Network Program: A report to Congress. Washington, DC: GPO. Retrieved from ftp://expres.cise.nsf.gov/pub/fnc /nrencongr.ascii

United States. White House. Information Infrastructure Task Force. (1992). The National Information Infrastructure: Agenda for Action. Washington, DC: GPO.

Whyte, W. F. (1991). *Participatory action research*. Newbury Park, CA: SAGE.

Willis, J. (1991). Computer mediated communication systems and intellectual teamwork: Social psychological issues in design and implementation. *Educational Technology, 31*(4), 10–20.

CHAPTER 11

AGENCY OF THE INSTRUCTIONAL DESIGNER

Moral Coherence and Transformative Social Practice

Katy Campbell, Richard Schwier, and Richard Kenny

In this chapter we propose a view of instructional design practice in which the instructional designer is an agent of social change at the personal, relational, and institutional levels. In this view designers are not journeymen workers directed by management, but act in purposeful, value based ways with ethical knowledge, in social relationships and contexts that have consequences in and for action. The paper is drawn from the data set of a three-year study of the personal meaning that instructional designers make of their work, in a world where identities rely less on institutionally "ascribed status or place" than on the spaces that we make as actors in the social world. Through the voices of two instructional designers in this study, we begin to make the case for instructional design practice as ethical knowledge in action, and for how agency emerges from the designer's validated sense of identity in institutions of higher learning.

Previously published in *Australasian Journal of Educational Technology (2005), 21*(2), 22–262.

243

INTRODUCTION

Practice is a good deal more than the technical things we do in classrooms —it relates to who we are, to our whole approach to life. (Goodson, 1994, p. 29)

The instructional design field has long debated the nature of instructional design practice. Is it a craft? Is it a science? Is it an art? In each view, the designer interacts with models and content—moving from a workbench to a laboratory to a studio—but in ways constrained by technical knowledge, cultural boundaries, and somewhat confounded identities. In these conceptions instructional designers act *on* content *to* externally defined expectations. We propose an alternative view, in which the instructional designer is an agent of social change at the personal, relational, and institutional levels. In this view designers are not journeymen workers directed by management but act in purposeful, value based ways with ethical knowledge, in social relationships and contexts that have *consequences in and for action.*

As instructional designers, and teachers and scholars of instructional design, we have become critical of our own understanding and practice and skeptical about the "grand narratives" of instructional design, for example, the traditional models of instructional design process that portray it as a rational, systematic, objective process based on the purposeful implementation of the principles of behavioral and cognitive science (cf. Braden, 1996; Dick, 1996; Gustafson & Branch, 1997; Kenny, Zhang, Schwier, & Campbell, 2004; Willis, 1998, 2000). From our own work as instructional designers, we suspect that instructional designers do not always, or even mostly, practice in ways that can be measured, quantified, and scientifically described (Wood, 1992) but that they act out of their own values and convictions about the social purposes of design. We think the instructional design process in higher education at least, in which faculty, designers, and others develop new ideas and understandings through conversation, may be a form of cultural learning or collaborative learning that leads to cultural change. In other words, we propose that rather than occupying the lowest rung of the professional ladder, that of the technicians who "do what they are told" by putting the theory into practice (Wood, 1992), instructional designers can, and do, challenge and shape the institutional "discourse" about the purposes and forms of learning. What do we mean by change agency, and what implications does this view have for the practice of instructional design?

At the core, change agency is a moral relationship with others. Fundamentally we believe that instructional design practice is not grounded in the rationality of behaviorism as much as in a "social morality in which caring values are central but contextualised in webs of relationships and constructed towards communities" (Christians, 2000, p. 142). Practice is embodied in the designer's core values and beliefs: Herda (1999) suggests that ethical knowledge, or moral judgment, can neither be learned nor forgotten; that it characterizes all authentic understanding and action. The consensual act of instructional design is a social, relational process created and shared through language, itself a form of action (Herda, 1999). For example, when we engage faculty in a conversation about the consequences of designing for active learning, including the development of critical thinking skills, we are "altering and changing (a) social context (and), those statements, themselves, contribute significantly to a basis for *personal and social change*" (Herda, 1999, p. 26). In this way design is a moral and political act, not merely a technical one.

Acting from ethical knowledge—with moral agency—implies a reflexive knowledge of self in action, an understanding of "one's biography, present circumstances, deep commitments, affective investments, social context and conflicting discourses" (Britzman, 1991, p. 8), about what it means to be an instructional designer in an institution of authoritative discourse about the monologic sources of knowledge and power, and one's role in it. Although she refers specifically to teacher education, we agree with Britzman (1991) that designing with moral authority "concerns coming to terms with one's intentions and values, as well as one's knowing, being and acting in a setting characterised by contradictory realities, negotiation, dependency, and struggle" (p. 8). To be agents of social change, designers must not only hold certain values, but also be conscious of them, and be able to articulate the choices for action that embody them. In this view, instructional design is purposeful and critical.

This chapter is drawn from the data set of a three-year study of the personal meaning that instructional designers make of their work, in a world where identities rely less on institutionally "ascribed status or place" than on the spaces that we make as actors in the social world. There are several purposes for this study and its products. As a reflexive project we share our stories as practitioners and, as conversational partners and in community, "reformulate our conceptions of identity and self-hood" (Goodson, 1995, p. 3). In this way the project itself is one of agency and transformation. Here, through the voices of two instructional designers in the study, we begin to make the case for instructional design practice as ethical knowledge in action, and for how agency emerges from the designer's validated sense of identity in institutions of higher learning. This is one of a series of completed (cf. Campbell, Gibson, & Gramlich, 2004; Schwier, Campbell, & Kenny, 2004) and developing papers that will explore identity, agency and community and challenge the grand narratives of instructional design.

THE THEORETICAL CHALLENGE

We use the following theoretical constructs to challenge the discourses that contextualized instructional design as a rational, technical, non-subjective process. The study this chapter represents is embedded in two theoretical constructs: instructional design as a social construct and critical pedagogy, in which designers act as agents of social change. In post-structuralist terms, we propose that instructional design practice is constituted by socially and culturally produced patterns of language, or discourse, with socially transformative power through the positioning of the self in explicit action (Francis, 1999). This construct is contrary to the idea of instructional design as decontextualized science. In other words, we view instructional design as a socially constructed practice rather than a technology to be employed.

In addition to the social implications of practice, we recognise that instructional design exists within a larger context of social change. Research on change and change management is multi-disciplinary, drawing on fields such as organizational behavior, sociology, psychology, economics, and anthropology, and the education literature is replete with useful advice for leaders who are part of the change process in large systems—and particularly school systems (Fullan, 2001, 2004). We argue that the process of change is particular to the context in which it occurs—in this case, instructional design in higher education, and it also has temporal characteristics. For example, Weick and Quinn (1999) suggested that change is either episodic or continuous. Episodic change refers to an infrequent and discrete kind of change, typically change that occurs once and is contained. Continuous change on the other hand refers to ongoing change, often change that occurs over time and that may resonate beyond the system within which the change initially occurs. While many products of instructional design are episodic (e.g., changes to a course are completed and implemented), the *process and influence* of instructional design is better understood as continuous.

Whether continuous or episodic, most change models fall into two broad categories: planned change and unplanned change. Planned change is deliberate, and it is normally the outcome of conscious reasoning based on some clear expectations. For example, the explicit practices of instructional design, and most models of instructional design, promote the idea that instructional design is a deliberate process that emphasizes planned approaches to development. But change has unplanned features that can introduce desirable or undesirable consequences, and instructional design similarly embraces tacit, creative and spontaneous elements that can influence the quality of outcomes. In order to maximise the benefits of change and avoid unintended consequences, change must be effectively managed, including social negotiation among individuals and groups, and larger transformational changes (Bolman & Deal, 1997). This is a particular challenge to instructional designers, given that professional programs in ID pay little or no attention to change or change management strategies, perhaps because it is an adolescent field of study and practice that is growing away from its original heritage (Hill, Bichelmeyer, Boling, Gibbons, Grabowski, Osguthorpe, Schwier, & Wager, 2004).

The roots of ID are well known. Instructional design as a field came of age after World War II, and was originally based on the behaviorist learning theories of Skinner, and Thorndike, among others (Saettler, 1992). That is, instructional design was based on the empiric assumption that behavior is predictable, and that educational design can occur in isolation from the contexts in which learning will take place (Koper, 2000). As a result, the language of traditional instructional design reflects a systematic approach based on social engineering and reflects the values of efficiency and effectiveness (cf. Braden, 1996; Dick, 1996; Dick & Carey, 1996; Merrill, 2002; Merrill, Drake, Lacy, Pratt, & ID2 Research Group, 1996). Traditional instructional design models describe an expressly linear, systematic, prescriptive approach to instructional design (Andrews & Goodsen, 1991; Braden, 1996; Wedman & Tessmer, 1993) and are strongly objectivist in nature (Jonassen, 1999). Although most authors have moved away from strict linearity and their approaches are less explicitly prescriptive, systematic models continue to thrive in various portrayals (e.g., Morrison, Ross, & Kemp, 2004; Seels & Glasgow, 1998; Smith & Ragan, 2005) and continue to be taught to thousands of graduate students (Willis, 1998).

These are supposedly value free ways of shaping and representing knowledge based on the assumption that educational technologies and environments are neutral and democratic, that knowledge can be codified and presented in templates or blueprints that describe what knowledge is in a "known world." Designers, programmers, and media developers emerging from this "scientific" field have learned models that value objective, rational, instrumental, and empirical approaches. Critics like Garrison (1993) and Vrasidas (2001) have described the products and environments they produce and deliver as too often prescriptive, formalistic, restrictive, and reductionist, due in no small way to the culture instructional designers have acquired within their areas of study and the training that they have received. Carter (2000) asks, "How aware and concerned are distance educators, instructional designers, and educational technologists about critical pedagogy, critical multiculturalism, and the powerful political nature of technological systems and their cultural practices?" (p. 28).

A cultural shift has been occurring over the past decade in education—a shift towards environments and approaches based on the ideas of social constructivism. In this worldview, learning is situated in rich contexts, and knowledge is constructed in *communities of practice* through social interactions. Cobb (1996) argues that knowledge is not held objectively, but is unique, wholly subjective, and passed on by establishing common ground between the knower and the learner. This common ground must embrace interests and personal values, which requires a sharing at both the socio-cultural and the cognitive levels (Ewing, Dowling, & Coutts, 1998, p. 10). In other words, the instructional designer's practice, to which self reflection is critical, will reflect his or her values and belief structures, understandings, prior experiences, and construction of new knowledge through social interaction and negotiation. Certainly, it is true that, over the past decade, the field of instructional design has experienced the strong influence of constructivist learning theory and a shift from teacher controlled to learner centred instruction (Reigeluth, 1996, 1999), and this movement has led to the emergence of a number of ID models based on constructivist learning principles (e.g., Cennamo, Abell, & Chung, 1996; Hannafin, Land, & Oliver, 1999; Jonassen, 1999; Mayer, 1999; Shambaugh & Magliaro, 2001; Willis, 2000).

What is less certain is whether instructional designers are being taught or are using such models in their practice. A recent review (Kenny, Zhang, Schwier, & Campbell, 2005) has indicated that instructional designers tend to follow the techniques delineated by traditional, process based models, although they do not follow them in a rigid fashion and also engage in a wide variety of other tasks that are not reflected in ID models, such as communications, editing, project management and team building.

In addition to these other activities, instructional designers also widely engage in faculty development (Kenny, Zhang, Schwier, & Campbell, 2005) through both formal and informal learning processes. In this regard, we believe that faculty working with instructional designers in development projects are actually engaging, as learners, in a process of professional and personal transformation that has the potential to transform the institution. Some theorists (cf. Glaser, 1991; Jonassen, Dyer, Peters, Robinson, Harvey, King, & Loughner, 1997; Tergan, 1997) believe that learning is most effective if it is embedded in social experience, and if it is situated in authentic problem solving contexts entailing cognitive demands relevant for coping with real life situations. Learning occurs through social intercourse, or design conversations. Even though much continues to be written about the effect of technology and computers on society, designers have not necessarily recognized their agency in the development of a knowledge economy that reflects culturally biased views of teaching, learning, and the construction of knowledge. We believe that instructional designers have not been encouraged to examine their cultural values and assumptions critically, and we challenge the idea that the expert knowledge of designers, gained through education, experience and interaction, should remain unexamined.

THE RESEARCH DESIGN

The stories reported in this chapter were drawn from a three-year (2002–2005) study involving, to date, twenty instructional designers at six Canadian universities. Initially, we selected participants from Medical Doctoral Universities, those with a broad range of PhD programs and research, as well as medical schools. The participating institutions also have an administrative and/or academic unit whose mandate is to support faculty developing (usually) technology enhanced, "blended," or online learning environments; and/or an "Extension Division," that employs at least two instructional designers.

Participation was elicited through a range of strategies including personal e-mail invitations, advertisements on lists and in institutional communications platforms, personal contacts at professional meetings and through collaborative projects, membership lists from professional associations, contacts through delegate lists from conferences, and visits to graduate classes. Sources of data include research conversations with instructional designers, e-mail, and group meetings and/or focus groups.

The main study is constructed as a narrative inquiry conducted mainly through the development of collaborative conversations. The "opening gambits"—designed to encourage designers to explore what they know, how they know it, and how this influences their actions in the particular sociocultural contexts in higher education—referred to their lives as learners and their memberships in social and professional communities, their career choices, their core values about the purposes of education and of design, and their design practices.

Narrative inquiry and the storying of experience are socially and contextually situated interpretive practices, starting from the personal as "personal knowledge has a practical function, not in a technical sense, or as an instrument for previously determined outcomes, but ... as a source for deliberation, intuitive decisions, daily action and moral wisdom" (Conle, 2000, p. 51). Narrative inquiry is transformative, because in defining how to become engaged as students of our own practice, the practice itself is examined and understood. In this way, thinking about and telling stories of practice requires a critical, reflective engagement leading to changed or transformed practice. Thus the methodological approach for the study mirrors a social constructivist framework for instructional design practice, which is one of social interaction and construction of meaning through conversation.

The two conversations included in this chapter were selected from the pool of data because these participants chose to discuss elements of their professional identities and performance that were tied to moral/ethical dimensions of their work and their roles as agents of social change. These conversations were not unique; other participants also discussed these dimensions of their work. But they were particularly clear, powerful and focused on these dimensions, probably more than the conversations with other people we interviewed. And the two participants approached the issues from very different directions—one more intimately/personally, and one more globally/politically. So, in other words, they were selected not because we were looking to generalize findings from the entire group, but rather because we felt these participants gave thoughtful, articulate, and divergent descriptions of their social agency, the moral stances that guide their work, and the transformations they have influenced and experienced.

Transcripts of the conversations are independently coded by two researchers. Transcripts are analyzed using Atlas ti® software, and as themes emerged, they are shared with the research team and the participants, and used to construct networks of meaning. This reflexive process is intended to further engage participants in identifying emerging personal and community issues related to instructional design by bringing the personal and community problems of practice into self awareness, leading to social action. In this way, narrative inquiry involves the "politics of identity construction and ongoing identity maintenance," where the lived experiences of instructional designers can "be used as the sites wherein and whereby we interrogate the social world theoretically and critically" (Goodson, 1995, p. 4).

Two Stories of Design: Ethical Knowledge in Practice

In this chapter we have chosen to exemplify the moral integrity of instructional design practice through the stories of two designers, illuminating how their values and beliefs embody the relational work they choose to do, and the ways in which they engage their own agency. Professionally prepared in unrelated disciplines, each of these designers came to instructional design through career paths and life and work experiences that have critical dimensions. Power is "the ability to take one's place in whatever discourse is essential to action and the right to have one's part matter" (Heilbrun, 1988, p. 18). In this excerpt from *Writing a Woman's Life,* Carolyn Heilbrun speaks of the power to be able to participate in culturally contextualized work of the higher education institution in ways that make a social and critical difference.

In some ways, making a critical difference demands subversion of the dominant discourse; in these stories subversion is a positive, generative power. We have chosen these stories for this chapter because they are imbued with the personal, moral strength key to a particular design ethics in practice. Each is a story of agency, yet they reveal what we think are somewhat different and complementary agencies, played out at different levels of personal, institutional, and societal engagement. Laura's agency is experienced and enacted at the level of the personal—agency in a web of personal relationships—that we have described as agency for social change. David's agency is less embedded in the personal and appears more externally directed: we speculate that this relates to agency for social justice. As our research progresses we are curious about how these two agencies develop and interact in the sociopolitical contexts of higher education. And we are curious about the agentic zones of designers and the faculty with whom they work.

Laura's Design Ethic in Practice

On the advice of a classroom teacher Laura's parents placed her in drama activities when she was ten years old, which "fundamentally shaped" her. She described an integrated approach to teaching writing during her teaching practice, in which she would use art, drama and physical activity to create a safe and caring environment for children who had "clinical depression, ADHD, severe learning problems;" given the economic status of the neighborhood she'd bring snacks from home because she worried about their nutrition. She described how, in the context of a novel study, she would take them outside for physical activity where she would, for example, have them "pretend that they were mice running in this field and this hawk was about to get them ... and they had to hide and crouch down." Laura characterizes her values as "holistic— learning involves the body and knowing is embodied." She is conflicted about the disembodiment of online learning. She says, "(We're) having a written discussion online ... we're not *talking* online. We are in actuality, typing; meanwhile all our body is learning is to sit stationary in front of a screen"

Laughing, she noted the family "theory of getting kids tired out by running a lot," and related her "philosophy of teaching stems a lot from their philosophy of raising me." A key event in shaping her worldview occurred when she was in high school and she observed a friend's encounter with racism: her struggles drove Laura to "look at human rights and to reexamine my own prejudices in life and that really drove me into the area of human rights (in graduate work)." With an emphasis on spiritual and physical well being as a way of coping, the same friend influenced Laura as a designer. While wanting to shape active, social, holistic and equitable learning environments, Laura is at the same time deeply conflicted about the "critical social issues surrounding the use of technology." Laura partially attributes the development of her identity as a critical instructional designer to a graduate supervisor who would engage in debates with her "about the socially constructed nature of technology and the inability to develop a technology based learning environment that didn't have a cultural bias."

Her values were challenged in a job developing materials for a company that delivered expensive training to hard to employ populations of young, single mothers, aboriginals, and new Canadians, "this was a very lucrative process for a company willing to exploit." Appalled at the cost of tuition paid by the clients and the company's lack of integrity, Laura resigned because she "just couldn't support their business practices." Laura achieves a degree of moral coherence when she is able to work with third world development projects where her commitment to human rights allows her to challenge the assumptions of what she terms "the neoconservative stance" to education. Her design ethic in practice is framed by the notion of conversation communities in which all members are morally accountable.

When we asked Laura to tell us a story of practice embodying her values of holistic learning that is socially responsible, she told one of professional and personal transformation for learners, faculty, and herself in a medical course designed for rural health care practitioners.

The Health Care Team: Challenging Cultural Myths

Laura is one of several designers in an academic support centre in a Western university for faculty who were developing blended learning environments. One of many programming initiatives, the unit offered up to $500 to faculty as seed funding for an instructional development project. One proposal requested a digital camera to support an information site for resources (texts, images; educational support) and emerging practices in end of life care. The faculty member, a new tenure stream appointment in the Faculty of Health Care, had had to cope with the complex physical, emotional, cognitive, social, and spiritual issues of the dying and their communities in his practice as a young rural physician. As he struggled to find the resources he needed he became determined to make them available to his colleagues around the world; he took up his faculty appointment at a time the Internet was becoming a more stable and accessible delivery platform.

The mission of Laura's unit was to support and enhance reflective practice in the development of transformative learning environments. Although many faculty clients would have preferred an arm's length production orientation, the unit took a faculty learning view through active inclusion on a collaborative instructional development team. Dr B. wanted a tool to develop a Web site; here was an opportunity to involve him in a project to transform the way in which medical professionals learn in their workplaces. Bemused, Dr B. was drawn into designing a case based, interdisciplinary, continuing professional medical education course. The design involved virtual health care teams composed of physicians, nurses, residents, and pharmacists working equitably and collaboratively to resolve end of life care issues with the support of a facilitative group of rural physicians and nurses, a spiritual advisor, a pharmacist, and a sociologist.

Coming from a family associated with the health professions, Laura's pedigree, or her "knowledge of the day to day life of a medical family," helped her establish an "immediate rapport" with Dr B. and an emotional commitment to his success. Since Laura was sensitive to the medical culture of authority, she realized that Dr B. would find it difficult, even risky, to trust her design expertise and accept advice that contradicted his professional enculturation into a moral and intellectual hierarchy with the physician at the head. Laura's design approach involved offering problem based alternatives to the instructor centred, text based presentation of evidence of the medical seminar, talking through the likely outcomes of his instructional ideas and implementing, with empathy, decisions that she knew would not result in successful learning experiences. Laura relates her practice to her goals of social emancipation rooted in the personal values of the community and felt that Dr B's increasing appreciation for instructional design as a valid process was partly grounded in the high value he placed in medical evidence based practice:

> I see ... the same parallel in working on a project in instructional design as doing development work in emerging countries ... this comes from my studies in global and human rights education and critical theory ... this has been fundamental in shaping my own philosophy of design and education. Any time an OECD country went in and said, 'This is the way we think you should develop...This is the right way, this is our way'... there has been no success.... Social change requires that people change how they are in the world—their thinking—their feelings—their actions—and this is extremely personal. Dr. B. could have come out of that (project) hating technology ... but the major change he experienced ... wasn't really his attitude towards

technology, but rather his view towards instructional design—it was like, "Wow, instructional design is an area of expertise that is necessary and important!"

She believes that change is realized in webs of personal commitments to others and enacts that through an instructional design practice in which

> you say, "We need to foster that change from within... from the grass roots ... and I am here to support that.... Change is a very emotional thing, and I think that is why in order for the course to be successful ... it needs to come from his heart."

By supporting Dr B's growing appreciation of inclusive, constructivist learning environments, Laura undergoes a personal transformation that leads her to the next stage of her moral and political growth. Technology can embody her values—she has to find out how; she can't work in any other context.

> I know that I'm not the person that I was when I started.... I am meant to go and have a life of adventure for a while and build and embody what I want my life to be ... the people who make social change happen are people like Dr B. .... He embodies what he preaches.... He is such a good, passionate man. He said to me one time, "I was working with a patient who was your age.... All I could do was go out in my car and cry," and I thought, "You are the person that I want beside me when I am dying."... Those are the people that make the important differences in the world .... I want to embody what I value

If she was criticized for working with technology, she responded

> Yes, but I am working on a Palliative Care project. There's meaning in this.... I don't think I would have stayed as long as I did ... If I couldn't find meaning in the project ... if I didn't find meaning in the people; if I didn't find meaning in supporting their success.

Growing up in a medical culture Laura was well aware of the claims to intellectual, moral and cultural authority, which she challenges through her own commitment to critical education. The personal values she embodies in her instructional design practice—framed by a critical ethic of care—are thus interposed within and shaped by "a broader linguistically and culturally determined weave of relationships, interactions, and possibilities making up an ideology or shared worldview" (Herda, 1999, p. 57). Laura enters into an ethical conversation with Dr B. and engages both his world view as a physician and his "disorienting" experience as a learner.

Through the familiar use of case studies and clinical knowledge Laura *acts* through her ethical knowledge when she asks Dr B. to confront the cultural myths that "structure(s) the individual's taken for granted views of power, authority, knowledge, and identity" (Britzman, 1991, pp. 6–7). By referencing her personal knowledge of physicians' beliefs and practices, she encourages Dr B.'s transformative thinking about power relationships in the health care teams that shape the learning experience in the course. In Herda's (1999) words, the design conversation is ontological in nature "because the cultural reference points that determine our own identity are reinterpreted in view of our personal expectations and singular circumstances" (p. 57). For Laura, this means working with Dr B. and his colleagues in the facilitators' group to explicitly value the knowledge of the "lower status" members of the learning teams, and to insist that the team members value each other.

> remember in the design when (the nurses) came back and said, "Well, you know, the doctors have said it all already. What is the point in (participating in the collaborative case study and offering our input) ... when the doctors have all the authority?' This was a form of invalidation. As developers and designers, we then went back and said, 'Ok, how can these learners feel valued? What can they bring to the learning that they feel is of value and how as a designer do you build on that?"

This significant challenge to the learners' personal and social investment in the cultural myth of Medical Authority led to a critical decision: at one point midway through the course the physicians resisted working in collaborative teams in which the voices of nurses and other non-physicians were accepted as equally valid, and demanded to be able to form their own closed team. In intense conversations with Laura in which he struggled with his discipline's expectations of higher status Dr B. (courageously) refused to privilege physician knowing by reorganizing the learning teams into separate and unequal disciplines.

Laura characterizes this transformative act as reflecting the mutuality of relational design practice. By standing by a learning design decision for social equity and inclusion, she risks her credibility with Dr B. to protect his credibility as a member of an academic and professional community. Dr B. acknowledges Laura's moral agency to affect social change in medical education, by placing his trust in her ethical design knowledge and rejecting certain long held cultural values of medical moral authority.

Ethical knowledge and morally coherent design practice requires an inner ear, tuned to the stories not told and the fears not expressed. A less mature instructional designer might not be aware of her ethical boundaries and where they interact with those of her client, where her zone of influence intrudes in harmful ways on other cultures and discourses, and where there is space for challenge and change. Laura's agentic zone is ultimately bounded and informed by her profound personal relationships, within the framework of her identity as a critical educator.

David and the Grand Politics of Design

As does Laura, David embodies his values in his design practice, values that derive from his early exposure to labour politics. While Laura acts in individual relationship and emotional connection, David's agency plays out in a broader institutional discourse.

David's career path, like many of the designers in this study, evolved through seemingly serendipitous events that have moral coherence when viewed retrospectively (Schwier, Campbell, & Kenny, 2004). In mid-career he joined a public educational organization that developed and delivered learning services to adults unable to attend a post-secondary institution. The "social mission" of the agency, related to access to education, appealed to him and, although he enjoyed teaching as an English professor he felt that there were "other people who also enjoy teaching the elite, which is who we were teaching." He describes his "epiphany" as a decision emerging from his personal politics, in that "increasing opportunities for the underprivileged in society has always been an interest and I've always felt, obviously, education is an important aspect of improving people's lives."

David's father was a radical labour leader who, in the mid 1930s, organized a pacifist "sit down" involving 1,500 unemployed workers, for which he was subsequently jailed. David's family fled to a remote part of the province where he grew up with a "working class, union based, multi-cultural mix of connections."

> One of the interesting things about (that) industry is that there are all sorts of races and interesting people that show up in those (camps) ... so the people we knew were the fellow workers, and they were Chinese, East Indian....
> A fair number of First Nations people and so the children of these workers were my playmates and friends as I grew up.

Attracted by the politics of Canadian literature David attended a university well known for its left leaning politics in the 1970s, where he became involved in student government. The opportunity to join the learning organization also reconnected him with marginalized communities such as the First Nations group who was one of the "forward looking nations and individual bands within that were looking to taking over or getting more control over their own education." David explicitly supports this project of appropriation as a cultural and moral issue by subverting the administrative practices of his organization: he assumes the authority to facilitate partnerships that are defined by the community stakeholders rather than the bureaucracy, and directly approaches policy makers at the highest levels of the provincial education ministry.

> (I) set up a number of sort of partnerships, collaborations, with the First Nations schools (who) were doing Adult Ed and college level programming and university transfer level programming.... Not all of them were remote ... but what they were short of was curriculum and accreditation and so with that kind of partnership then allowed them to have our accreditation. They would use our courses but they would use them most often in the face to face situation, delivered by the teachers in the First Nations college or school. So they were sort of transforming distance education materials and methodology to the local sites. It's been sort of a practice that gets me in hot water now and then, but everybody needs hot water now and then...once the idea sort of struck, I started contacting other groups and they networked fairly well themselves, so other people would contact me as well saying, "Hey, this is interesting. We would like to get onboard too." I'd say, "Fine! Let's do it." There were about 23 different First Nations groups.

Like Laura, David reflects moral coherence, or "thoughtful agency" (London Feminist Salon Collective, 2004) in that he is conscious of and acts out of a consistent ideology in the organizations and institutions in which he works; while Laura's practice seems grounded in the gentle reciprocity of personal relationships, David's language reflects his political enculturation in disruption of the dominant discourse. He sees the potential of instructional design in "subverting the traditional system" depending on the context of which the designer is part. Although he acknowledges that an "instructional designer in a very highly regarded research university might do very good work in helping transform teaching in better ways for the elite" and is "valuable," he prefers to spend his agentic capital on "quality education that's equal to what the elite are offered." He is critical of the "second class" status of distance education, knowing that "there's all sorts of reasons why people don't get on that sort of traditional high class, upper class route" and that excluded communities of learners, such as First Nations and women, continue to be marginalized by the irony of technologies that simultaneously increase and limit access while maintaining the lower status of its users. Although an apparent contradiction, his move from an organization that was responding directly to these issues to a large, research university reflects a moral coherence because "it was intriguing to think, 'Well, how can we make it better for students at (this traditional university) and possibly even open things up a bit (here)?" because the lifelong learning aspect has always been really important to me." Proud of the organization's learner demographics—60-70% women with an average age of 35 to 46—he thought,

> This could be interesting with the university's (cultural) mix and also with (a colleague's) international work, because one of the advantages, if you can get out into the rest of the world and influence that, then you can bring that back to your own institution and kind of leverage that to make some changes there.

David raises issues of knowledge representation in dominant forms that are "institutionally controlled" but represented as neutral: Britzman's (1991) monological knowledge.

> That was one of the issues I had with broadcast ... 75% of the people had VCRs and we should be designing our media pieces for the VCR.... I tried to explain that broadcast is institutionally controlled and VCR can be learner controlled.... Because what was always an issue for me with broadcast was that as a student you're certainly going to learn something but you just don't have any sort of control over what's gone zinging by and how are you going to apply it and what sort of deeper learning activities opportunities do you have.

And he underlines his ethical knowledge in action by relating an example of his practice for which he typically addresses a larger social issue, that is, animal research. One of his instructional design projects involved the development of a curriculum based on the Canada Council ethical guidelines for animal care. His moral entree was that "researchers aren't doing more harm than they need to do to animals." But he found it difficult to reconcile his personal value for animal life with "a kind of animal slavery and ... the larger philosophical issues of what are beings." In the end, David saw an opportunity to subvert the discourse through design activism: he engaged the subject matter expert throughout the project in a critical conversation about the morality of research that privileges human needs over the "right of every living thing ... to be untouched and left alone."

Whereas Laura's focus is on personal transformation of self and faculty with whom she works, David's goal is the "development of the social citizen" by confronting the "corporatisation" of universities that "have very little to do with education any more" and that feed students' expectations that if they study hard they will "learn what they need to be successful at the job and get up the corporate ladder." David's commitment, to increasing access to curriculum associated with cultural power through distance delivery, is a political ideological choice that he anticipates will address inherent "inequities and substandard" access for both undergraduate and alternate learning communities.

CONCLUSION

Implications of Agency for Instructional Design Practice and Education

Although the field is evolving, the dominant discourse of instructional design—that it is a set of scientific principles embedded in a rational, technical process operating outside of, or in spite of, social, political, cultural, and personal contexts—deskills the instructional designer in higher education institutions in fundamental human ways. We maintain that instructional design is a moral practice that embodies the "relationship between self concept and cultural norms, between what we value and what others value, between how we are told to act and how we feel about ourselves when we do or do not do act that way" (Anderson & Jack, 1991, p. 18). Instructional design involves the ethical knowledge of the designer acting in moral relationship with others in a dialogue among curriculum, the sources and forms of knowledge and power, and the social world. As ethical actors in that world we use the language of design in collaborative conversation with our colleagues, our clients, and our institutions to create an alternate social world of access, equity, inclusion, personal agency and critical action. Herda (1999) captures this notion of transformative social change when she credits language with a "generative role in enabling us to create and acknowledge meaning as we engage in discourse and fulfill social obligations ... (that) are characterised as moral activities" (p. 24).

Agency refers to doing and implies power (Hartman, 1991). Designer agency develops into a positive social force when designers have the moral space and authority for the reflexive practice that makes available critical relationships to knowledge (Britzman, 1991).

Referring to the coupling of knowledge and power, James Donald (1979) argues that agency can only be understood and become a directed source of personal and political power if "it is conceived not just as a source of social change, but above all, as an effect of particular social and institutional practices" (in Britzman, 1991, p. 37). What then are the implications for instructional design practice that is transformational; that contributes in significant ways to the public good?

We believe that designers are not technicians who simply implement techniques and principles, although when challenged they can certainly use that language to describe what they 'do, but are principled actors whose practices embody core values, and are represented by moral language and political acts. What could we achieve if we were thoughtful, deliberate, and unapologetic in aligning design projects with the ethical knowledge of designers? If we developed a community in which the moral dimensions of practice were explicitly developed through reflexive dialogue? If we publicly explored the "conscious and unconscious influences on (our) practice and personal resistances to change" (Kugelmass, 2000, p. 179) by asking ourselves: Who am I, why am I practising this way, and what effect does this have on others?

Goodson and Cole (1994) described the developmental circle of novice teachers as they participated in a reflexive project of sustained, critical conversation with their peers. Over time the teachers talked less about the technical aspects of their practice, and more about effecting substantive change in their schools and communities: "Broadly stated, the teachers expanded their conceptions of teaching and themselves as teachers from an early image of teacher as classroom technician to one of teacher as agent of change" (p. 96). We need to articulate our experiences, make connections with ourselves, challenge theory and theorise our practice before we can influence institutional change. We must move beyond lamenting the failure of designers to faithfully implement the theoretical models of design at a micro-level, and inquire into "the epistemology of practice" that is complex, ill-structured, situational, and value laden.

How might we redefine the curriculum in graduate programs of instructional design? Several possibilities exist within the framework of critical, moral practice. For example, engaging pre-service designers early in identity work through approaches such as autobiographical writing, providing more situated experiences that are then deconstructed in group conversations, working with cases based on ethical dilemmas, developing international links and project teams that challenge cultural assumptions about learning, internships—these are a few of the activities embodied in the change management process. Further, the focus in many course on the mastery of tools should be re-examined.

In the meantime, since most graduate programs of professional preparation in educational technology are silent on these issues, narrative communities seem the best sites for this inquiry as designers rehabilitate their identities and "emplot" new narratives that effect structural changes in their institutions (Hartman, 1991). We are listening closely to the stories of the designers for hints for harnessing the transformational power of community.

ACKNOWLEDGMENT

This research was supported by a grant from the Social Sciences and Humanities Council of Canada.

REFERENCES

Anderson, K., & Jack, D. C. (1991). Learning to listen: Interview techniques and analyses. In S. B. Gluck & D. Patai (Eds.), *Women's words: The feminist practice of oral history* (pp. 11–26). New Yoek: Routledge.

Andrews, D. H., & Goodson, L. A. (1991). A comparative analysis of models of instructional design. In G. J. Anglin (Ed.), *Instructional technology, past, present, and future* (pp. 133–155). Eaglewood, CO: Libraries Unlimited. (Reprinted from *Journal of Instructional Development, 3*(4), 2–16).

Bolman, L. G., & Deal, T. E. (1997). *Reframing organisations: Artistry, choice and leadership.* San Francisco: Jossey-Bass.

Braden, R.A. (1996). The case for linear instructional design and development: A commentary on models, challenges and myths. *Educational Technology, 36*(2), 5–23.

Britzman, D. P. (1991). *Practice makes practice: A critical study of learning to teach.* Albany, NY: SUNY.

Carter, V. K. (2000). Virtual shades of pale: Educational technologies and the electronic "other." In N. M. Rodriguez & L. E. Villaverde (Eds.), *Dismantling white privilege: Pedagogy, politics and whiteness (Counterpoints,* Vol. 73) (pp. 25-40). New York: Peter Lang.

Cennamo, K., Abell, S., & Chung, M. (1996). A "layers of negotiation" model for designing constructivist learning materials. *Educational Technology, 36*(4), 39–48.

Christians, C. G. (2000). Ethics and politics in qualitative research. In N. K. Denzin & Y. S. Lincoln (Eds.), *Handbook of qualitative research* (2nd ed., pp. 133–155). London: SAGE.

Cobb, P. (1996). Where is the mind? A coordination of sociocultural and cognitive constructivist perspectives. In C.W. Fosnot (Ed.), *Constructivism: Theory, perspectives and practice.* New York: Teachers College Press.

Conle, C. (2000). Narrative inquiry: Research tool and medium for professional development. *European Journal of Teacher Education, 23*(1), 49–63.

Dick, W. (1996). The Dick and Carey model: Will it survive the decade? *Educational Technology Research and Development, 44*(3), 55-63.

Dick, W., & Carey, L. (1996). *The systematic design of instruction* (4th ed.). Glenview, IL: Scott, Foresman.

Donald, J. (1979). Green paper: Noise of crises. *Screen Education, 30,* 13–49.

Ewing, J. M., Dowling, J. D., & Coutts, N. (1998). Learning using the World Wide Web: A collaborative learning event. *Journal of Educational Multimedia and Hypermedia, 8*(1), 3–22.

Francis, B. (1999). Modernist reductionism or post-structuralist relativism: Can we move on? An evaluation of the arguments in relation to feminist educational research. *Gender and Education, 11*(4), 381–393.

Fullan, M. (2001). *Leading in a culture of change.* San Francisco: Jossey-Bass.

Fullan, M. (2004). *Leadership and sustainability: System thinkers in action.* Thousand Oaks, CA: Corwin Press.

Garrison, D. R. (1993). A cognitivist constructivist view of distance education: An analysis of teaching-learning assumptions. *Distance Education, 14*(2), 199–211.

Glaser, R. (1991). The maturing of the relationship between the science of learning and cognition and educational practice. *Learning and Instruction, 1*(2), 129–144.

Goodson, I. (1994). Studying the teacher's life and work. *Teaching and Teacher Education, 10*(1), 29–37.

Goodson, I. (1995). *Storying the self: Life politics and the study of the teacher's life and work.* Paper presented at the annual meeting of the American Educational Research Association, San Francisco, CA.

Goodson, I. K., & Cole, A. L. (1994). Exploring the teacher's professional knowledge: Constructing identity and community. *Teacher Education Quarterly, 21*(1), 85–105.

Gustafson, K. L., & Branch, R. M. (1997). *Survey of instructional development models* (3rd ed.). Syracuse, NY: ERIC Clearinghouse on Information and Technology.

Hannafin, M., Land, S., & Oliver, K. (1999). Open learning environments: Foundations, methods and models. In C. M. Reigeluth (Ed.). *Instructional-design theories and models, Volume II: A new paradigm of instructional theory.* Mahwah, NJ: Erlbaum.

Hartman, J. E. (1991). Telling stories: The construction of women's agency. In Joan Hartman & Ellen Messer-Davidow (Eds.), *(En)gendering knowledge: Feminists in academe* (pp. 11–31). Knoxville: University of Tennessee Press.

Herda, E. A. (1999). *Research conversations and narrative: A critical hermeneutic orientation in participatory inquiry.* London: Praeger.

Heilbrun, C. (1988). *Writing a woman's life.* New York: Ballantine Books.

Hill, J. R., Bichelmeyer, B. A., Boling, E., Gibbons, A. S., Grabowski, B. L., Osguthorpe, R. T., Schwier, R. A., & Wager, W. (2004). Perspectives on significant issues facing instructional design and technology. In M. Orey (Ed.), *Educational media and technology yearbook, 29,* 23–43.

Jonassen, D. H. (1999). Designing constructivist learning environments. In C.M. Reigeluth (Ed.), *Instructional design theories and models, Volume II: A new paradigm of instructional theory* (pp. 215–239). Mahwah, NJ: Erlbaum.

Jonassen, D., Dyer, D., Peters, K., Robinson, T., Harvey, D., King, M., & Loughner, P. (1997). Cognitive flexibility hypertexts on the Web: Engaging learners in making meaning. In B. H. Khan (Ed.), *Web-based instruction* (pp. 119–133). Englewood Cliffs, NJ: Educational Technology.

Kenny, R. F., Zhang, Z., Schwier, R. A., & Campbell, K. (2005). A review of what instructional designers do: Questions answered and questions not asked. *Canadian Journal of Learning and Technology, 31*(1), 9–16.

Koper, R. (2000). *From change to renewal: Educational technology foundations to electronic learning environments*. Inaugural address of the Educational Technology Expertise Center, Open University of the Netherlands.

Kugelmass, J. W. (2000). Subjective experience and the preparation of activist teachers: Confronting the mean old snapping turtle and the great big bear. *Teaching and Teacher Education, 16*, 179–194.

London Feminist Salon Collective. (2004). The problematization of agency in postmodern theory: As feminist educational researchers, where do we go from here? *Gender and Education, 16*(1), 25–33.

Mayer, R. H. (1999). Designing instruction for constructivist learning. In C. M. Reigeluth (Ed), *Instructional design theories and models, Volume II: A new paradigm of instructional theory* (pp. 141–159). Mahwah, NJ: Erlbaum.

Merrill, M. D. (2002). First principles of instruction. *Educational Technology, Research and Development, 50*(3), 43–60.

Merrill, M. D., Drake, L., Lacy, M. J., Pratt, J., & the ID2 Research Group (1996, Sept/Oct.). Reclaiming instructional design. *Educational Technology*, 5–7.

Morrison, G. R., Ross, S. M., & Kemp, J. E. (2004). *Designing effective instruction*. Hoboken, NJ: Wiley.

Reigeluth, C. M. (1996). A new paradigm of ISD? *Educational Technology, 36*(3), 13–20.

Reigeluth, C. M. (1999). What is instructional-design theory and how is it changing? In C. M. Reigeluth (Ed), *Instructional design theories and models, Volume II: A new paradigm of instructional theory* (pp. 5-29). Mahwah, NJ: Erlbaum.

Saettler, P. (1992). *A history of instructional technology*. New York: McGraw-Hill.

Shambaugh, R. N., & Magliaro, S. G. (2001). A reflexive model for teaching instructional design. *Educational Technology Research and Development, 49*(2), 69–91.

Schwier, R. A., Campbell, K., & Kenny, R. (2003). Instructional designers' interpretations of their communities of practice. In M. Simonson (E.), *Proceedings of Selected Research and Development Paper Presentations*, Association for Educational Communications and Technology (AECT), Anaheim, California.

Schwier, R. A., Campbell, K., & Kenny, R. F. (2004). Instructional designers' observations about identity, communities of practice and change agency. *Australasian Journal of Educational Technology, 20*(1), 69–100. Retrieved from http://www.ascilite.org.au/ajet/ajet20/schwier.html

Seels, B., & Glasgow, Z. (1998). *Making instructional design decisions* (2nd ed.). Upper Saddle River, NJ: Merrill Prentice Hall.

Smith, P. L., & Ragan, T. J. (2005). *Instructional design* (3rd ed.). Hoboken, NJ: Wiley.

Tergan, S. O. (1997). Misleading theoretical assumptions in hypertext/hypermedia research. *Journal of Educational Multimedia and Hypermedia, 6*(3/4), 257–283.

Vrasidas, C. (2001). Constructivism versus objectivism: Implications for interaction, course design, and evaluation in distance education. *International Journal of Educational Telecommunications, 6*(4), 339–362.

Weick, K. E. & Quinn, R.E. (1999). Organizational change and development. *Annual Review of Psychology, 50*, 361–386

Willis, J. (1998). Alternative instructional design paradigms: What's worth discussing and what isn't? *Educational Technology, 38*(3), 5–16.

Willis, J. (2000). The maturing of constructivist instructional design: Some basic principles that can guide practice. *Educational Technology, 40*(1), 5–16.

Wood, D.R. (1992). Teaching narratives: A source for faculty development and evaluation. *Harvard Educational Review, 62*(4), 535–550.

CHAPTER 12

CONSTRUCTIVIST UNDERPINNINGS IN DONALD SCHÖN'S THEORY OF REFLECTIVE PRACTICE

Echoes of Nelson Goodman

Elizabeth Anne Kinsella

Donald Schön's theory of reflective practice has garnered unprecedented attention in the field of continuing professional education. In this chapter, I examine Schön's writing on reflective practice and argue that a constructivist orientation is a central, although largely unexplored, underpinning of his work. In particular I consider the work of philosopher Nelson Goodman and suggest that he is a major constructivist influence within Schön's theory. Secondary constructivist influences such as George Kelly, Jean Piaget and Ernst von Glasersfeld are also highlighted. Given this link, one of the avenues for understanding Schön's theory of reflective practice may well lie in deepening one's understanding of constructivist thought.

Previoulsy published in *Reflective Practice* (2006), 7(3), 277–286.

Constructivist Instructional Design (C-ID): Foundations, Models, and Examples, pp. 265–276

Donald Schön's (1983, 1987) theory of reflective practice has captured the imagination of educators and has garnered unprecedented attention in the field of continuing professional education. Yet, the constructivist underpinnings of Schön's theory have remained unexamined in the literature to date. In this chapter, I consider constructivist underpinnings in Schön's theory; I argue that Schön's conception of reflective practice is infused with constructivist roots and that one of its defining characteristics is this constructivist orientation. I suggest that those interested in understanding Schön's work would do well to consider constructivism as it relates to reflective practice and continuing professional education.

INFLUENCES OF NELSON GOODMAN

Schön appears to have been influenced to a significant extent by Nelson Goodman's (1978) constructivist views as expressed in his seminal book *Ways of Worldmaking*. Goodman is an American philosopher who has written important works in metaphysics, aesthetics and epistemology. His work is sometimes regarded as dense and impenetrable; however it has in recent years been rendered more accessible by one of his students, Catherine Elgin. Elgin (2000) notes that throughout Goodman's work there is a concern with the ways that worlds are made by symbols, and an argument that the symbols people construct inform the facts they find and how they structure their understanding of these facts. In this chapter, I consider the evolution of constructivist terminology as it relates to reflective practice, examine constructivist influences in Schön's work, and raise consequent implications for scholarship that seeks to foster understanding of reflective practice and its applications.

Use of Constructivist and Constructionist Terminology

In his earlier work Schön (1987) uses the term constructionism, and later switches to the term constructivism (Schön, 1992; Schön & Rein, 1994). Likewise, Goodman (1978) adopts the term constructionism; however, later his work is referred to in terms of constructivism (Elgin, 1997). Both men's writing largely preceded the vigorous discussions about constructivism that have recently emerged in fields such as philosophy, social science, psychology and education (see Fosnet, 1996; Golinski, 1998, Gould, 2003; Kukla, 2000; Neimeyer & Mahoney, 1995; Phillips, 1995, 2000; Steffe & Gale, 1995; von Glasersfeld, 1995; Zehfuss, 2002). As discursive understandings of these terms have evolved over time, I suggest that Schön and Goodman's intentions when they speak of constructionism in their early work are analogous to what many people today refer to as constructivism.

In recent years the word constructionist has taken on new meaning. It has frequently been invoked by those who focus on the social construction of knowledge (Berger & Luckmann, 1967; Burr, 1995; Gergen, 1994, 1999; McNamee & Gergen, 1992; Potter, 1996; Shotter, 1993). In contrast the term constructivist is more frequently adopted by those who focus on the active manner in which individuals construct knowledge (see Fosnot, 1996; Kelly, 1955; von Glasersfeld, 1995). In this chapter, I use the terminology adopted by Schön and Goodman in its original form—constructionism—when quoting their work. However, as mentioned above I believe the intentions of both men when they use this term are closer to what is currently referred to as a constructivist approach.

Definitions of Constructivism

Defining constructivism is not a straightforward affair, as there are epistemic debates and conceptual differences in current constructivist views (see Geelan, 1997; Mathews, 1997; Noddings, 1990; Phillips, 1995; Steffe, 1995). Indeed, von Glasersfeld (1995), in recounting his experience of the first symposium of leading constructivist thinkers around the world, stated that rather than trying to formulate basic constructivist principles the group spent most of the time "arguing about relatively small individual discrepancies" (p. 18). Noddings (1990) suggests that despite such differences, constructivists generally agree on three points:

- all knowledge is constructed, at least in part, through a process of reflection
- there exist cognitive structures that are activated in the process of construction, and
- cognitive structures are under continual development (purposive activity induces transformation of those structures, and the environment presses the organism to adapt).

Meichenbaum (1995), a former behaviorist turned constructivist, gives a succinct description: "The constructivist perspective is founded on the idea that humans actively construct their personal realities and create their own representational models of the world" (p. 23). Furthermore, common to various proponents of constructivism is the tenet that the human mind is a product of constructive symbolic activity and that reality is a product of the personal meanings that individuals create. It is not as if there is one reality and individuals distort that reality; rather there are multiple realities, and the task is to help individuals to become aware of how they create these realities and of the consequences of such constructions (Meichenbaum).

SCHÖN'S CONSTRUCTIVISM

With respect to Schön's constructivism, the influences appear to have evolved over time (Argyris & Schön, 1974/1992; Schön, 1983, 1987, 1992; Schön & Rein, 1994), with Goodman as the central and most explicit. In his early writing with Argyris (Argyris & Schön, 1974/1992), there appears one reference to a well-known psychologist and pioneer of constructivist learning, George Kelly; however Kelly's (1955) influence is neither elaborated, nor cited in future discussions. In 1983, in his classic text *The Reflective Practitioner*, Schön writes in a manner that suggests a constructivist influence; however, he does not explicitly use the term or cite others from the field. In 1987, in *Educating the Reflective Practitioner*, Schön links Nelson Goodman to what he then refers to as constructionism. And, in 1992, he adds Jean Piaget, and in 1994, Ernst von Glasersfeld, as references for his constructivist orientation. Table 12.1 illustrates these citations.

That Schön (1992) values a constructivist approach is clear from his critique of John Dewey, the philosopher whose notion of reflective thinking he says he reworked into his own version of reflective practice. However, Schön (1994) has referred to Piaget and von Glasersfeld's influence as well. Therefore, a brief overview of their position is offered prior to the more extensive consideration of Goodman's thinking.

Piaget's and Von Glasersfeld's Constructivism

The corpus of Piaget's work is significant; he published 88 books and hundreds of articles, and he has been interpreted in numerous, often conflicting, ways (von Glasersfeld, 1995). Piaget is well known as a

Table 12.1.   Schön's constructivist in?uences

Influences	Citation
Kelly	Argyris & Schön (1974)
Goodman	Schön (1987)
Goodman + Piaget	Schön (1992)
Goodman + Piaget + von Glasersfeld	Schön & Rein (1994)

constructivist, and traces his constructivism to Kantian philosophy. He credits Kant with the first description of an epistemological subject and a description of the structures by which a competent subject generates knowledge. Piaget followed Kant in distinguishing between empirical knowledge and logico-mathematical knowledge, but he differed from Kant by describing cognitive structures as products of development rather than as innate structures (Noddings, 1990). Noddings points out that in a Piagetian view:

> the objects play a role in reflective abstraction; that is, epistemological subjects and objects are indissociably linked in operational events. We cannot force certain results onto the objects we operate on. Our operations are somehow constrained. (Noddings, 1990, p. 9)

Thus, according to Piaget, one's actions are constrained by the objects one operates on.

Von Glasersfeld (1995) has built his version of radical constructivism on the teachings of Piaget. He suggests that Piaget has often been misinterpreted in North America due to inaccurate translations of his original texts. He offers a distinct interpretation, built upon reading Piaget in the original French versions and upon a study of the corpus of Piaget's work. The distinguishing assumption of von Glasersfeld's (1995) radical constructivism is his belief that cognition serves the subject's organization of the experiential world, not the discovery of an objective ontological reality. In this sense radical constructivism replaces the notion of truth with the notion of *viability within the subject's experiential world. Von Glasersfeld acknowledges that* his version of constructivism is difficult for many to accept, and argues that this theory of knowing is intended as a tool that should be tested for its usefulness rather than as a metaphysical proposal. In this way, von Glasersfeld sidesteps metaphysical questions about the nature of reality and proposes a pragmatic view that supports a pragmatic focus on the practitioner's organization of his or her experiential world. Such a view focuses on viability or fit within the subject's experiential world, as opposed to a fixed notion of truth.

This discussion is relevant, as Schön refers to Piaget and von Glasersfeld's position in his later work; however, one can only infer the significance with respect to Schön's view. In particular, the notions of practitioners as active organizers of their experiential worlds, of practitioner actions being constrained by the environments in which they practice, and the viability or fit within the subject's experiential world are relevant considerations with respect to (a) how reflective practitioners construct their practice worlds, and (b) how practitioners test out their constructions.

Goodman and Schön: Relativism With Constraints

Goodman's ideas clearly have the most significant influence on Schön's constructivist thinking. From a philosophical position, both Goodman and Schön appear to advocate a type of relativism with constraints. Goodman (1978) suggests that what emerges in his work can perhaps be described as a radical relativism under rigorous constraints. He presents his position as at odds with empiricism, rationalism, materialism, idealism and dualism, essentialism and existentialism, mechanism and vitalism, mysticism and scientism, and indeed at odds with most ardent doctrines. Goodman (1978) describes his general orientation as skeptical, analytical and constructionalist. He proposes a diversity of right and even conflicting versions or worlds in the making, as opposed to a single world that is fixed or found. Yet, his relativism is not unlimited. He notes that willingness to accept countless alternative true or right world versions does not mean that everything goes, but only that truth must be otherwise conceived than as correspondence with a ready-made world. In this regard he writes:

> Though we make worlds by making versions, we no more make a world by putting symbols together at random than a carpenter makes a chair by putting pieces of wood together at random. The multiple worlds I countenance are just the actual worlds made by and answering to true or right versions. (Goodman, 1978, p. 94)

Thus, although Goodman proposes that worlds are made, he also puts constraints on acceptable versions based on their "fit." Schön's position appears quite similar to Goodman's, although he tends to avoid discussions of relativism per se. Schön (1987) considers himself a constructionist, yet notes that a constructionist point of view need not lead to relativism and the abandonment of every claim of knowledge. Within a particular created world, he suggests, it is possible to discover the consequences of one's moves, make inferences, and establish by experiment whether one's way of framing the situation is indeed appropriate. This is similar to Piaget's position as articulated above: the notion that one cannot force certain results onto the objects one operates on, but rather that one's operations are somehow constrained. Such a perspective is also consistent with von Glasersfeld's (1995) description of viability within the subject's experiential world. Drawing on Spence's (1982) discussion of historical versus narrative truth in psychoanalysis, Schön (1987) suggests that all interpretations are essentially creative; and that any number of different ones, equally coherent and complete, might be provided for any particular clinical event. In this view, right interpretations have a power to persuade grounded in their aesthetic appeal. They may also acquire pragmatic usefulness, grounded in the expectation that they will lead to additional clarifying clinical material (Schön, 1987, p. 229).

Schön appears to be putting forth a type of relativism with "practical" or "practice" constraints, in which different versions of the world of practice are tenable, yet are judged according to their aesthetic appeal, persuasiveness, and pragmatic utility. Such a view is similar to Goodman's notion of 'radical relativism with rigorous restraints', in which versions are judged according to fit, while recognizing that more than one coherent version can exist. As mentioned above, this view is also similar to that of Piaget, who notes that one's operations are constrained by the objects one operates on, and to that of von Glasersfeld, who looks for viability within the subjects' experiential world.

CONSTRUCTIVISM AND WORLDMAKING

Schön (1987) contrasts the objectivist view, in which he sees an underlying technical rationality, with the constructionist view, which underlies the notion of the reflective practitioner. According to the objectivist view, facts are what they are, and the truth of beliefs is strictly testable by reference to them. All meaningful disagreements are resolvable by reference to facts, and professional knowledge rests on a foundation of facts. Professional competence is seen as technical expertise, in which skilled professionals have accurate models of their objects and powerful techniques for manipulating them to achieve professionally sanctioned ends. Schön contrasts this objectivist view with the constructionist view, in which the practitioner is seen as constructing the situations of practice. The constructionist perspective encompasses 'the way people frame and shape their worlds' and the "kinds of reality individuals shape for themselves" (p. 322). Schön (1987) notes that constructive activity gives coherence to more or less indeterminate situations, and that individuals test their frames through a web of moves, consequences and implications. He writes:

> A constructionist view of a profession leads us to see its practitioners as worldmakers whose armamentarium gives them frames with which to envisage coherence and tools with which to impose their images on situations of their practice. A professional practitioner is, in this view, like an artist, a maker of things. (Schön, 1987, p. 218)

Likewise, Goodman (1978) has critiqued the idea that people can discover a ready-made world grounded in facts. He seeks to "irritate" fundamentalists who believe that such facts constitute the one and only real world, and that knowledge consists in believing such facts. Goodman argues that faith in facts so firmly possesses most of us that it binds and blinds us. Rather than finding a ready-made world, he argues that we continually make and remake versions of the world using words, numerals, pictures, sounds and other symbols. It is this process of making and remaking versions of the world that Goodman calls "worldmaking." Goodman (1978) suggests that universes of worlds as well as worlds themselves are built in many ways, and that "worldmaking as we know it always starts from worlds already on hand; the making is a remaking" (p. 6). Thus, both Goodman and Schön posit a world that is constructed through versions, as opposed to a world that is "found" or "ready-made." Whereas Goodman considers ways in which worlds are made, Schön considers the implications of a constructionist view for professional practice.

Although Schön (1983) alludes to constructionism, frames and ways of seeing in his book, The reflective practitioner, it is later—in *Educating the Reflective Practitioner* (1987)—that he elaborates to a greater extent on these ideas and appears to first use the word worldmaking. In particular, it is in writing about constructionism that Schön invokes this notion. Schön (1987) writes:

> In the constructionist view, our perceptions, appreciations, and beliefs are rooted in worlds of our own making that we come to accept as reality. Communities of practitioners are continually engaged in what Nelson Goodman (1978) calls "worldmaking." (Schön, 1987, p. 36)

Whereas Goodman attempts to tease out specific processes whereby worldmaking occurs, Schön adopts a more general and applied approach. Schön does not elaborate on or tease out specific worldmaking processes, but they are perhaps implicit in his descriptions of how practitioners construct situations of practice, and for this reason are highlighted below. Without attempting to put forth a rigid or exhaustive system, Goodman (1978) illustrates and comments on five processes involved in worldmaking. He is careful to say that these are 'ways' worlds are made, rather than 'the ways' worlds are made, and that there may be others not yet illuminated. Goodman notes that such processes often occur in combination, and that some illustrative examples may fit equally well under more than one heading.

First, Goodman (1978) suggests that much, but by no means all, worldmaking consists of taking apart and putting together; this he calls composition and decomposi*tion. Composition and decomposition is normally assisted or consolidated by the* application of labels: names, gestures, pictures and so forth. Second, Goodman notes that worldmaking is a matter of emphasis, and that different material is *weighted differently to render different worlds. Here he gives the example of* the difference between two histories of the Renaissance. One, without excluding the battles, emphasizes the arts; the other, without excluding the arts, stresses the battles. This difference in style is a difference in weighting that gives us two different Renaissance worlds. Third, Goodman suggests that ordering influences the process by which worlds are made. He gives the example of daily time, noting that it provides a means of marking the day into 24 hours, and of ordering each hour into 60 minutes and each minute into 60 seconds. This mode of organization, he notes, is not "found" in the world, but rather is built into a world. Fourth, Goodman highlights deletion and supplementation, suggesting that the making of one world out of another usually involves some extensive weeding out and filling, excision of old and supply of some new material. Finally, Goodman suggests that deformation can be involved in worldmaking. As an example, artists frequently use distortion as a means of making a world: Picasso for instance has worked with distortions that amount to revelation. Thus, Goodman highlights composition and decomposition, weighting, ordering, deletion and supplementation, and deformation as ways of making a world.

Schön describes some of the processes of worldmaking as he perceives it within the world of professional practitioners. He does not elaborate on various processes in the manner that Goodman does, but such processes appear to underlie the practitioner's construction of the world of practice. Schön (1987) writes of practitioners:

> Through countless acts of attention and inattention, naming, sensemaking, boundary setting, and control, they make and maintain the worlds matched to their professional knowledge and know-how. They are in transaction with their practice worlds, framing the problems that arise in practice situations and shaping the situations to fit the frames, framing their roles and constructing practice situations to make their role-frames operational. They have, in short, particular, professional ways of seeing their world and a way of constructing and maintaining the world as they see it. When practitioners respond to the indeterminate zones of practice by holding a reflective conversation with the materials of their situations, they remake a part of their practice world and thereby reveal the usually tacit processes of worldmaking that underlie all of their practice. (Schön, 1987, p. 36)

One can see in this description that the practitioner builds and takes apart the world of practice (as in Goodman's composition and decomposition), emphasizes and makes sense of different dimensions (as in weighing and ordering), and pays attention to certain facts and not to others (as in selection and deletion). According to Schön, when a situation occurs that is disorienting for the practitioner, the practitioner begins the process of remaking the practice world, trying to find a way of framing the world that is coherent and makes sense. In this view the problem setting that practitioners undertake is an "ontological process ... a form of worldmaking"

(Schön, 1987, p. 4). Schön notes that debates in professional practice often involve conflicting frames in which practitioners pay attention to different facts and make different sense of the facts they notice. For instance, in a road-building situation, civil engineers may see drainage, soil stability, and ease of maintenance issues, however they may not see the effects of the road on the economies of the towns that lie along its route. Through acts of naming and framing, the practitioner selects things for attention and organizes them, guided by an appreciation of the situation that gives it coherence and sets a direction for action (p. 4). In this way the worlds of professional practice are made and remade. Implicit within these acts of naming and framing are the processes of worldmaking: composition and decomposition, weighting, ordering, deletion and supplementation, and deformation, as described by Goodman.

The assumption that worlds are made has important implications for professional practice, and for approaches to education in professional schools. As Schön (1987) writes, "It is not by technical problem solving that we convert problematic situations to well-formed problems; rather, it is through naming and framing that technical problem solving becomes possible" (p. 5). Practitioners set the problems they go about solving, and such problem setting is a form of worldmaking that often falls outside the realm of the technical knowledge learned in professional schools. Problem setting often begins when one's usual understanding of the world bumps up against a disorienting dilemma or problematic situation that falls outside of one's usual frames.

> Because the unique case falls outside the categories of existing theory or technique, the practitioner cannot treat it as an instrumental problem to be solved by applying one of the rules in her store of professional knowledge. The case is not "in the book." If she is to deal with it competently, she must do so by a kind of improvisation, inventing and testing in the situation strategies of her own devising. (Schön, 1987, p. 5)

In this way the practitioner is viewed as setting the problem within a world of his or her own making. Goodman's (1978) suggestion that rather than one world there are a multiplicity of worlds has important implications for Schön's view, and by extension for practitioners and educators who adopt his theory.

CONCLUSION

A constructivist orientation appears to underlie Schön's theory of reflective practice, though surprisingly little literature has illuminated this link. Individual practitioners are seen as constructing viable worlds of their own making. I suggest that the thinker to which Schön most clearly relies in this respect is Nelson Goodman; however Schön also acknowledges the influence of George Kelly, Jean Piaget and Ernst von Glasersfeld. Schön has invoked Goodman's notion of worldmaking, a constructivist perspective, within his theory of reflective practice. From this perspective worlds are made, not found, and right interpretations are constrained by their fit with the world. Both men appear to adopt a similar philosophical position with respect to an underlying argument for relativism with constraints. Schön, however has not described the ways in which worlds are made using Goodman's language. Rather he has applied such processes to the world of professional practice, and through examples discusses the ways in which worlds are made in practice.

The contentious literature using the word constructivism has been prolific in the time following Schön's and Goodman's writing. It would indeed be interesting to hear their perspectives as well as their own locations within the current debates. Although this is not possible, I suggest that an examination of Donald Schön's writing on reflective practice clearly demonstrates constructivist underpinnings rooted in the work of Nelson Goodman; one of the relatively unexplored avenues for understanding reflective practice may well lie in deepening our understanding of constructivist thought.

REFERENCES

Argyris, C., & Schön, D. (1992). *Theory in practice: increasing professional effectiveness.* San Francisco, Jossey-Bass. (Originally published in 1974)

Berger, P., & Luckmann, T. (1967). *The social construction of reality.* London: Allen Lane.

Burr, V. (1995). *An introduction to social constructionism.* New York: Routledge.

Elgin, C. (Ed.). (1997). *The philosophy of Nelson Goodman: Nominalism, constructivism, and relativism in the work of Nelson Goodman.* New York: Garland.

Elgin, C. (2000). Goodman. In *Concise Routledge encyclopedia of philosophy.* London: Routledge.

Fosnet, C. T. (1996) Constructivism: A psychological theory of learning. In C. T. Fosnet (Ed.), *Constructivism: theory, perspectives and practice.* New York, Teachers College Press.

Geelan, D. (1997) Epistemological anarchy and the many forms of constructivism. *Science & Education, 6,* 15–28.

Gergen, K. J. (1994). *Realities and relationships: soundings in social construction.* Cambridge, MA: Harvard University Press.

Gergen, K. (1999). *An invitation to social construction.* London: SAGE.

Golinski, J. (1998). *Making natural knowledge: constructivism and the history of science.* Cambridge, England: Cambridge University Press.

Goodman, N. (1978). *Ways of worldmaking.* Sussex, England: The Harvester Press.

Gould, C. (2003). *Constructivism and practice: Toward a historical epistemology.* Lanham: Rowman & Littlefield.

Kelly, G. A. (1955). *A theory of personality: The psychology of personal constructs.* New York: Norton.

Kukla, A. (2000). *Social constructivism and the philosophy of science.* London: Routledge.

Mathews, M. (1997). Introductory comments on philosophy and constructivism. *Science & Education, 6*(1/2), 1–4.

McNamee, S., & Gergen, K. (Eds.). (1992). *Therapy as social construction.* London: SAGE.

Meichenbaum, D. (1995). Changing conceptions of cognitive behaviour modification: retrospect and prospect. In M. Mahoney (Ed.), *Cognitive and constructive psychotherapies.* New York: Springer.

Neimeyer, R., & Mahoney, M. (1995). *Constructivism in psychology.* Washington, DC: American Psychological Association.

Noddings, N. (1990) Constructivism in mathematics education. In R. B. Davis, C. A. Maber, & N. Noddings (Eds.), *Constructivist views on the teaching and learning of mathematics.* Reston, VA: National Council of Teachers of Mathematics.

Phillips, D. C. (1995). The good, the bad, and the ugly: The many faces of constructivism. *Educational Researcher, 24*(7), 5–12.

Phillips, D. C. (Ed.). (2000). *Constructivism in education: Opinions and second opinions on controversial issues: Ninety-ninth yearbook of the National Society for the Study of Education* (Part 1) Chicago: National Society for the Study of Education.

Potter, J. (1996). *Representing reality: Discourse, rhetoric and social construction.* London: SAGE.

Schön, D. (1983). *The reflective practitioner.* New York: Basic Books.

Schön, D. (1987). *Educating the reflective practitioner.* San Francisco: Jossey-Bass.

Schön, D. (1992). The theory of inquiry: Dewey's legacy to education. *Curriculum Inquiry, 22*(2), 119–139.

Schön, D., & Rein, M. (1994). *Frame reflection: toward the resolution of intractable policy controversies.* New York: Basic Books.

Shotter, J. (1993). *Cultural politics of everyday life: social constructionism, rhetoric and knowing of the third kind.* Buckingham: Open University Press.

Steffe, L. (1995) Alternative epistemologies: an educator's perspective. In L. Steffe & J. Gale (Eds.), *Constructivism in education.* Hillsdale, NJ: Erlbaum.

Steffe, L., & Gale, J. (Eds.). (1995). *Constructivism in education.* Hillsdale, NJ: Erlbaum.

Von Glasersfeld, E. (1995). *Radical constructivism: A way of knowing and learning.* London: Falmer Press.

Zehfuss, M. (2002). *Constructivism in international relations: The politics of reality.* Cambridge, England: Cambridge University Press.

SECTION III

R2D2 and Other C-ID Models

Jerry Willis

Interest in Constructivist-ID models was first demonstrated in the educational technology literature through articles that discussed how constructivist principles and methods of teaching might be integrated into the instructional design process. The ID literature was, until the 1990s, dominated by behavioral theories of learning and associated movements such as the drive to tie instruction to specific, predetermined objectives, and the use of objective assessments (usually tests) of instruction. In that context papers began to appear that suggested we could think about teaching and learning from a very different viewpoint.

The early papers on constructivism and instructional design tended to focus on (1) the constructivist theory of learning and (2) encouraging designers to use instructional strategies that were rapidly emerging from the fertile minds of constructivist educators. For example, in his paper on constructivism and ID, Lebow (1993) discussed seven constructivist values (collaboration, personal autonomy, generativity, reflectivity, active engagement, personal relevance, and pluralism) that point designers to the use of certain types of instructional strategies. In addition, his five principles of constructivist ID all emphasized constructivist instructional strategies (e.g., "Embed the reasons for learning into the learning activity itself." Other early C-ID papers (e.g., Hannafin, Hannafin, Land, & Oliver, 1997) also emphasized the need to consider different instructional strategies such as problem-based learning, anchored instruction, and collaborative learning. And, Jonassen (1996) detailed a set of "implications" of constructivism for ID that focused on characteristics of an appropriate teaching/learning environment:

1. Provide multiple representations of reality;
2. Represent the natural complexity of the real world;
3. Focus on knowledge construction, not reproduction;
4. Present authentic tasks (contextualizing rather than abstracting instruction);
5. Provide real-world, case-based learning environments, rather than pre-determined instructional sequences;
6. Foster reflective practice;
7. Enable context-and content dependent knowledge construction;
8. Support collaborative construction of knowledge through social negotiation. (p. 35)

The tendency to equate C-ID with using constructivist teaching and learning strategies is an example of what I called the Pedagogical approach to ID in Section I of this book. It was a common framework for papers about C-ID written in the 1980s and 1990s. C-ID is presented from a Pedagogical ID perspective in more recent papers as well. Maureen Tam (2000), for example, reviewed constructivist theory and then summarized the implications for ID this way:

> The constructivist propositions ... suggest a set of instructional principles that can guide the practice of teaching and the design of learning environments. It is important that design practices must do more than merely accommodate the constructivist perspectives, they should also support the creation of powerful learning environments that optimize the value of the underlying epistemological principles.

The Pedagogical approach to ID even produced several full C-ID models that focus almost entirely on the instructional strategies that should be used. On such Pedagogical ID model was proposed by Sun and Williams (2004).

Articles on C-ID that put the emphasis on using constructivist instructional strategies are still being published, but in the 1990s another type of C-ID paper began to appear. In these papers authors offered sets of "principles" or "guidelines" for designing instruction based on constructivist theory. These papers differed from the Pedagogical ID approach because at least some of the principles or guidelines offered included a consideration of what the process of ID should look like. For example, Karagiorgi and Sumenu's (2005) exploration of C-ID includes consideration of a number of process questions. However, their focus is still primarily on instructional strategies based on constructivist theory. Puntai (2007), on the other hand, offers an extended discussion of the impact and implications of constructivist theory on the process of instructional design.

This brings me to a third type of scholarship on C-ID – the creation of models that focus on the process of ID from a constructivist perspective. This aspect of the C-ID research literature is relatively small and mostly recent, but there is now a small but growing number of ID models based on constructivist theory. This type of work differs from the Pedagogical ID papers because the process of design is the focus. It is similar to the papers on guidelines and principles, but the work on C-ID models represents a further step toward professional practice because the scholarship on ID models is more closely tied to work in the field and is generally easier to apply to practice.

This last type of research, the development of C-ID models, is the focus of this section. Five different C-ID models are presented. The first two chapters detail the basic principles and suggested procedures of the R2D2 model, which is my own model of ID. R2D2 is one way of designing instruction through a process that is collaborative, recursive, and emergent. A second C-ID model is the focus of the third chapter. Katherine Cennamo's Layers of Negotiation Model emphasizes the need for collaboration across the design and development process. In her own words the Layers of Negotiation Model is based on the idea that "instructional design is at process of knowledge construction, involving reflection, examining information at multiple times for multiple purposes, and social negotiations of shared meanings."

Today there is no C-ID model that has been through decades of development and refinement, but the R2D2 and Layers of Negotiation Models have both gone through several phases of refinement. Three other very appealing C-ID models, are introduced in the final three chapters in this section. An Australian scholar, Rod Sims (who now works in the United States) presents a C-ID model. It was created specifically for the development of online learning resources. More specifically, the Sims model offers a new framework to guide the work of college and university distance education specialists as they work with faculty to develop and refine distance education services.

The fourth ID model presented in this section was created by Feng Wang and Michael Hannafin. It is not, strictly speaking, a C-ID model. Instead, Wang and Hannafin explore the concept of design-based research (DBR) and develop a framework for doing research that is also instructional design. DBR was discussed in Section I and I made a suggestion that there is enough overlap between C-ID and DBR that cross-fertilization would be beneficial. In this chapter Wang and Hannafin propose an "interactive, iterative, and flexible" approach to DBR that is also integrative and contextual. While I do not agree with everything proposed by Wang and Hannafin, the approach has much in common with the C-ID models. The chapter also provides a good overview of both the history of DBR development and the several models for doing DBR.

The final chapter in this section introduces Karen Norum's Appreciative Instructional Design (AiD) model. Based on the broader concept of Appreciative Inquiry, the AiD model is not a C-ID model. Instead, it is an ID model based on another theoretical foundation. However, it is a "compatible" model that shares with C-ID models a rejection of the top-down, linear, behaviorist characteristics of traditional ID models. It has a number of interesting charactistics including a focus at the organizational level. AiD also adds an innovative twist to the design process by emphasizing a novel way to begin ID. It is not with searching for what is wrong or ineffective about current instructional alternatives. Instead it proposes that we begin by identifying what is good about current alternatives. As Norum put it:

> The process begins with Discovery: questions are crafted to discover the "best of" performance level in the organization. Stakeholders (those who will be the audience for the training and/or those who need to support it) are interviewed to elicit stories about what the "best of" performance level looks like as well as their "best" instructional or training experiences. Questions are also asked about the "ideal" system and "ideal" instruction or training. Drawing inspiration from what is working, people are encouraged to dream about what could be. They are dared to expand the realm of the possible. The information gathered in the interviews is relevant to the Discovery and Dream phases of AiD. The goal is to discover what is already working well in the system's performance and to understand why it is working well. What life-giving factors are present at this "best of" level of performance and what does the system want "more of"

The traditional or positivist approach to ID begins, explicitly or implicitly, with the assumption that there is a Right way to do ID and it is our job to find it and tell others about it. Thus, many books and papers in this tradition tend to present an ID model as the current Right way to do things. C-ID developers, on the other hand, have a different task. They must tell others about their models and provide enough details about both the model and the context of application, that readers can make their own minds up about what warrants further consideration. Thus, the goal of C-ID development is not to gradually narrow our choices down until we arrive at the *One Best C-ID model*. Instead, the goal is to develop a rich and expanding set of models that reflect both the diversity of perspectives and the diversity of application contexts where ID is practiced. I commend these five models to you for you consideration and hope that should a second (or more) edition of this book be published, that your C-ID model will be included.

REFERENCES

Hannafin, M. J., Hannafin, K. M., Land, S. M., & Oliver, K. (1997). Grounded practice and the design of constructivist learning environments. *Educational Technology Research and Development, 45*(3), 101–117.

Jonassen, D. H. (1996). *Computers in the classroom: Mindtools for critical thinking.* Englewood Cliffs, New Jersey: Prentice-Hall.

Karagiorgi, Y., & Symeou, L. (2005). Translating constructivism into instructional design: Potential and limitations. *Educational Technology & Society, 8*(1), 17–27.

Lebow, D. (1993). Constructivist values for systems design: five principles toward a new mindset. *Educational Technology Research and Development, 41,* 4-16.

Puntai, W. (December, 2007). Integrating constructivism into instructional design. *Journal of Language and Communication, 12*(12), 101–112. Retrieved October 1, 2008, from http://www4.nida.ac.th/lc/journal2007/wiwat.pdf

Sun, L., & Williams, S. (2004). *An instructional design model for constructivist learning.* Retrieved October 1, 2008, from http://www.ais.rdg.ac.uk/papers/con50-An Intructional design.pdf

Tam, M. (2000). Constructivism, instructional design, and technology: Implications for transforming distance learning. *Educational Technology & Society, 3*(2), 50–60.

CHAPTER 13

BASIC PRINCIPLES OF A RECURSIVE, REFLECTIVE INSTRUCTIONAL DESIGN MODEL

R2D2

Jerry Willis

As the foundations for instructional design expand to beyond the traditional behavioral, information processing, and cognitive science theories that guided the first few generations of instructional design (ID) models, a new family of ID models has emerged, based on social and cognitive constructivist theory, and on nonlinear systems theories. We now have several "families" of ID models. Two of those families are compared in this chapter and three "guidelines" suggested as the foundation for one family of ID models—Constructivist-ID. Those three guidelines are (1) non-linear design, (2) reflective practice, and (3) participatory design.

This chapter is a major revision of Jerry Willis (2000). The maturing of constructivist instructional design: Some basic principles that can guide practice. *Educational Technology, 40*(1), 5–16

Constructivist Instructional Design (C-ID): Foundations, Models, and Examples, pp. 283–312

In 1995, I described the first version of an instructional design model based on constructivist learning theories and an interpretivist philosophy of science (Willis, 1995). It was developed during work at NASA's Johnson Space Center in Houston and at the Center for Information Technology in Education at the University of Houston. The model, named Recursive and Reflective Design and Development (R2D2), was one of the first to lay out in some detail a process for creating instructional materials that was based on constructivist theory. The 1995 paper emphasized the philosophical and epistemological underpinnings of the R2D2 model. That paper also dealt with the "family characteristics" of ID models based on the modernist perspective that held sway in both psychology and educational technology for most of the twentieth century versus the characteristics of ID models based on postmodern epistemologies, particularly interpretive and constructive world views (see Tables 13.1 and 13.2).

Tables 13.1 and 13.2 highlight the major differences between the ID models based on behavioral theories of learning and empirical epistemologies of knowing versus ID models based on constructivist theories of learning and interpretive epistemologies of knowing. The two tables focus on the practical implications of the two foundations for ID and in this chapter, I will not address the broader theoretical issues in any detail. Instead, I would like to briefly discuss some of the emerging constructivist instructional design (C-ID) models. Then I will suggest a set of general principles or guidelines for instructional design that are based on an interpretivist philosophy of science and constructivist theories of learning. In the following chapter, I will provide a more detailed guide to ID practice using the R2D2 model.

Until recently, the great majority of instructional design (ID) models have been based on modernist philosophies of science and learning theories from the behavioral and information processing families (Dick, 1996). There is, however, a growing body of literature on the practice of instructional design from a constructivist perspective (Winn, 1992). Much of this literature has emphasized the types of learning environments that can be developed, such as anchored instruction or problem-based learning (Cognition and Technology Group at Vanderbilt, 1993; Lin, Bransford, Hmelo, Kantor, Hickey, Secules, Petrosino, Goldman, and the Cognition and Technology Group at Vanderbilt, 1995; Reigeluth, 1996a, 1996b; Wilson, 1996). Some very useful papers have also offered general guidelines for constructivist instructional design (Wilson, 1997; Winn, 1992). Some have emphasized the differences between constructivist and other ID approaches at the philosophical level. Wilson (1997), for example, delineates four issues that separate constructivism from traditional ID:

Table 13.1.   Family Characteristics of
Objective-Rational Instructional Design (ID) Models

1. The Process is Sequential and Linear

The design process proceeds through a series of steps or stages; it is an objective process, and it is focused on the work of experts who have special knowledge. The process is like a waterfall; the results of one step are the data needed to begin the next step.

2. Planning Top Down and "Systematic"

Design begins with a precise plan of action including clear behavioral objectives. Work proceeds through the ID process in a systematic, orderly, preplanned manner.

3. Objectives Guide Development

Precise behavioral objectives are essential. Considerable effort should be invested in creating instructional objectives and objective assessment instruments.

4. Experts, Who Have Special Knowledge, Are Critical to ID Work

Experts, who know a great deal about the general, universally applicable principles of instructional design and pedagogy, are needed to produce good instructional materials and resources.

5. Careful Sequencing of Content and the Teaching of Subskills Are Important

In the ID process you must break complex tasks down into subcomponents and teach the subcomponents separately. Pay particular attention to the sequence of the subskills taught as well as the events of instruction.

6. The Goal is Delivery of Preselected Knowledge

There is an emphasis on the delivery of basic facts and enhancement of skills selected by experts, which favors direct instruction methods. When computer technology is used the technology tends to take on the roles of a traditional teacher—information deliverer, evaluator, record keeper.

7. Summative Evaluation is Critical

Invest the most assessment effort in the summative evaluation because it will establish whether the material developed is effective or not.

8. Objective Data Are Critical

The more data the better, and the more objective the data the better. From identifying entry bnehaviors to task and concept analysis, pretests, embedded tests, and posttests, the model emphasizes collection and analysis of objective data.

- the nature of reality,
- the nature of knowledge,
- the nature of human interaction, and
- the nature of science.

Others have presented principles that should guide design (Lebow, 1993). In addition to general guidelines for C-ID, a number of more specific frameworks for "doing" constructivist ID have been proposed. These are constructivist instructional design (C-ID) models.

Table 13.2.   Family Characteristics of
Interpretivist-Constructivist Instructional Design (ID) Models

- **The ID Process is Recursive, Iterative, Non-linear, and Sometimes Chaotic**

The ID process is recursive or iterative; you will address the same issues many times, such as what should be taught and how to best involve learners in the educational experience. Development is also non-linear. There is no required beginning task that *must* be completed before all others. Some problems, improvements, and needed changes will only be discovered in the context of designing and developing instruction. Plan for recursive evaluations by users and by experts. Plan for false starts and redesigns as well as revisions. The scholarship on iterative software development (Cusumano, 2007; Thumboo et al., 2006) provides some guidance on iterative ID as does the work on *Agile Software Development* and Extreme Programming (Ambler, 2008; Beck & Andres, 2005; Vanderburg, 2005; Northover et al., 2007).

- **Planning is Organic, Developmental, Reflective, and Collaborative**

Begin with a vague plan and fill in the details as you progress. Vision and strategic planning come later. Premature visions and planning can "blind" (Fullan, 1993). Development should be collaborative. The design team, which should include many who will use the instructional material, must work together to create a shared vision. That vision may emerge over the process of development. It cannot be "established" at the beginning. "Today, 'vision' is a familiar concept in corporate leadership. But when you look carefully, you find that most 'visions' are one person's (or one group's) vision imposed on an organization. Such visions, at best, command compliance- not commitment.... If people don't have their own vision, all they can do is 'sign up' for someone else's." (Senge, 1990, pp. 206–211)

- **Objectives Emerge from Design and Development Work.**

Pre-selected objectives do not guide development. Instead, during the process of collaborative development, objectives emerge and gradually become clearer.

- **General ID Experts Don't Exist; Participatory Approaches are Preferred**

General ID specialists, who can work with subject matter experts from any discipline, are a myth. You must understand the 'game' being played before you can help develop instruction. Specialists in ID cana make important contributions to the ID process, but they should play roles closer to that of a facilitator than an "expert" and they should be immersed in the environment of use either before or during the design process. Participatory (McIntyre, 2008) and emancipatory (Zuber-Skerritt, 1996) models of research and participatory models of design (Winters & Moy, 2008; Baek & Lee, 2008) provide considerable guidance here.

- **The Instructional Resources Developed Tend to Emphasize Learning in Meaningful Contexts. (The Goal is Personal Understanding Within Meaningful Contexts)**

Standard direct-instruction approaches that focus on teching content outside a meaningful context often result in "inert" knowledge that is not useful outside the classroom. Students cannot apply what they have learned to real world tasks as opposed to classroom tasks (e.g., taking tests). The instructional emphasis should be on developing understanding and skill in context. This approach favors strategies such as anchored instruction, situated cognition, cognitive apprenticeships, and cognitive flexibility hypertext. It also favors instructional approaches that pose problems and provide students with access to knowledge and tools needed to solve the problems. This encourages the development of hypermedia/multimedia information resources, electronic encyclopedias, and a wide range of accessible, navigable electronic information resources.

- **Formative Evaluation is Critical**

Designers should invest the most assessment effort in formative evaluations because they are the type that provides feedback you can use to improve the product. Summative evaluation does nothing to help you improve the product and you cannot "prove" that an instructional package "works" in any general and universal sense. Whether it works or not depends heavily on how it is used in a particular, local, context.

- **Subjective Data May Be the Most Valuable**

Many important goals and objectives cannot be adequately assessed with multiple choice exams, and excessive reliance on such measures often limits the vision and value of instruction. Some important outcomes of learning can be shown or observed but they cannot be quantified. Many types of assessment, including authentic assessment, portfolios, ethnographic studies, and professional or expert critiques, should be considered both as methods of formative evaluation in the design process and as formative evaluation methods in the instructional resources being developed. Accepting subjective data also means behavioral objectives become much less important as either starting or ending points for ID and instruction. "Reliance on behavioral objectives has not benefitted mathematics education, nor has stimulus-response psychology proved capable of describing insightful mathematics learning effectively. Often, the exclusively behavioral characterization of desirable learning outcomes leads educators to rely on the teaching of discrete, disconnected skills in mathematics, rather than on developing meaningful patterns, principles, and insights. Entire public school mathematics departments have devoted their summers to rewriting the objectives of their textbooks in behavioral terms, replacing non-operationally-verifiable words like 'understand' with operationally verifiable words like 'solve correctly' " (Goldin, 1990, p. 36)

C-ID MODELS

Constructivist theory, and particularly the epistemology upon which it is based, is not a comfortable framework within which to proclaim that your new and improved ID model is a step closer to some Platonic *ideal form* that represents perfection and Truth. Constructivist theories of knowledge view truth and knowing as local events, and highlight the importance of context in making meaning. Multiple perspectives, the importance of context, and the social construction of meaning are all values that push the ID theorist away from any tendency to view his or her creation as THE way to think about design. Thus, it is important to view the R2D2 model that will be presented in this chapter, and any other C-ID model, as just one of many possible constructivist ID models.

R2D2 is one of many possible "truths" when it comes to ID models. When I taught the introductory and advanced instructional design courses at Iowa State University and Lousiana State University, I did not present the R2D2 model as the "best" or "only" model students should consider. As students construct their own understanding of design, I expect them to explore a number of different models and to consider what the impact of using various models would have on their own design work. Fortunately, there are a number of other C-ID models in use today as well as several models that, while not based specifically on constructivist theory alone, have much in common with C-ID models.

Layers of Negotiation

Catherine Cennamo, Sandra Abell, and Mi-Lee Chung (1996) constructed the Layers of Negotiation ID model while they were creating a series of case-based interactive videodiscs for use in constructivist teacher education programs. Cennamo has also recently updated this model (see a later chapter in this section). Cennamo and her colleagues see C-ID as a process that involves "five basic actions:

1. Embrace the complexity of the design process.
2. Provide for social negotiations as an integral part of designing the materials.
3. Examine information relevant to the design of instruction at multiple times from multiple perspectives.
4. Nurture reflexivity in the design process.
5. Emphasize client—centered design."

As they analyzed their work on the video cases, they concluded the Layers of Negotiation ID model was different from most other ID models in three ways:

- It is process-based rather than procedure based. "Whereas traditional instructional design models prescribe a set of procedures to be followed to design instruction, we found our emphasis shifted to the process of decision-making that is involved in designing instruction" (p. 42).
- It is question driven rather than task driven. Instructional designers should ask good questions instead of following a linear sequence of prescribed steps that are part of traditional ID models.
- Spiral cycles were used instead of discrete stages. "Whereas traditional instructional design models often include discrete stages for analysis, design, development and evaluation activities, we addressed the questions of design in a spiral fashion, progressing through a series of stages at one level, then spiraling back and adding more detail within" (p. 43).

The Layers of Negotiation ID model has much in common with the R2D2 model that will be discussed in detail later in this chapter.

Chaos Theory ID

One of the most interesting ID models to be described in the literature in recent years is the Chaos Theory ID model of Yeongmahn You (1993). You contrasts traditional ISD (instructional systems design) models with a model based on chaos theory. You summarizes chaos theory in three key elements:

1. sensitive dependence on initial conditions;
2. fractals; and
3. strange attractors (see his paper for an explanation of these concepts).

You then points out four weaknesses of traditional ISD models and suggests that chaos theory points to what we should move away from as well as appropriate alternatives we should move toward:

1. Away from linear design and toward nonlinear design.
2. Away from determinism and expected predictability toward "indeterministic unpredictability."
3. Away from closed systems and toward open systems.
4. Away from negative feedback (input from users is considered a "problem" if it suggests major changes) and toward positive feedback (input that suggests the need for change presents opportunities).

You's paper is one of the most thoughtful treatments of the theory-to-design question. His discussion of the implications of chaos theory for ID calls into question many of the basic assumptions underlying traditional ID models.

Other C-ID Models

There are also a number of other C-ID or compatible models currently in development and I believe most of them deserve attention and consideration. Each has strengths and desirable characteristics that may make them a good match for certain types of instructional design work. Many are based on interpretivist/constructivist theory but some contribute diversity by using other theoretical frameworks such as chaos theory.

In the remainder of this chapter I will present the current revision of the R2D2 C-ID model. The guiding principles of R2D2 will be discussed here, and in the next chapter the focus will be on procedures and practices. Before turning to R2D2, however, I want to note that the guiding principles to be discussed should not be considered laws or required approaches. They are guides that may be profitably used or ignored, depending on the particular context of practice.

An Overview of the Revised R2D2 Model

The original R2D2 model has received some attention in the literature. It is now covered in a number of instructional technology courses around the country, and the discussions of constructivism and ID often discuss R2D2. R2D2 has also generated some criticism. Apparently because the original 1995 paper on R2D2 challenged some of the long-held assumptions of traditional instructional design theory, Braden (1996) commented that the author (me) was "much like an arrogant adolescent with a chip on his shoulder" (p. 19). For what I consider a more thoughtful comparison of the Dick and Carey ISD model and the R2D2 model by one of the creators of the ISD model (see Dick, 1996).

I have now had an opportunity to work on a number of additional projects that used the R2D2 model. That experience, plus discussions with graduate students and colleagues, my own reflections, and the useful criticisms of designers such as Braden (1996) and Dick (1996), have all influenced the evolution of the R2D2 model. In the remaining sections of this chapter, I would like to discuss some potential general principles for all C-ID models, including R2D2. I will also link those principles to work in other fields, including educational and social science research, software engineering, and industrial design.

Educational technology can be a somewhat isolated profession that has only weak links to other fields in education, such as curriculum and instruction, educational research, educational foundations, and educational administration. Educational technology also has limited contact with other, related fields, such as industrial design and computer software development. That isolation is a barrier to full participation in the discussions about ID, especially C-ID. Few of the principles of C-ID are completely new; they have been used in many other fields for years. However, they are sometimes treated as odd and unsupported ideas by critics within the field of educational technology, who do not seem to be aware that many professionals outside the field of educational technology take them as established guides to practice. For example, in software development and software engineering there are movements such as iterative software development (Cusumano, 2007; Thumboo et al., 2006). And, a major shift to what has come to be called *Agile Software Development* or *Extreme Programming* (Beck & Andres, 2005; Vanderburg, 2005; Northover et al., 2007) have much in common with the effort to create and use C-ID models in the field of educational technology.

General Principles Versus Detailed Steps

One of the issues when teaching instructional design, or creating an ID model, has to do with whether design work is a process of applying tried and true techniques derived from research to solve well-defined problems or an artistic and professional process of creating possible solutions for fuzzy and often ill-defined problems. Roberts Braden (1996) is a good example of a theorist who adopts the first approach. He calls designers who take the second position "Pablo Picasso" designers. Another designer, Merrill (1996, 1997), represents that first approach, sometimes called the "technical-rational approach." For Merrill, the problems ID addresses can and should be well defined. And, for a well-defined problem there is already a collection of solutions that have been proven to work in research studies. As Merrill (1996), put it,

> There are known instructional strategies. The acquisition of different types of knowledge and skill require different conditions for learning.... If an instructional experience or environment does not include the instructional strategies required for the acquisition of the desired knowledge or skill, ... the desired outcome will not occur. (p. 1)

Even more to the point:

> There are different kinds of knowledge and skill (Gagné assumption). The different kinds of knowledge and skill each require different conditions (strategies) for learning. IF an instructional strategy does not include presentation, practice, and learner guidance that is consistent with, and appropriate for, the type of knowledge or skill to be taught, THEN it will not teach. IF A PRODUCT DOES NOT TEACH, IT HAS NO VALUE. (Merrill, 1997, emphasis in original)

This approach is in stark contrast to the "reflective practice" approach of Donald Schön (1987). His famous comment about high ground and swamps expresses this view.

In the varied topography of professional practice, there is a high, hard ground overlooking a swamp. On the high ground, manageable problems lend themselves to solution through the application of research-based theory and technique. In the swampy lowland, messy, confusing problems defy technical solution. The irony of this situation is that the problems of the high ground tend to be relatively unimportant to individuals or society at large, however great their technical interest may be, while in the swamp lie the problems of greatest human concern. The practitioner must choose. (p. 3)

For Schön much of what makes a real difference in professional practice cannot be reduced to the technical formulas that Merrill seeks. However, most of the available ID models accept, to one degree or another, the technical-rational approach. They assume a problem can be well defined and that well-defined solutions are already available to solve that problem. The problem can be anything from how to organize a course on American history to preparing technicians to operate a nuclear power plant.

The technical-rational approach seems most suited to the creation of materials that use direct instruction approaches. And ID models such as the instructional systems design (ISD) approach of Dick and Carey (1996) seem best suited to the task of creating direct instruction. But what if the design team adopts an alternative theory of learning? What if the assumptions about how people learn come from a constructivist theory (Brooks & Brooks, 2001; Wilson, 1996)? The principles of constructivist learning have been simply but eloquently stated by Jonassen (1996) in a three-page article. (More detailed coverage is available in books such as Brooks & Brooks, 2001; Duffy & Jonassen, 1992; Jonassen, Peck, & Wilson, 1999; Marlowe & Page, 2005; Martin, 2008; and Wilson, 1996). Most of the information resources on constructivist learning environments emphasize that such environments are based on some guiding principles rather than specific recipes that precisely specify what the teacher is to do and how students are to perform. Jonassen's (1996) article, for example, includes a discussion of three core ideas of constructivist learning:

- context,
- collaboration, and
- construction of knowledge.

He also suggests that a constructivist design process should be concerned with designing environments which support learning that involves students in the collaborative construction of knowledge in meaningful, authentic environments. For constructivists, such general principles are not concepts that are to be fleshed out by specific, detailed, how-to guide-lines that become available when research in a field matures and provides real directions and prescriptions to practitioners. The best we can do, say constructivists, is to provide guidance, not specific rules that invariably apply to all situations. This, as with most of the issues discussed in this chapter, is not an isolated issue that is relevant only to the field of educational technology. The April, 1996 issue of *Educational Researcher* contained several papers on the role of research in guiding practice. Gage (1996) made an eloquent plea for the use of social science research to guide practice from a technical-rational perspective. Much of the defense offered by Gage is based on the assumption that social science research can lead us to "long lasting generalizations." Gage concludes that there are "universals" that "hold up over considerable variation across the individuals or other units studied, across different ways of describing and measuring those individuals or other units, across varied settings, and across decades" (p. 14). Therefore, Gage proposes that we can, through good empirical research, come to know how things are and thus base our practice on that knowledge. He refers to those who disagree with that view of social science and education research, and there are many today, as counsels of despair who

> assert that, whereas the behavioral sciences once promised to reveal universal relationships between phenomena in the social world—generalizations that would hold true everywhere and forever—we should now realize that they have failed, and must inevitably continue to fail to produce such, generalizations. (p. 5)

The view expressed by Gage of the link between research and practice is the framework many prescriptive ID models adopt. Researchers discover s universal rules, and practitioners must learn to apply those rules in specific situations. But there are alternatives. In the same special issue of *Educational Researcher* that carried the Gage paper, two scholars from the University of Utrecht in the Netherlands, Kessels and Korthagen (1996), also addressed the question of how theory relates to practice. They pointed out that for much of this century theoretical, abstract knowledge has been considered superior to "concrete skills or the tacit knowledge of good performance." Kessels and Korthagen compare the current situation, especially views on how research and practice interact, to the differences between the ancient Greek concepts of *episteme* and *phronesis*. They believe these two concepts are relevant to the issue of how research/theory interacts with professional practice.

> Plato's version of knowledge, episteme, is general, abstract, and universal. The propositions or assertions of epistemic knowledge "are of a general

nature; they apply to many different situations and problems, not only to this particular one. Consequently, they are formulated in abstract terms. Of course, these propositions are claimed to be true; preferably their truth is even provable.... Because they are true, they are also fixed, timeless, and objective.... It is this knowledge that is considered of major importance, the specific situation and context being only an instance for the application of the knowledge." (p. 18)

Episteme is the type of knowledge that Merrill counts on for his form of instructional design, and it is what Gage is talking about when he speaks of knowledge that allows us to make generalizations. Aristotle's knowledge, *phronesis*, is situated in a context and is dependent on that context for its meaning. It is *practical wisdom* rather than abstract universal wisdom. This type of knowledge is

knowledge of a different kind, not abstract and theoretical, but its very opposite: knowledge of concrete particulars.... In practical prudence, certitude arises from knowledge of particulars. All practical knowledge is context-related, allowing the contingent features of the case at hand to be, ultimately, authoritative over principle. (p. 19)

For Aristotle, it is not possible to capture good practice "in a system of rules." Kessels and Korthagen quote Aristotle, "Let this be agreed from the start, that every statement (*logos*) concerning matters of practice ought to be said in outline and not with precision" (p. 19). Kessels and Korthagen believe that knowledge about professional practice is best thought of as *phronesis* rather than *episteme*. Trying to precisely apply universal rules is doomed to failure. As Donmoyer (1996) put it,

Aristotle's practical problems are as much about framing questions as finding answers and as much about values as facts. Problems, from a phronesis frame of reference, are also, by definition, about idiosyncrasy and uniqueness, about particular students or particular clients rather than about seeing particular students or clients merely as exemplars of a general category or ideal type. (p. 4)

With Platonic knowledge we can provide detailed steps about how to design because we have what Gage calls universal or generalizable knowledge. With Aristotelian, phronetic knowledge we cannot do that because the context is so important to meaning. The R2D2 model is based on the assumption that social and behavioral science knowledge is more Aristotelian than Platonic. There is much that is idiosyncratic and unique about a given design project and context. Understanding that context is an important aspect of successful design work. Some of the reasons for that position will become obvious later in this chapter. However, Winn's (1992) explanation of the difference between technical-rational and reflective practice approaches to instructional technology highlights the differences in perspective. (Winn does not use these terms but his paper suggests the categories he uses are roughly equivalent.) For one approach, the most important thing is applying tried and true solutions to new but familiar problems. For the other, the most important thing is understanding theory and the implications of theory because the problems you will face are new and unique—thus there are no prefab solutions stored in some cognitive warehouse. In their plea for including is "a sound learning theory base" in courses on instructional design, Gentry and Csete (1995) make a similar distinction when they point out that "more often than not, students in courses are taught to design instruction using a cookbook approach." Gentry and Csete feel, however, that

> while this method may work in test cases developed for the course, this kind of training is insufficient to deal with unexpected constraints that inevitably arise in real situations. Educational technologists need a sound understanding of the learning theory base from which they work, so I that they can make creative decisions that take the constraints and assets of a particular situation into account. (p. 24)

I would go further and say that instructional designers need to know more than learning theory; they need to know theories about change, about diffusion, about what knowing and understanding is, and about how to work collaboratively in groups.

This section began with a question about whether specific rules and recipes can be the foundation for ID or whether general principles are desirable (and with specific decisions made by the participants in the design process in the context of practice). The R2D2 model is based on the assumption that general guidelines can help us as we come to understand the specific context of design and deployment. Trying to follow detailed, specific rules of design is discouraged because each context is unique. There is no Platonic ideal form of ID that can be used as a guide in all situations. The best we can do is work from flexible guidelines or principles that are subject to change and to being overruled, to work from cases and exemplars shared with us by others, and from our base of experience.

Three Flexible Guidelines

I believe there are three important but flexible guidelines for Constructivist Instructional Design Models. They are, to use Winn's terminology, three *first order principles*. They are Recursion, Reflection, and Participation. These are all concepts that flow naturally from constructivist theory, but I have learned over these past few years that the implications of these three principles for ID practice are not always obvious. They are also not universally accepted. Indeed, all three are vigorously opposed by segments of the design community. As you will see, however, they are also championed by scholars and practitioners in many fields.

The First Flexible Guideline: Recursive (Iterative), Non-Linear Design

The great majority of instructional design models are based on the assumption that a somewhat linear approach is best. That is, there are steps in the model and those steps are best carried out in a particular sequence. Braden (1996) is one of the clearest proponents of linear ID. "Linear instructional design and development (LDD) is what we commonly call basic instructional design, basic instructional development, or just basic ID" (p. 5). While Braden does not believe that linear design is the "solution to every performance problem," he does believe linear ID is a powerful tool. "Adherence to a procedure such as the one described herein will result in the creation of instructional products that achieve their intended, purpose" (p. 21). Braden's (1996) detailed description of his ID model is one of the best presentations of a design model based on the assumption that linear is better. Even where a linear, *Step 1 comes before Step 2* approach is not absolutely required, many ID models were designed so that it is very convenient to do the steps in order. Dick (1996) has pointed out that linearity is not an absolute requirement of the Dick and Carey ISD model and that many designers do indeed complete steps in a non-linear sequence.

> This point has served as a straw man for more than one critic who has observed that design is just not practiced that way—that designers move back and forth in the model and do not always get to start at the beginning.... When the model is used to create instruction, the flow of information is always two-way and changes are made to various components of the process based upon the new information. (p. 59)

However, after discussing linearity as a straw man argument, Dick (1996) commented that:

> The output of one step is the input for the next step. Ultimately there must be a connection between the boxes, a consistency in the flow, from box to box. Similarly, how can novices be told, just start anywhere you like and try to cover as many of the steps as you can in any sequence that seems appealing to you? That would be chaos; the frustration level would be extremely

high. It is likely that little would be learned and the result would be unskilled designers with bad attitudes. Novice designers are encouraged to learn the process by beginning at the beginning and working through the model in an orderly fashion. (p. 59)

It seems clear that Dick believes there are steps in his ID model that should come first and be followed by others in an orderly fashion. If that is not done, chaos will be the result, the designers will be out of sorts, and learning will not occur. For Dick, non-linearity is the occasional exception to the general rule of linearity. Step 1 usually comes before Step 2, unless you have a very good reason for doing it differently. The basic structure of the Dick and Carey model encourages a linear approach. For example, writing performance objectives is one of the steps in the ISD model. Those objectives are necessary for the next step, developing criterion-referenced test items. Controversy over whether a process should be linear or non-linear is not limited, however, to instructional design. Carpenter (2006) has argued that in curricular areas such as the humanities there is much to be said for nonlinear learning:

Non-linear, interactive curriculum differs from traditional linear curriculum in that it does not follow a finite set of prescribed paths as determined by a single author. Rather, nonlinear, interactive curriculum enables the reader to pursue paths of content for presentation to students that make the most sense at that particular reading

Carpenter has even developed a set of resources to help teachers design nonlinear curricula (Carpenter & Taylor, 2003).

Hlynka (1995), in his discussion of postmodern approaches to educational technology, points out that it pervades education:

Modern education is linear education. Linear models abound. Textbooks are linear and produced chapter by chapter. Classes are linear, following an exact number, an exact timeline, and an exact schedule. A lecture is linear. Linear teaching is comfortable, effective, and efficient. Give the students a statement of objectives and teach to the objectives. The student learns. Modernist theorists argue that this model is necessary if we are to have order.... If only teaching and learning were really like that! Postmodern education, trying to come to grips with the information explosion, finds content everywhere and all at once. Of course, it would be just fine if first things came first, but, in fact, things come as they come. (p. 116)

Non-linearity is one of six characteristics, or indicators, of postmodernism, as proposed by Hlynka. Programmed instruction, drill and practice, and tutorials are all suited to a modernist conception of teaching and learning. They are all linear. Other approaches, such as problem-based learning and anchored instruction, are more suited to postmodern, nonlinear conceptions of teaching and learning. If nonlinearity is a component of new forms of teaching and learning, it makes sense that the design and development of nonlinear curricular materials should be accomplished through a process that is also nonlinear. In the R2D2 model, the suggested procedures can be completed in any order that makes sense; there is no single starting or ending place. If non-linear ID seems chaotic to many, it is perhaps not surprising that another source of support for it is an emerging theory that is influencing many areas of the social sciences today: chaos theory. You (1994) points out that

> the conventional model of ISD takes the form of a straight line through a relatively linear sequence of procedures.... Within the linear ISD model, the second step cannot be implemented without carrying out the first step because the first step is antecedent to the second. (p. 20)

You believes this "is one of the major shortcomings of traditional ISD models" (p. 20) and he argues that

> a linear approach is not sufficiently flexible for working with environmental turbulence or sophisticated educational systems.... The linear ISD process imposed upon a dynamic system typically overlooks one or more messy variables that interfere at each stage of design and development. (p. 20)

You proposes the use of nonlinear ID models because they "can represent the dynamic interrelationship and interdependence among their components" (p. 20).

Non-linear ID is recursive or iterative. The same issues will be addressed over and over across the entire design and development process. For some, the approach seems chaotic. It is chaotic in the sense that it does not prescribe in advance a specific pattern. It suggests instead that you let the project guide your decisions. What you should do next is dependent on the situation. Liam Bannon (1991), in his discussion of the ways in which human factors researchers must change if they are to inform designers, comments that research must shift from a product to a process focus.

> By this I mean that more attention needs to be paid to the process of design, that is, working with users in all stages of design, to see the iterative nature of design and the changing conception of what one is designing as a result of the process itself. This is in contrast to a view of design that proceeds from a set of fixed requirements without iteration and without involvement of the users. (p. 209)

Fundamentally, iteration is the process of developing instructional material in a way that allows both users and experts to fully participate in the process of revision and reformulation. This may happen over and over again in a complex project. It may also happen at many levels. Members of the design team may initially look at little more than scribbles on a flip chart that represent a rough and rapidly developed prototype. More accurately, they create the scribbles on the flip chart through collaborative development of the concept. Later they may go through a scenario that tells the story of how the material would be used in a class and then revise their conceptions of what it is that they will create. Still later they might work with rough but functional prototypes, and then progressively more complete, prototypes of the material. Finally, relatively finished alpha and beta versions of the material would be used.

Few ID models avoid all forms of recursion, but the structure of many models does not encourage or facilitate frequent recursion. Many identify specific points where experts and end users are to evaluate material, for example, but input at other times is not encouraged. The concept of recursion (iteration) that I have in mind here is more, however, than simply giving end users a chance to have their say about a product. That level of recursion is in most ID models. Somewhere in the process, formative evaluations generate data from end users. Revisions can then be based on that data.

In my view, more and smaller formative evaluations throughout the design process are better than the one or two big ones that are typical of ID today. But *strong recursion,* as I see it, involves even more than that. Bodker, Gronbaek, and Kyng (1995) embody the richer and more inclusive concept of recursion in their discussion of cooperative prototyping:

> The way we do prototyping—cooperative prototyping —is different from traditional prototyping in that traditional prototyping approaches mainly take the perspective of the developers, analyst/designers, conduct investigations in the user organization and develop prototypes on their own. Such prototypes are tested by or demonstrated to users to give the developers feedback on their solution. It has a superficial resemblance to our use of prototypes for illustrating new technological possibilities, but whereas we see this use of prototypes as part of the users' learning, traditional approaches view it as part of the feedback to developers. Traditional approaches put little emphasis on active user involvement in the actual design process.... [Cooperative prototyping] is an exploratory approach ... where prototyping is viewed as a cooperative activity between users and designers, rather than an activity of designers utilizing users' more or less articulated requirements.... The cooperative prototyping approach establishes a design process where both users and designers are participating actively and creatively with their different qualifications. (p. 220)

For additional suggestions on prototyping, see Nixon and Lee (2001) and Borenstein (1991). Michael Muller (1991) has also described an interesting approach to prototyping that he calls PICTIVE: Plastic Interface for Collaborative Technology Initiatives through Video Exploration. It is a prototyping approach that emphasizes early and continued involvement of end users in the design process. PICTIVE treats design as a partnership between the designers and end users, and goes to considerable lengths to make the process open and available to end users. Muller, for example, is critical of rapid prototyping because it is generally done in a programming or authoring environment that is not easy to learn. And,

> if the user lacks the time or inclination to be trained in the prototyping software, then the user is dependent upon a software professional whose personal or organizational agenda may be quite different from the user's.... The user may thus be alienated from the design process and from the artifact produced by the process. (p. 213)

Muller's PICTIVE process uses low-tech prototyping media-such as paper-and-pencil, markers, plastic icons, and other easily used and understood media to encourage and facilitate participation by all team members. In addition, design sessions are videotaped and the video is used as a guide to design and redesign. Muller's (1991) paper describes the use of PICTIVE for a real-world design project and to teach students the process of design.

Another area that has experienced a major shift in perspective, relative to the desirability of recursion, or iteration, is social science research. In the standard scientific method that many of us were taught in graduate school, the hypotheses and methodology of the research had to be clearly and precisely defined before the study began. The process is much like the procedure in ISD that insists detailed objectives and such must be created before design and development begin. If a researcher discovered something else that was interesting as the data were gathered, it could only be treated on a post-hoc basis, which meant that it was suspect. To really study that new phenomenon, which was not covered in the prepared hypotheses, the researcher had to begin again with a new study. This approach is in contrast to a method now known as *grounded theory* (Willis, 2007). In this method, the purpose

> is to develop theory, through an iterative process of data analysis and theoretical analysis, with verification of hypotheses ongoing throughout the study. A grounded theory perspective leads the researcher to begin a study without completely preconceived notions about what the research questions should be, assuming that the theory on which the study is based will be tested and refined as the research is conducted. (Savenye & Robinson, 1996, p. 1177)

What a difference! In grounded theory what was wrong in traditional research is perfectly acceptable, even desirable and required. Grounded theory research is a corollary in research to a recursive or iterative approach in instructional design. Over the course of the design process, things like the objectives, the content, the teaching and learning activities, and much more, gradually emerge rather than being precisely specified early in the process.

The Second Flexible Guideline: Reflective Design

Reflection is perhaps the most difficult of the three basic principles of the R2D2 model to explain. The opposite approach, technical rationality, is relatively easy to explain. "Technical rationality is an epistemology of practice derived from positivist philosophy, built into the very foundations of the modern research university" (Schön, 1987, p. 1). Good practice from this perspective involves carefully and precisely defining the problem and then applying clear, well-formed solutions derived from good research. A reflective model of practice assumes that many, if not most, important problems in professional practice cannot be well-formed and solved with pre-formed solutions. In such a situation,

> the terrain of professional practice, applied science and research- based technique occupy a critically important though limited territory, bounded on several sides by artistry. There are an art of problem framing, an art of implementation, and an art of improvisation—all necessary to mediate the use in practice of applied science and technique. (Schön, 1987, p. 13)

Those arts, of problem framing, implementation, and improvisation, make up reflective practice. They are more artistic than scientific, and they call for thoughtful and careful attention to, as well as understanding of, the context in which the professional work occurs. Schön uses the terms *reflection-in-action* and *reflection-on-action* to refer to the type of thoughtful cognitive work in context that I believe should be an important principle in design. He also refers to this as *artistry* to contrast it with technical approaches. Schön (1987) describes reflection-on-action as a recursive process in which each effort to solve a problem that has not yielded to routine solutions is a trial that presents a reflective opportunity.

> But the trials are not randomly related to one another; reflection on each trial and its results sets the stage for the next trail. Such a pattern of inquiry is better described as a sequence of moments in a process of reflection-in-action. Thinking reflectively about what we have done, and are doing, leads to

reformulations of the problem as well as to experimentation. New approaches are tried, sometimes discarded, sometimes adopted or revised again. (p. 36)

I have contrasted constructivist ID models with what I considered to be traditional ISD models that are, in my opinion, based on a positivist, objectivist foundation. Schön (1987) makes the same point about reflective-practice. His reflective practice is based on a constructivist view of reality—in this case the real world of professional practice. That "real world" is not "out there" in some concrete form the practitioner must address and deal with. Instead, it is constructed by the practitioner, who then makes professional practice decisions from within that constructed reality. The process of making those decisions is artistic and reflective. It is not based on technical rationality that assumes there is a knowable external world governed by universal laws and regulations the practitioner must learn and use. The debate is not over whether the social world of humans operates on laws and regulations. Perhaps it does, perhaps it does not. The debate is over whether we, as humans ourselves, can know those laws and regulations so confidently that we can adopt an objectivist, technical-rational approach to professional practice. C-ID and Schön's concepts of reflective practice are based on the assumption that we cannot access those laws and regulations, if they exist at all, while traditional ISD models and technical-rational models of practice assume we can. Schön's books (1983, 1987, 1990, 1995) are excellent guides to reflective practice. His chapter on architecture in the 1987 book is particularly recommended, since it deals with another design profession.

The Third Flexible Guideline: Participatory Design

In their classic and influential book, *Participatory Design: Principles and Practices*, Douglas Schuler and Aki Namoika (1993) described participatory design as "a new approach towards computer systems design in which the people destined to use the system play a critical role in designing it" (p. xi). They contrast it with other ways of designing.

Participation stands in contrast to the cult of the specialist. In the specialist model, an expert is sought out. The question is presented to the Expert who will eventually produce the Answer. With this approach, those most affected by the conclusion must sit idly by, waiting patiently for enlightenment. [Participatory Design] PD, of course, demands active participation. PD, however, is not against expertise. There is no reason or motivation to belittle the role of expertise. Specialized training and experience, both technical and interpersonal, are important. In the participative model, however, this special expertise becomes yet another resource to be drawn on—not a source of unchallenged power and authority. A partnership between implementers and users must be formed and both must take responsibility for the success of the project. (pp. xi—xii)

That, in a nutshell, is the essence of participatory design. Involve the users in design as participants, not as observers from the sidelines or objects to be studied. PD began in Scandanavia in design fields such as software design and industrial design, but it has spread around the world and into many different design fields. User involvement has been promoted and studied in fields such as software engineering and industrial and systems design for over 20 years, but instructional design has been slower to consider the full possibilities of user involvement, perhaps because behavioral theories have dominated the field longer and more deeply than they have in other design disciplines. Participatory design is one of the more controversial aspects of alternative ID models. The idea has been criticized by traditional ID proponents on several counts. The heart of the criticisms, however, is that this just won't work. Merrill (1996) had this to say in his response to Reigeluth's (1996a,1996b) proposal that users be heavily involved in the ID process:

Reigeluth's next suggestion is the really frightening section in his paper. A visioning activity is a recipe for disaster in the real world of instructional development. It is a dream of academics who value collaborative approaches to knowledge; but, in practice, it often leads to disaster. There is no doubt that stakeholders must have a role in determining ends (how the learners will be different as a result of instruction), but when stakeholders play a significant role in determining means (how those changes in the learners will be fostered), then the result is often ineffective instruction that does not teach. (p. 58)

Merrill goes on to say that "Theconsensus of stakeholders often equals poor Iearning" and he raises doubts about whether students can play a meaningful participatory role in ID since "students are, for the most part, lazy." Merrill's basic criticism can be packaged in several frameworks. Customers don't have the expertise to make good decisions, and end users make decision based on factors other than the best way to teach something.

There are instructional design principles. There are principles based on the way students learn and these are not subject to collaborative agreement. No amount of argument to the contrary will negate these principles. Learning will be negatively affected if these principles are violated.... Stakeholders (and too many so-called instructional designers) in recommending means (how those changes in the learners will be fostered) often violate these principles. (Merrill, 1996, p. 58)

Thus, teachers, and students, are ignorant, fickle, lazy, or otherwise unfit to fully participate in the process of making design decisions. Proponents of participatory design (Bodker, Kensing, & Simonsen, 2004; Anthopoulos, Siozos, & Tsoukalas, 2007; Ross & Ross, 2008) base their argument for it on the assumption that understanding the context in which the resulting product will be used is critical to success, and that context can be understood, represented, and interpreted by users. There is actually some research on this question. Spinuzzi (2005), for example, reported that participatory design approaches produce better results because the design process is informed by the "tacit" knowledge of users. The concept of tacit knowledge was developed by Michael Polanyi, and is knowledge people use to make decisions in specific contexts. It is not codified and is not written down in a logical form such as "If X happens, then do Y." The person who is using tacit knowledge may not even be able to express it or write it down. A frequently quoted statement of Polanyi is

> We know more than we can tell" which nicely summarizes what tacit knowledge is. Tacit knowledge is difficult to share in a formal way, such as writing a paper or delivering a lecture. An outstanding teacher, for example, cannot pass on all her or his expertise through lectures, for example. However, observing, working with, and being mentored by that same teacher will bring more of that tacit knowledge into the open where a novice educator or colleague can learn about it. When teachers, students, workers, staff, and other stakeholders are members of participatory design teams that important but difficult-to-access tacit knowledge can be brought into play throughout the design process. Another way to explain the power of participatory design is *collective intelligence*, which is "a shared insight that comes about through the process of group interaction, particularly where the outcome is more insightful and powerful than the sum of individual perspectives. (Sanoff, 2007, p. 213)

The concepts of tacit knowledge and collective intelligence are similar, and both help us see how participatory teams can facilitate instructional design work. However, success must be based on a respect for the participants and the belief that they can make significant contributions. This is virtually the opposite of the "designer as expert" approach which is based on the assumption that instructional designers have special expertise and knowledge about pedagogy that must be given precedence in decisions about how material should be designed. Sanoff (2007) succinctly expressed the appropriate foundation for participatory design:

[Participatory design] practitioners share the view that every participant in a PD project is an expert in what they do, whose voice needs to be heard; that design ideas arise in collaboration with participants from diverse backgrounds; that PD practitioners prefer to spend time with users in their environment rather than "test" them in laboratories. Participatory design professionals share the position that group participation in decision-making is the most obvious. They stress the importance of individual and group empowerment. Participation is not only for the purposes of achieving agreement. It is also to engage people in meaningful and purposive adaptation and change to their daily environment. (p. 213)

Although the concept of participatory design began the factories, hospitals, and offices of Scandanavia where it was supported by more open and flexible worker-supervisor and worker-company relationships, Sanoff (2007) has highlighted a number of uniquely American trends that have had an influence on the way participatory design is practice in the United States. Sanoff began his discussion of American influences with a comment about life on the American frontier as "the shaping force in grass roots democracy" which established "the people's right to participate" (p. 214) but he also noted a number of contemporary American influences (See Figure 13.1).

The choice between expert-based design versus participatory design thus centers on this point—are experts who have extensive training on the universal knowns (e.g., Merrill's principles) of ID and pedagogy) the best people to make decisions even if they do not know the local context well? Or, is it better to comprehensively involve end users and other experts in the decision making processes that occur over the cycles of the ID process? One position assumes there are universals that apply to all contexts and that those universal are known and understood primarily by design experts. The other assumes there are very few universals and those that do exist must be interpreted and reinterpreted based on an understanding of context. R2D2 is based on the latter position. Participatory design reflects a shift from the perspective that the-designer-knows-best to one in which the designer is part of a team that, collectively, can accomplish much more when each person is a full participant. This shift from an Expert-Object to Expert-Expert model has its parallel in research methodology as well. As qualitative research methods, and the paradigms that underlie many of them, have become more popular, so has the research equivalent of participatory design. Participant observation, or ethnographic research, might be called a weak form of participatory design because the researcher still tends to take somewhat of an expert role.

The hallmark of participant observation is interaction among the researcher and the participants. The main subjects take part in the study to varying

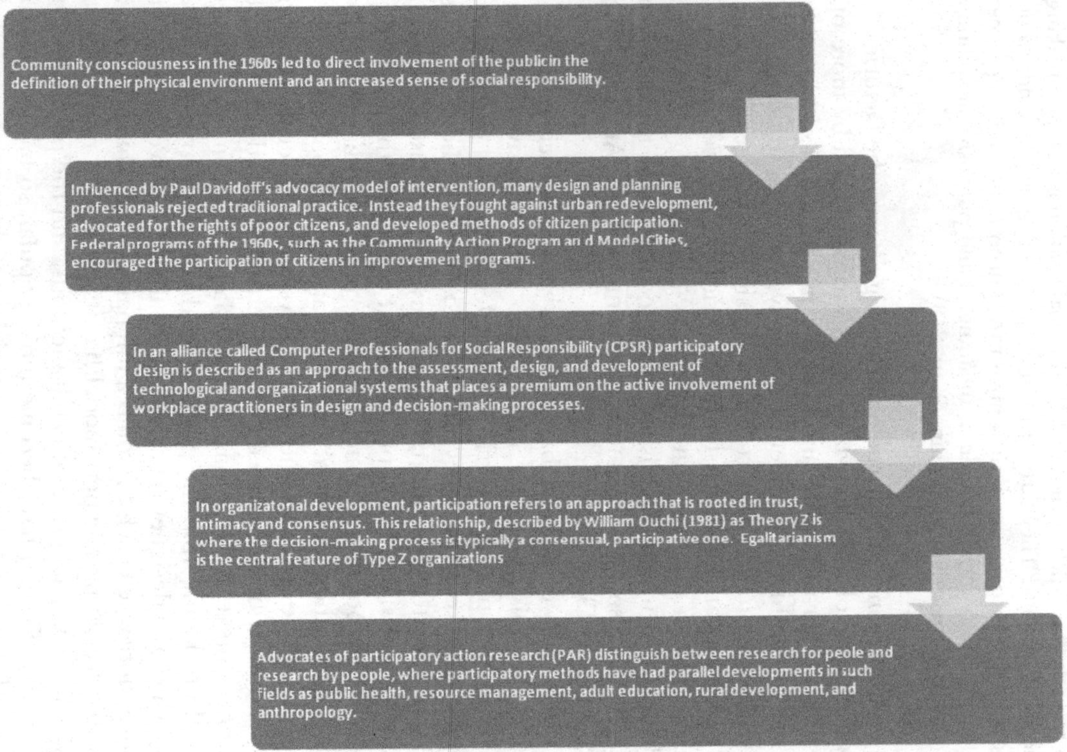

Community consciousness in the 1960s led to direct involvement of the public in the definition of their physical environment and an increased sense of social responsibility.

Influenced by Paul Davidoff's advocacy model of intervention, many design and planning professionals rejected traditional practice. Instead they fought against urban redevelopment, advocated for the rights of poor citizens, and developed methods of citizen participation. Federal programs of the 1960s, such as the Community Action Program and Model Cities, encouraged the participation of citizens in improvement programs.

In an alliance called Computer Professionals for Social Responsibility (CPSR) participatory design is described as an approach to the assessment, design, and development of technological and organizational systems that places a premium on the active involvement of workplace practitioners in design and decision-making processes.

In organizational development, participation refers to an approach that is rooted in trust, intimacy and consensus. This relationship, described by William Ouchi (1981) as Theory Z is where the decision-making process is typically a consensual, participative one. Egalitarianism is the central feature of Type Z organizations

Advocates of participatory action research (PAR) distinguish between research for peole and research by people, where participatory methods have had parallel developments in such fields as public health, resource management, adult education, rural development, and anthropology.

Figure 13.1.   Influences on American adoption of Participatory Design (Sanoff, 2007)

degrees, but the researcher interacts with them continually. For instance, the study may involve periodic interviews interspersed with observations so that the researcher can question the subjects and verify perceptions and patterns. (Savenye & Robinson, 1996, p. 1177)

In participant observation, as generally practiced, the researcher tends to draw the conclusions and seek verification from the participants. The same is true of some forms of action research in education. The researcher is the expert and the participating teachers tend to play less central roles. I would still term this participatory, though a weak form, because there is some sharing of the responsibility for the conclusions. Certain types of action research are much more participatory—the researcher is part of a team in which responsibility is much more evenly distributed (McIntyre, 2007; Whyte, 1991). This form of action research is often referred to as Participatory Action Research or PAR. A related approach, *cooperative inquiry* (Heron, 1996), seems to be a full research equivalent of participatory design:

> Cooperative inquiry is a form of participative, person-centered inquiry that involves research with people not on them or about them. It breaks down the old paradigm separation between the roles of researcher and subject. In traditional research in the human sciences these roles are mutually exclusive: [only] the researcher ... contributes the thinking that goes into the project—conceiving it, designing it, managing it and drawing knowledge from it—and the subjects only contribute the action to be studied. In cooperative inquiry this division is replaced by a participative relationship among all those involved. This participation can be of different kinds and degrees. In its most complete form, the inquirers engage fully in both roles, moving in cycling fashion between phases of reflection as co-researchers and of action as co-subjects. In this way they use reflection and action to refine and deepen each other. They also adopt various other procedures to enhance the validity of the process and its outcomes. (p. 19)

Heron's (1996) book is an excellent introduction to the conduct of research from a participatory or cooperative perspective. It represents in research methodology what I am proposing here in instructional design. For a detailed justification for using participatory design in the development of K–12 classroom Internet applications (see Silva & Breuleux, 1994).

Finally, although this discussion has focused on an either-or choice between participatory versus expert-dominated design procedures, there are actually many levels of participation. There can be, for example, "Weak Participation" in which stakeholders have an opportunity to provide input on design decisions but they play advisory roles and have no real power in the process. At the other end of the continuum, Very Strong

Participation" would mean the stakeholders have independent decision making power that would trump the recommendations of others on the design team.

IN SUMMARY

To summarize, there are three flexible guidelines that can support C-ID: reflection, recursion (iteration), and participation. All three are general principles that have been widely adopted and used in other design fields. They are not as widely practiced in ID for a variety of reasons, but the paradigm shift that seems to be occurring now in the field provides a much more hospitable environment for them. Constructivist theories of learning and interpretivist theories of epistemology are successfully competing for attention today and they provide a strong theoretical foundation for C-ID. Also, the emerging epistemologies that support alternative views of what research is and how, as well as why, it is done, also support these three principles. Constructivist learning theory is a more compatible framework for reflective, recursive design based on participatory involvement than are behavioral or information processing theories. When it comes to reflective, recursive, participatory research, the same is true for interpretivist paradigms when compared to positivist and post-positivist paradigms. I should point out, however, that these three general principles are not the same type of concept as the principles Merrill (1996) talks about. For Merrill, a principle is a much more precise and prescriptive thing that has a great deal in common with Gage's (1996) *long-lasting generalizations* and Plato's *epistemic knowledge*. The principles I have discussed are much less precise and require considerable interpretation and adaptation on the part of a practicing professional. Merrill's principles are the foundation for technical-rational practice. Mine are much more in the tradition of Aristotle's *phronesis* and Schön's *reflective practice*.

With the basic guiding principles dealt with in this chapter as a flexible foundation, I will explore the process and practices of R2D2 in the next chapter.

REFERENCES

Ambler, S. W. (June, 2008). Has agile peaked? *Dr. Dobbs Journal: The World of Software Development, 33*(6), 52–54.

Anthopoulos, L., Siozos, P., & Tsoukalas, I. (2007, April 1). Applying participatory design and digital for discovering and re-designing e-government services.

Government Information Quarterly, 24(7), 353–376. Retrieved from http://www.sciencedirect.com/science?_ob=ArticleURL&_udi=B6W4G-4MSR8NF-1&_user=10&_rdoc=1&_fmt=&_orig=search&_sort=d&view=c&_acct=C0 00050221&_version=1&_urlVersion=0&_userid=10&md5=7cd2ead2bf767 d65ca9afb2863e74cc1

Baek, J., & Lee, K. (September, 2008). A participatory design approach to information architecture design for children. *CoDesign, 4*(3), 173–191.

Beck, K., & Andres, C. (2005). *Extreme programming explained: Embrace change.* Boston: Addison-Wesley.

Bannon, L. (1991). From human factors to human actors: The role of psychology and human-computer interaction studies in system design. In R. Baecker, J. Grudin, W. Buxton, & S. Greenberg (Eds.), Readings in human- computer interaction: Toward the year 2000 (pp. 205- 214). San Francisco: Morgan Kaufmann.

Bodker, K., Kensing, F., & Simonsen, J. (2004). *Participatory IT design: Designing for business and workplace realities.* Boston: MIT Press.

Bodker, S., Gronbaek, J., & Kyng, M. (1995). Cooperative design: Techniques and experiences from the , Scandinavian scene. In R. Baecker,). Grudin, W. Buxton, & S. Greenberg (Eds.), *Readings in human-computer interaction: Toward the year 2000.* San Francisco: Morgan Kaufmann.

Borenstein, N. (1991). *Programming as if people mattered.* Princeton, NJ: Princeton University Press.

Braden, R. (1996, March-April). The case for linear instructional design and development: A commentary on models, challenges, and myths. *Educational Technology, 36*(2), 5–23.

Brooks, J., & Brooks, M. (2001). *The case for constructivist classrooms* (2nd Ed.). Alexandria, VA: Association for Supervision and Curriculum Development.

Carpenter, B. (2006, April). *Toward non-linear curriculum design: Hypertext and the preparation of teachers in art education.* Retrieved from http://www.ts.vcu.edu/etech/etday/papers/carpenter.pdf

Carpenter, B. S., & Taylor, P. G. (2003). *The intentionally tangled curriculum.* Proceedings of the Annual Conference of the Society for Technology and Teacher Education. Albuquerque, New Mexico. Charlottesville, VA: AACE.

Cennamo, K., Abell, S., & Chung, M. (1996, July–August). A "layers of negotiation" model for designing constructivist learning materials. *Educational Technology, 36*(4), 39–48.

Cognition and Technology Group at Vanderbilt. (1993, March). Anchored instruction and situated cognition revisited. *Educational Technology, 33*(3), 52–69 .

Cusumano, M. (2007, October). Extreme programming compared with Microsoft-style iterative development. *Communications of the ACM, 50*(10), 15–18. Retrieved October 13, 2008, from Academic Search Premier database.

Dick, W. (1996). The Dick and Carey model: Will it survive the decade? *Educational Technology Research and Development, 44*(3), 55–63.

Dick, W., & Carey, L. (1996). *The systematic design of instruction* (4th ed.). New York: HarperCollins.

Donmoyer, R. (1996, April). Juxtaposing articles/posing questions: An introduction and an invitation. *Educational Researchen, 25*(3), 4, 23.

Duffy, T., & Jonassen, D. (Ed.). (1992). *Constructivism and the technology of instruction: A conversation.* Hillsdale, NJ: Erlbaum.

Fullan, M. (1993, March). Why teachers must become change agents. *Educational Leadership, 50*(6), 1–13.

Gage, N. (1996, April). Confronting counsels of despair for the behavioral sciences. *Educational Researchen, 25*(3), 5–15, 23.

Gentry, C., & Csete, 1. (1995). Educational technology in the 1990s. In G. Anglin (Ed,), *Instructional technology: Past, present, and future* (pp. 20–33). Englewood, CO: Libraries Unlimited.

Goldin, G. (1990). Chapter 3. Epistemology, constructivism, and discovery learning in mathematics. *Journal for Research in Mathematics Education.* Monograph, Vol. 4, Constructivist Views on the Teaching and Learning of Mathematics (pp. 31-210).

Heron, J. (1996). *Cooperative inquiry: Research into the human condition.* Thousand Oaks, CA: SAGE.

Hlynka, D. (1995). Six postmodernisms in search of an author. ln G. Anglin (Ed), *Instructional technology: Past, present, and future* (pp. 113–118). Englewood, CO: Libraries Unlimited.

Jonassen, D. A. (1996, April). Thinking technology: Toward a constructivist design model. *Educational Technology, 34*(4), 34–37.

Jonassen, D., Peck, K., & Wilson, B. (1999). *Learning with technology: A constructivist perspective.* Columbus, OH: Merrill.

Kessels, J., & Korthagen, F. (1996, April). The relationship between theory and practice: Back to the classics. *Educational Researcher, 25*(3), 17–22.

Lebow, D. (1993). Constructivist values for instructional systems design. Five principles toward a new mindset. *Educational Technology Research and Development, 47*(3), 4–16.

Lin, X., Bransford, L., Hmelo, C., Kantor, R., Hickey, D., Secules, T., Petrosino, A., Goldman, S., & the Cognition and Technology Group at Vanderbilt. (1995, September/October). Instructional design and development of learning communities: An invitation to dialogue. *Educational Technology, 35*(5), 53–63.

Marlowe, B., & Page, M. (2005). *Creating and sustaining the constructivist classroom* (2nd ed.). Thousand Oaks, CA: Corwin Press.

Martin, D. J. (2008). *Elementary science methods: A constructivist approach* (5th ed.). Belmont, CA: Wadsworth.

McIntyre, A. (2007). *Participatory action research.* Thousand Oaks, CA: SAGE.

Merrill, M. (1996, July–August). What new paradigm of ISD? *Educational Technology, 36*(6L), 57–58.

Merrill, M. (1997, January). Onlinstructional strategies. *Dr. Dave's Corner 1*(2) 1–5. Retrieved from http://www.coe.usu/edu/it/id2

Muller, M., (1991). PICTIVE—An exploration in participatory design. In *Proceedings of the ACM CHI 91 Human Factors in Computing Systems Conference* April 28–June 5, 1991, New Orleans, Louisiana, 225-231.

Nixon, E., & Lee, D. (2001, January 1). Rapid Prototyping in the Instructional Design Process. *Performance Improvement Quarterly, 14*(3), 95–116. (ERIC Document Reproduction Service No. EJ643438) Retrieved October 13, 2008, from ERIC database.

Northover, M., Northover, A., Gruner, S., Kourie, D., & Boake, A. (2007). Agile software development: A contemporary philosophical perspective. In ACM International Conference Proceeding Series; Vol. 226 *Proceedings of the 2007 annual research conference of the South African Institute of computer scientists and information technologists on IT research in developing countries*, Port Elizabeth, South Africa. 106-115. Available through the ACM Digital Library.

Reigeluth, C. (1996a, May/June). A new paradigm of ISD? *Educational Technology*, *6*(3), 13–20.

Reigeluth, C. (1996b, July/August). Of paradigms lost and gained. *Educational Technology*, *36*(4), 58–59.

Ross, R., & Ross, D. (2008). *Walking to New Orleans: Ethics and the concept of participatory design in post-disaster reconstruction.* Eugene, OR: Wipf and Stock.

Sanoff, H. (2007). Special issue on participatory design. *Design Studies*, *28*(3), 213–215.

Savenye, W., & Robinson, R. (1996). Qualitative research issues and methods: An introduction for instructional technologists. In D. Jonassen, (Ed.), *Handbook of research on educational communications and technology* (pp. 1171–1195). Mahwah, NJ: Erlbaum.

Senge, P. (1990). *The fifth discipline.* New York: Doubleday.

Schön, D. (1983). *The reflective practitioner.* New York: Basic Books.

Schön, D. (1987). *Educating the reflective practitioner.* San Francisco: Jossey-Bass.

Schön, D. (1995). *The reflective practitioner.* New York: Ashgate.

Schuler D. & A. Namoika, A. (Eds). (1993). *Participatory design: Principles and practices.* Hillsdale, NJ: Erlbaum.

Silva, M. & Breuleux, A. (1994). The use of participatory design in the implementation of internet-based collaborative learning activities in K–12 classrooms. *Interpersonal Computing and Technology: An Electronic Journal for the 21st Century*, *2*, 99–128. Retrived from http://www.helsinki.fi/science/optek/1994/n3/silva.txt

Spinuzzi, C. (2005). Lost in the translation: Shifting claims in the migration of a research technique. *Technical Communication Quarterly 14*(4), 411–446.

Thumboo, J., Wee, H., Cheung, Y., Machin, D., Luo, N., & Fong, K. (2006, September). Development of a smiling touchscreen multimedia program for HRQoL assessment in subjects with varying levels of literacy. *Value in Health*, *9*(5), 312–319.

Vanderburg, G. (October, 2005). A simple model of agile software processes–or–Extreme programming annealed. *ACM SIGPLAN Notices*, *40*(10), 539–545. Proceedings of the 20th annual ACM SIGPLAN conference on Object oriented programming systems languages and applications. Retrieved from the ACM Digital Library.

Whyte, W. (Ed.). (1991). *Participatory action research.* Thousand Oaks, CA: Sage.

Willis, J. (1995). A recursive, reflective instructional design model based on constructivist-interpretivist theory. *Educational Technology* , *35*(6), 5–23.

Willis, J. (2007). *Foundations of qualitative research.* Thousand Oaks, CA: Sage.

Wilson, B. (1996). (Ed.), *Constructivist learning environments.* Englewood Cliffs, NJ: Educational Technology.

Wilson, B. (1997). Reflections on Constructivist ID. In C. R. Dills and A. A. Romiszowski (Eds.), *Instructional Development Paradigms*. Englewood Cliffs NJ: Educational Technology Publications. Retrieved from http://carbon.cudenver.edu/~bwilson/construct.html

Winn, W. (1991, September). The assumptions of constructivism and instructional design. *Educational Technology, 31*(9), 38–40.

Winn, W. (1992). The assumptions of constructivism and instructional design. In T. Duffy & Ionassen, D. (Eds.), *Constructivism and the technology of instruction: A conversation* (pp. 177-182). Hillsdale, NJ: Erlbaum.

Winters, N., & Mory Y. (2008, February). IDR: A participatory methodology for interdisciplinary design in technology enhanced learning. *Computers & Education, 50*(2), 579-600.

You, Y. (1993). What can we learn from chaos theory? An alternative approach to instructional systems design. *Educational Technology, Research and Development, 43*(3), 17–32.

Zuber-Skerritt, O. (1996). *New directions in action research.* New York: Routledge-Falmer.

CHAPTER 14

A GENERAL SET OF PROCEDURES FOR C-ID

R2D2

Jerry Willis

The previous chapter presented a set of general or guiding principles for a constructivist approach to instructional design (C-ID). The ID model I have developed with colleagues working on projects at the NASA Johnson Space Center, the University of Houston, Iowa State University, Louisiana State University, and Manhattanville College is called the Reflective Recursive Design and Development (R2D2) model. In this chapter the focus is on the way guidelines might be put into practice. This is not, however, a set of required steps or essential ingredients for C-ID. It is, instead, an example of one way the basic principles of C-ID might be implemented. In addition, the suggestions are illustrative and flexible, not prescriptive, in the spirit of constructivist learning

This chapter is a major revision of Jerry Willis and Kristen Wright Willis (2000). A general set of procedures for constructivist instructional design: The New R2D2. *Educational Technology, 40*(2), 5–20.

PROCEDURES FOR A REVISED R2D2 MODEL

Three principles discussed in the previous chapter are the most important part of the R2D2 mode. Those three principles—recursion, reflection, and participation—were explored in detail in the previous chapter.
Here I will present an overview of the activities and procedures typical of
design work based on R2D2, but I must emphasize that these are suggestions, guidelines, and ideas rather than "set in stone" procedures that
"must" be followed. ID at this point is much more art than science. And it
is much more art than it is the correct application of technical recipes that
require you to match well-defined conditions to well-defined solutions. If
someone followed the three principles discussed in the previous chapter
but did not use the procedures described in this chapter, the work would
be more in the spirit of the R2D2 model that work that followed the procedures but did not use the principles.

Before explaining some of the details of the revised R2D2 model, I
want to offer one more caveat. In the first version of the R2D2 model
(Willis, 1995), I used existing instructional design terms like *front-end analysis* that were recycled from traditional ID models. However, these terms
had new and quite different meanings in the R2D2 model. I used the
terms because they were familiar to many instructional designers and they
seemed to provide a bridge from traditional to C-ID models. That the
terms had radically different meaning in the R2D2 model was picked up
by some authors (e.g., Seels & Glasgow, 1998). Others, such as Dick
(1996), saw that the 1995 version used terms that were common in traditional linear ID models and seemed to conclude that they must have the
same meaning in the R2D2 model. In part because of conclusions like the
ones drawn by Dick, I decided that recycling terms that have established
meanings in the traditional ID literature was probably a mistake. The
revised version of R2D2 presented here generally uses terms from constructivist and related theories instead of recycling existing terms from
traditional ID models.

THE FOCAL POINTS

There are three focal points in the R2D2 model: Define, Design and
Development, and Dissemination. The graphic representation of the
model is deliverately non-linear and flexible. There is no required beginning step and there are no second, third, and fourth step that follow one
after the other. Instead, there is an open, flexible process that results in an

instructional resource. The model is based on the assumption that designers will work on all three focal points of the process in an intermittent and recursive pattern that is neither completely predictable nor prescribable. The focal points are, in essence, a convenient way of organizing our thoughts about the work. The type of work a design team might participate in for each of the focal points is described in the following sections.

Define Focus

Traditional Instructional Systems Design (ISD) begins with a range of preparatory activities that must precede the design of instruction. Learner analysis, task and concept analysis, and the specification of instructional objectives are all common components of ISD models. These tasks are generally conducted in a technical-rational and an expert-led framework. That is, the "expert" designer studies the students as well as the content to be taught, much as an architect might begin work on the design and construction of a building by studying the site and the proposed uses of the building.

A principle of R2D2 is that design work can best be accomplished by creating a participatory team. Thus, instead of the expert designer collecting data on the flora and fauna of the design task, he or she becomes a facilitator of the design process and shares decision-making and exploration of issues with other members of the design team. This might be called the move toward "participatory design" in the field of ID and it is a reflection of a general move toward democratic participation on the part of consumers and other stakeholders in many types of design. For example, the term "participatory architecture" refers to an approach to designing buildings that involves stakeholders such as those who will live in or use the buildings in the process of designing and building them (Marschall, 1998). In architecture Roodt (2002) has even distinguished between two types of "community participation." An example of one type is the British colonial approach "where decisions are made at the top [by British administrators] and implementation is done by the people on the ground" (p. 104). The other type of participation "is based on the concept of a transformation of people's consciousness which leads to a process of self-actualization and empowerment" (p. 104). It is this second type of participation, which emphasizes empowerment of participants, and includes them in the decision making process, that is an important component of R2D2.

Figure 14.1.   A graphical representation of the R2D2 ID model.

This type of participatory approach eliminates, radically changes, or shifts the time frame of all the common "up front" activities of ISD: learner analysis, task and concept analysis, and creation of instructional objectives. The beginning of a project is probably the worst time to create specific detailed objectives. That is when there is the least agreement about what should be learned. The beginning is also the worst time to complete learner, task and concept analyses. Such work can begin there, but that understanding will emerge across the design process and will be of much higher quality than information and perceptions gained primarily at the beginning. Knowledge and understanding will emerge across the design and development process.

This idea of letting understanding emerge is strongly supported by chaos theory as well as constructivist theory. You (1994) pointed out that the chaos theory concept of fractals, which involve repeated patterns at different levels of examination, calls into question the linearity of ISD models and the requirement that much of the analysis of context occurs at the beginning of design work.

> ISD's linearity gives the impression that analysis occurs at the beginning, design at the middle and evaluation at the end of the process.... These fixed, separate steps are problematic in terms of fractal scaling, since fractal structure, in which the whole is present in the part, and the part in the whole, results from a process of iteration. (p. 31)

> When seen as a fractal structure, the process prescribed by ISD models are not mechanical, sequential parts independent of other components, but are a dynamically shifting web of parts that all connect together. This suggests that a complete analysis of task and learner performed prior to implementing instruction does not guarantee that either the task or the knowledge and skill of the learner are unchanging. Thus, iteration of all ISD operations such as analysis, design, and evaluation at every step become imperative.... Further, no single IS operation has preeminence over any other, for each operation plays its own unique role and together they form the distinctive character of the whole. (pp. 24–25)

From the perspective chaos theory, You (1994) asserts that two aspect of ISD are problematic. One is linearity. It is a problem because linearity means a job such as learner analysis will be scheduled for one point in the process of ID (such as at the beginning) and the results of that analysis will then be used in the remaining steps in the ID process. He objects to that because the information and perceptions we develop at one point in the process may be inaccurate or incomplete. You proposes, instead, that the analyses done in the beginning of ID when a traditional or ISD model is use, be done recursively, over the entire ID process. A recursive or iterative process replaces the linear process.

The other objection to linear ID models is that they ignore the dynamic nature of systems. This idea is based on a fundamental discovery of quantum physicist Werner Heisenberg that has come to be called the *Principle of Indeterminancy* or the *Heisenberg Uncertainty Principle*. Because he developed it at the Copenhagen laboratory of Niels Bohr, it is also referred to as the *Copenhagen Interpretation* of quantum mechanics. There are several aspects to the principle but the most significant one for this discussion is that you cannot accurately and simultaneously determine both the position of an electron *and* the speed of the electron. You can determine one or the other, but you cannot measure both at the same time. Thus, there is an aspect of uncertainty built into the nature of the world. The reason for this is that the act of measuring actually interferes with the behavior of the electron. That is, if you measure speed, you influence position and vice versa. Heisenberg was talking about the movement of an electron within an atom, but the idea has been applied to the social sciences as well. In ID, it suggests that both the context of assessment and the process of measurement and assessment, will have an impact on what is being measured. Thus, efforts to assess things like learner characteristics should be done in context rather than out of context or in isolation. Further, they should be conducted cross the design process rather than at only one point.

Linear ID models are sometimes referred to as "waterfall" models (Jarche, 2007) because the results of one phase in the ID process becomes the data needed to do the next phase. Waterfall models have been criticized on a number of points including the lack of flexibility. As Alex Iskold (2007) put it in his criticism of the waterfall model when used for software development:

> The Waterfall Model is now considered a flawed method because it is so rigid and unrealistic. In the real world, software projects have ill-defined and constantly evolving requirements, making it impossible to think everything through at once. Instead, the best software today is created and evolved using agile methods. These techniques allow engineers to continuously re-align software.

> The accelerating pace of business requires constant changes to software. Older development methods completely fail to address business needs. Using the Waterfall Model, these changes were impossible, the development cycle was too long, systems were over engineered and ended up costing a fortune, and often did not work right.

The problem was that the Waterfall Model was arrogant. The arrogance came from the fact that we believed that we could always engineer the perfect system on the first try. The second problem with it was that in nature, dynamic systems are not engineered, they evolve. It is the evolutionary idea that lead to the development of agile methods.

Jarche and Iskold (2007) both recommend an approach to software development they refer to as *Agile Software Development.* It has much in common with iterative or recursive approaches to ID. Iskold (2007) presents 9 principles for Agile Software Development:

- Have fun
- Embrace change
- Communicate
- Focus on simplicity
- Get feedback
- Release new versions of software often
- Adapt the code rather than continuously writing new programming code
- Refactor: "improve the design of existing code without changing how it works"
- Do developer testing regularly

Iskold believes Agile Software Development is preferred over waterfall models because:

The software systems created using agile methods are much more successful because they are evolved and adapted to the problem. Like living organisms, these systems are continuously reshaped to fit the dynamic landscape of changing requirements. Without a doubt, agile methods made a major impact on how we think about building software today - dynamically and continuously.

Although the Agile Software Development model was created to produce software for businesses that change and evolve continuously and therefore need frequent software changes and updating, there is much in common between C-ID models like R2D2 and Agile Software Development models.

In the revised R2D2 ID model described in this chapter, the Define Focus includes three activities. Creating and Supporting a Participatory Team, Progressive Problem Solution, and Developing Phronesis (Contextual Understanding). All three activities are important in the beginning of

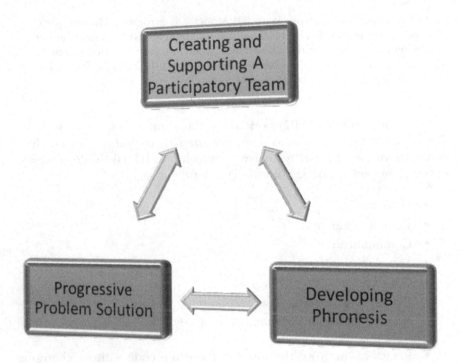

Figure 14.2.   Activities of the design focus.

a design project, but they must be attended to across the entire process (see Figure 14.2). They are part of a holistic design process, not distinct activities that occur at the beginning of the work.

Creating and Supporting a Participatory Team

Henry Sanoff (2007) began the special issue of *Design Science* (Volume 28, 2007) with this comment:

> Participatory design is an attitude about a force for change in the creation and management of environments for people Its strength lies in being a movement that cuts across traditional professional boundaries and cultures. Its roots lie in the ideals of a participatory democracy where collective decision-making is highly decenralised throughout all sectors of society, so that all individuals learn participatory skills and can effectively participate in various ways in the making of all decisions that affect them. (p. 213)

The idea of organizing design work around a participatory team was discussed earlier. Our experience suggests that this is one of the most difficult tasks facing a constructivist designer. Creating a team is not difficult. But supporting one, encouraging the team, and facilitating participation, is not easy. There are two basic ways of creating a team. One is to enlist the cooperation of a small group of individuals who represent different stakeholder groups, such as teachers, students, graphic artists, designers, and so on. This team then constitutes the participatory group that takes the project from start to finish. We have found that this approach works well when the team members are all heavily invested in the project and have time to participate. When some members are less invested, or when time and availability are a problem, this approach can be frustrating to all concerned because it is difficult to get and maintain full team participation. The result can be a project that has a participatory team in name only. In fact, only one or two people may end up making all the decisions, while the other team members do not provide significant input.

An alternative is to organize a small core team, perhaps only two or three people, and then involve various people at different points in the process. For example, a designer and a teacher might form a core, and different students as well as other teachers and specialists might provide input at different points. One small group of students might, for example, discuss the basic idea at the onset. Another group might look at early versions of the material, and yet another small group of students might critique the material when it is closer to finished form.

Regardless of the form the participatory team takes, it will require thoughtful support, and careful arrangement of the opportunities to participate, if the members are to do their best as collaborators. There are a number of examples of participatory design projects in the literature that may be helpful to instructional designers who create a participatory team. Several are reported in the classic book edited by Schuler and Namoika (1993). Greenbaum and Madsen (1993), for example, describe what I would call a "weak participatory design" approach that puts the designers in the work environment for an extended period. They organized a series of formal activities such as storytelling workshops (each worker prepared two short oral stories about his or her worst or most successful use of computers at work) and fantasy workshops (workers brainstormed ways of improving the computer support available in the workplace and then discussed the potential of the ideas generated.) I call this weak participatory design because much of the decision-making is still in the hands of the designer, even though users have significant input.

This point was made by Bannon (1995) in a discussion of user-centered versus user-involved design:

Exactly what the term user-centered system design means, or how it can be achieved, is far from clear. In some cases it dissolves into platitudes such as "Know the User." Such kinds of general guidelines are of little use in practical situations of design due to their lack of specificity. Gould (1988) discusses how important it is to have an early and continuous focus on users, to develop iterative designs, and to have early and continuous user testing. This is a step in the right direction. Users are being given a larger role in the design process, but it is still a relatively passive role. Although actual participation by users on the design team is mentioned, it does not figure prominently in this user-centered approach. A more radical departure from current thinking within the mainstream HCI (human-computer interaction) world is to look at users not simply as objects of study, but as active agents within the design process itself. This involvement of users in design is both a process and a way to ensure that the resulting computer system adequately meets the needs of the users. (p. 211)

In his introduction to the special issue of *Design Studies*, Sanoff goes on to describe the work on participatory design in areas like urban design, planning, industrial design, and information technology. I agree with many other instructional designers that we should strive for full user participation in instructional design as well, and that we should practice "strong participatory design." Other authors such as Bodker, Gronbaek, and Kyng (1991), have described "strong participatory design" in the area of software design. They believe "computer applications that are created for the workplace need to be developed with full participation from the users—both from a democratic point of view and to insure that competencies central to the design are represented in the design group" (p. 215). They point out, however, that users may require help to take on the new role of participant, and designers may need help in their new roles as facilitators. "Full participation," of course,

> requires training and active cooperation, not just token representation in meetings or on committees. We use the term *cooperative design* to designate such cooperation between users and designers.... Designers should know how to set up the process and need to make sure that everyone gets something out of the interaction. (pp. 215-216)

There are, fortunately, a number of resources for instructional designers interested in developing and supporting participatory design teams. For example, two Swedish participatory designers, Pilemalm and Timpka (2007), describe how "third generation" participatory design can be used in large-scale projects as well as the small scale projects "with homogenous user groups in local settings" where they have flourished in the past. Pilemalm and Timpka also pay some attention to the criticisms and difficulties of doing successful participatory design and they frame their discussion of these issues around the development of different approaches to PD over the last 30 years.

There are also papers on specific PD projects that provide enough information on the way a participatory team was developed and supported to be helpful to other participatory designers. Caitlin Cahill (2007), for example, describes in detail how she used a participatory action research to support a team of young women on the Lower East Side of New York City who were concerned about the sexist and discriminatory advertisements on New York public transportation. The *Makes Me Mad: Stereotypes of Young Urban Women of Color* project developed and disseminated counteradvertisements that mocked real ads but were designed to "plant a seed in the minds of society" (p. 325). The article (Cahill, 2007) and the Web site (www.fed-up-honeys.org) both provide helpful insights into how to work in participatory design teams.

Finally, in a study of PD in architecture, Rachael Luck (2007) at the University of Reading in the UK studied the patterns of a successful and experienced PD architect and a less experienced architect as they worked with a participatory team. Luck looked specifically at the communication and conversation patterns of the two architects and offers a number of suggestions about the important task of supporting and encouraging participation in the design process. Although I recommend you read the entire article, here are some of her suggestions:

- The facilitator should make it easy for users to become involved in the discussion about design.
- The conversations about design should be seamless "where the users quickly understand the subject being discussed, its relevance to their lives" and "quickly join the discussion of these issues" (pp. 233–234).

There are also a number of resources written for participatory action researchers that are also very helpful for instructional designers who want to use participatory design. Sara Kindon's (2008) book, *Participatory Action Research Methods: Connecting People, Participation, and Place*, is one of several books that cover all aspects of the PAR process including methods and procedures for supporting the participatory team. There are also other good resources such as Shirley White's (2000) edited book, *The Art of Facilitating Participation* which has both examples of how participation was supported in a particular project and discussions of particular approaches to support. Kaner's (2007) book, *Facilitator's Guide to Participatory Decision Making*, is a comprehensive guide for facilitators that will be very helpful to designers using participatory methods.

There are also a great many papers and web resources on participatory methods of research, design, and community change. One is a paper about work in the highlands of Uganda with local farmers (Sanginga, Kamugisha, Martin, Kakuru, & Stroud (2004) that both tells the story of a particular participatory process to change policies and also provides one framework for thinking about how to support participation. Another useful paper describes methods used to sustain community-based participatory research partnerships in Detroit, New York, and Seattle (Israel et al., 2006).

Progressive Problem Solution

The second component of the Define Focus is progressive problem solution. It is an overgeneralization to say that traditional ID models break the general design problem down into pieces and then attack the pieces one at a time. But it is not *too* much of an overgeneralization. Objectives are defined and finished first, then the criterion-referenced tests are written based on the objectives, and so on. This approach, which involves finishing each task before moving on to the next, was termed the *stagewise* model by Boehm (1995). Problems with the stagewise model led to the *waterfall* model. Water (completed tasks) flows out of one pool (a particular stage in the ID process) and into the next, where it is used to complete the subsequent task in the sequence. The waterfall model added a feedback loop at the end of each stage that involved getting feedback on the product of that stage and then revising it before passing the results to the next stage. Braden's (1996) ID model is probably the clearest example of a modern waterfall ID model. For a detailed analysis of the problems with waterfall models (see Hunter & Ellis, 2000; Leffingwell, 2007, chapter 2; and Boehm, 1995).

Stagewise and waterfall models are not the only ways to think about the work of design. R2D2 views it as a process of progressively solving multiple problems in context. That is, "solutions," such as the set of objectives for a project, progressively emerge across the entire design process instead of being completed early and then used to guide design and development work. The fuzzy objectives you begin with will influence your design and development work, but conversely, design and development work will also influence the objectives. R2D2 views design work as a richly interactive and dynamic process, in which solutions emerge across a process, and in which work on many different parts of the whole influence each other (and the whole). It is an example of what You (1994) described as an "open system" that assumes initial conceptions and frameworks will change and evolve across the process. The idea of progressively solving the problems of instructional design is one of the most difficult for many designers to accept because it means they must keep the process open and accept the possibility of making major changes even late in the process. Another C-ID model, Layers of Negotiation (Cennamo, 2003), is a non-traditional ID model and Boehm's (1995) Spiral Model of Software Development and Enhancement also emphasize the emergent aspects of design. As Cennamo, Abell, and Chung (1996) put it,

> Whereas traditional instructional design models often include discrete stages of analysis, design, development and evaluation activities ... we addressed the question of design in a spiral fashion, progressing through a series of stages at one level, then spiraling back and adding more detail.

This approach is

> consistent with the holistic manner in which experienced designers often apply traditional models ... we initially made decisions across stages based on the data that was relevant, then, as more information became apparent or relevant, we spiraled back and added more detail across stages. (p. 43)

The idea of progressively solving the problems of professional practice is incorporated into a number of research as well as design methods. For

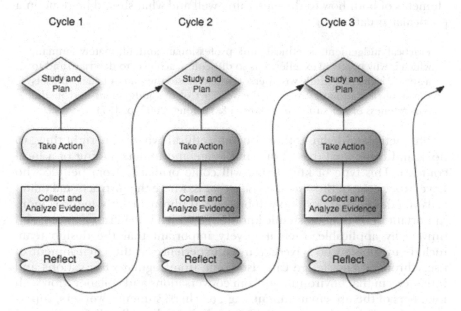

Figure 14.3.   Riel's (2007) Model of action research as progressive problem solving. Used with permission of the author.

example, Margaret Riel (2007) incorporates progressive problem solving in her vision of how to do action research (see Figure 14.3).

Developing Phronesis or Contextual Understanding

Are most design contexts like other design contexts? That is, can we develop generalizations, or rules, that can be easily applied across different design projects? Many ISD models are based on the assumption that there is much that can be generalized, and thus they emphasize commonalities across contexts. They are based on the idea that Plato's *epistemic knowledge* is the foundation for ID work. Epistemic knowledge is universal and applicable to many contexts.

R2D2 and other C-ID are based on the assumption that there is much that is unique in each design context and thus a designer (as well as other members of the design team) must have, or develop, a sophisticated understanding of the particular context in which design work will take place. Aristotle's type of practical knowledge, *phronesis,* is dominant here. There is no word in English that is an exact translation for phronesis, but "practical wisdom" comes close. Phronesis is practical wisdom developed through experience in a context (Gibbs, 2007). However, phronesis has elements of both how to do something well and what should be done in a particular situation.

> Practical judgement is ethical and professional, and ultimately human, which is why practical excellence is so difficult to achieve, to describe, and to assess, it is the practically wise person who chooses the salient issues and sets the implicit standards through the very act of her judgement and in the concreteness of her situation. (Oancea & Furlong, 2007, p. 127)

Phronetic knowledge is thus concrete, ethical (what is the right thing to do?), and contextual (based on knowledge and understanding of a local context). This type of knowledge will come primarily from people who have experience in the *context of use*—that is, where the instructional materials you design and develop will be used. If phronetic knowledge is more important in ID than epistemic knowledge, which is general, abstract, and universally applicable, then it is very important that the design team include members who have extensive experience in the environment of use. Phronetic knowledge can also come from regular observations and interviews in that environment, from conversations and collaboration with members of the use environment (e.g., teachers, students, workers, supervisors, parents, and so on). Different design jobs will call for different approaches, but the core issue is that there is no such thing as a "general

ID expert" who can take his or her skills from one setting to another, and design well without understanding the context. A good example of the importance of understanding context is the case study reported by Thoresen (1993). The case concerned a project to make changes in the support software for a large hospital. Several attempts to make constructive changes actually created major problems for nursing supervisors because the prototype software did not fit their work pattern. However, because the supervisors had input into the design and development process, the problems were identified and changes were made. The type of knowledge the nursing supervisors had, phronesis, is difficult, if not impossible, for external experts to know. It is only by being immersed in the context, and by involving people from that context in the design process, that it can be known and used.

DESIGN AND DEVELOPMENT FOCUS

The second focal point is where much of the action in the ID model happens. The R2D2 model does not separate design from development. They are integrated activities in this model, for a number of reasons that were detailed in the original article on R2D2 (Willis, 1995). Perhaps the most important justification for integrating these two activities is that the reasons for separating them in the first place were technical. In the past, most development environments for creating computer-based educational resources were not change-friendly. For example, writing a major section of code for a multimedia program in a language like C was not easy, and making major revisions was expensive in terms of both time and money. Thus design was separated from development because it was cheap to create flow charts or storyboards of the material to be developed, and expensive to make a rough "trial version" to try out and experiment with. A storyboard is not a very good representation of the final product. A mockup or trial version would be much better, but until recently, trial versions were much more expensive to create.

Fortunately, things have changed. Today there are powerful authoring environments for everything from relatively plain electronic documents to movies. When creating online courses and resources there are several commercial design platforms such as Blackboard and a growing number of open source programs such as Moodle. Some design and creation environments for web-based materials are powerful but complex, but there are also relatively versatile programs that are also relatively easy to use. In fact, the creation of virtually every type of material—from textbooks to videos to multimedia to Web sites—has changed over the past 20 years and become easier and more flexible because new technology and software can be used. Thus, in the R2D2 model, design and development are a single, integrated activity. Because new methods of producing both mockups and prototypes as well as relatively finished materials also make it easier and less expensive to make changes and revisions, design teams can ask "what if" questions. What if we add a video here that illustrates this point? What if we include an animation here that students can use if they click this button? What if we change the user interface and put all the navigation options at the bottom of the screen? What if we make this lesson a problem-based collaborative learning activity instead of a tutorial? In the past, asking, and answering, such *what if* questions was so expensive they were rarely addressed. Today, such questions, and many more, can often be asked and answered without breaking the financial or time budget for a project.

In the original version of R2D2, the design and development focus included four activities:

- Selection of a Development Environment
- Media and Format Selection
- Evaluation Procedures
- Product Design and Development

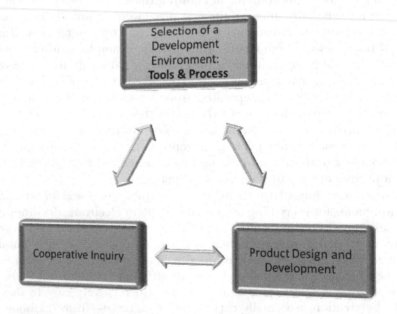

Figure 14.4.   Activities of the design and development focus.

In this revision, the first activity, selection of a Development Environment, has the same name, but the terminology for the other three is different.

Selection of a Development Environment

As noted previously, very powerful authoring and development environments and programs are available today that are also flexible in terms of their ability to handle changes easily. *Power and flexibility* are two important characteristics of a development environment. I would add a third, *accessibility*. The R2D2 model encourages experimentation and exploration of alternatives. For that, you need power and flexibility. R2D2 also encourages participation. For that you need accessibility That is, the development environment should be usable by team members other than a professional programmer. Otherwise, when creating instructional materials involving information technologies, the programmer is in a privileged position and other members of the participatory team are to some extent disenfranchised.

As a practical matter, the most powerful environment is rarely the most flexible, and it is rare for the most powerful and the most flexible to also be the most accessible. Therefore, the selection of a development environment is often a process of compromise that considers the makeup of the participatory team, the type of work being done, and the degree of involvement to be expected from different team members. For example, suppose you are a designer who is leading a team that will create a new social studies curriculum for the middle grades in a school district. If the end result will be a document that details the new proposed curriculum, you could select Word as the development environment because virtually everyone on your team knows how to use it, and you could use the Track Changes command so that all the changes each team member makes can be viewed and accepted or rejected. However, using Word with the Track Changes command gets complicated when several people are trying to make changes at the same time. A common problem is that you end up with two or three versions of the document, each with different suggested changes. Word is a powerful and flexible way of creating a standard document but accessibility is only average because of the problem of keeping track of changes. An option is to set up an online Wiki where all participants can access the developing document, add to and edit it at will, and read the changes other participants have made. The word processing features of most Wikis are not equal to Microsoft Word, but Wikis are more accessible. Thus, you might sacrifice some power and flexibility (e.g., the ability to incorporate all sorts of illustrations and graphics into the document) for the increased accessibility of a Wiki.

The development environment actually consists of two components: the *tools of design* and the *process of design*.

Tools of Design

Tools include everything from a flip chart to a $200,000 computerbased video editing and authoring environment. The tools will vary radically from project to project, but the three criteria for evaluating tools should always be kept in mind: power, flexibility, and accessibility. When graphics are being created, Photoshop is one of the most powerful and flexible programs available today, but there are other programs with less power and flexibility that are more accessible. For example, the 2007 versions of Microsoft Word added a Smart Art command that allows the spectacularly untalented, such as me, to create a wide range of charts and figures. The relative importance of each of these factors—power, flexibility, and accessibility—will vary from project to project, but all three will influence the way the team works and the quality of the final product.

The Process of Design

Originally, I advised designers using the R2D2 model to follow a specific pattern as their team worked on the design and development of materials. This process had four levels:

- Component Design
- Single Path Prototype
- Alpha Version
- Beta Version

Wok began with components such as the interface. Then, after a number of components were finished in some rough form, a single-path prototype was created. A single-path prototype is a prototype that has one relatively complete path through the instructional resource. As long as certain choices are made, the user can proceed from beginning to end as if the entire program were operational. There may be rough versions of artwork and other media, but one path is essentially usable. Feedback and exploration of that prototype can be used to create an alpha version of the software. An alpha version may still have rough versions of some material and there may be "placeholders" for some items like photographs or supplementary material. Feedback and exploration of the alpha version can be used to create the beta version.

Of course, there is usually more than one single path-prototype, and a component such as the user interface may evolve through several iterations, even after an alpha or beta version has been completed. Each prototype, each alpha, and each beta version is gradually revised and improved (or abandoned when it is necessary to start over from scratch). The result is that the final product gradually comes into focus as it comes closer and closer to the version that will be used for instruction. When revisions on the beta version of the material are complete, you have Version 1.0.

There is some linearity to this part of the model because, obviously, you cannot do a beta version before completing an alpha version. However, there is still quite a bit of flexibility and recursion. The team can, for example, decide to substantially revise a component even after a beta version has been created.

This way of thinking about the ID process—components, single path prototype, alpha version, and then beta version—has worked well for us. However, there are a number of other frameworks for thinking about the ID process that would also be quite appropriate. Bodker, Gronbaek, and Kyng (1995), for example, described an approach that includes four major phases. Their original phases were workplace-focused and I have modified them to emphasize an instructional focus:

- *Designers Learn About the Learning Context.* This is accomplished through school and classroom visits, interviews with teachers and students, observations, examination of existing curricular materials and plans as well as student work, and demonstrations.

- *Futures Workshop.* Team members use a variety of techniques to think about how the future might be different from the past. Often this can be done in group meetings where a facilitator helps the team brainstorm possibilities. One format involves three phases: critique, fantasy, and implementation. The critique phase focuses on developing a shared perspective on the problem and need. The fantasy phase allows team members to think through "what if" fantasies about future possibilities, and the implementation phase is a discussion of what resources would actually make realistic changes possible in the teaching and learning context.

- *Organizational Games.* Team members look at the current instructional activities and curricula. Then they broaden their perspective of how teaching and learning might occur as well as what content might be taught. Mockups and prototypes may be used to explore possibilities. Then the team comes to consensus about what might be done in the future and an action plan is developed.

- *Embodying Ideas.* Using mockups and cooperative prototypes, the team iteratively creates new instruction.

Bodker, Gronbaek, and Kyng (1995) and Greenbaum and Kyng (1991) provide much more detail on this framework for design. I consider it to be one alternative to the process I have used that would be more appropriate for C-ID projects. There are, however, many other models of the process of ID that would be appropriate. Cennamo's (2003) Layers of Negotiation Model has already been mentioned. Another is a participatory design process (Good, 1995) developed at Digital Equipment Corporation that has five phases. Good has described the application of his model to the creation of a portable torque-feedback device. His participatory design team included designers, chemists, and computer engineers. His steps are described here, but again, I have modified the steps to reflect an instructional design emphasis instead of equipment/computer design:

1. *Building relationships.* Create a working team that includes end users (e.g., teachers and students) as well as designers, other stakeholders, and other relevant experts.
2. *Contextual Inquiry.* Through interviews and observations, become familiar with the context in which the instructional material will be used.
3. *Brainstorming.* In sessions where criticism is forbidden, the group generates many ideas about the instructional material to be developed.
4. *Storyboarding and Scenarios.* The team develops some of the most promising brainstorming ideas into illustrated scrips of a "day in the life" of a classroom where the proposed instructional would be in use.
5. *Iterative Design.* Using the storyboards, members of the team create prototypes of the instructional materials. The prototypes are tested and evaluated by teachers and students. Their input is used to iteratively revise the prototype.

Another alternative process for C-ID was proposed by Borenstein (1991) who detailed a framework for software development that is quite similar to Good's approach. In his very readable book, Bornstein divided development into four phases: Definition, prototyping, production, and maintenance.

Salomon (1991) described a somewhat different approach in her case study of the iterative design of an information kiosk. Salomon's iterations were across progressively improved prototypes. She divided the design and development work into three phases:

1. *Initial Design Specification Phase.* In this phase the team created rough paper sketches and screen mock-ups that were critiqued and modified.

2. *Storytelling Prototype Phase.* The most promising designs "were refined and used to 'tell stories' about the system's functionality to others. Feedback helped pinpoint areas where the interface design was potentially unclear or nonintuitive" (p. 27).

3. *Functional Prototype Phase.* Semi-functional prototypes were tested by users and revised as well as improved based on the users' input.

Salomon's paper provides details on this approach to participatory design. Other interesting ways of organizing participatory design work have been described by Karat and Bennett (1991) as well as Baecker, Nastos, Posner, and Mawby (1991).

Most of the models and processes for ID discussed thus far have been general models that can are flexible enough, and general enough, to be adapted to many different contexts. In fact, most of the developers of these models do not expect them to be used as rigid recipes for practice. Instead, the models can be revised, reformed and adapted in many ways. Thus, the work on C-ID models in the last twenty years of the twentieth century tended to focus on general models of C-ID.

The type of scholarly work being done on C-ID models shifted somewhat after the turn of the century, and a number of newer models have been proposed that I would call "custom C-ID models." They are customized in two different ways. Some were developed to emphasize the application of a particular theory or framework that is compatible with general constructivist theory. Others have been developed to help designers create a particular type of instructional resource. One new model that does both is the work of Irlbeck, Kays, Jones, and Sims (2006). These authors, who work in American and Australian universities, argue that ID models based on the principles of emergence theory can "best harness the power and potential of design and development for online distance education" (p. 171).

Emergence Theory and the Three-Phase Design (3PT) Model

Emergence theory, which draws its name from the title of a seminal book by Steven Johnson (2001), *Emergence: The Connected Lives of Ants, Brains, and Software,* is a variation of chaos theory that emphasizes the interaction that occurs in complex systems that does not involve what some have called an "executive function"—that is, a top down directive function that organizes and controls the system. Instead, systems are viewed as "self organizing" from the bottom up. Some equate the lack of top down control as equal to a random chaos in which each element of a system goes in unpredictable and often counterproductive directions. Emergence Theory takes the opposing view: that self-organizing systems do not disintegrate. Instead they will exhibit order, obvious systematic behavior, and interrelated patterns of interaction. The subtitle of Johnson's book mentions ants, brains, and software because he uses them to illustrate how self-organizing systems can operate in the real world: ant colonies, brains, and self-learning software. Another commonly used example is a flight of birds that seem to change direction as a single unit. In all these examples Emergence Theory suggests that what seem to be group decisions, perhaps directed by a "boss" who makes the decision and transmits it to the followers, is not that at all. Instead, it is the result of an emerging system of behavior that comes from the bottom of a system rather than from the top. Emergence Theory has implications for many fields of human interaction and it has become the foundation for new conceptual and practical theories in many areas. In interviews on Sunday talk shows, for example, economists will sometimes use the term "Emergent Economy" and activists trying to support change in a political system will sometimes use the slogan "Think Local, Act Local" to express their view that change can be accomplished without an overall and global plan (e.g., Think Globally, Act Locally).

Irlbeck and her colleagues (2006) used Emergence Theory as the foundation for building an instructional design model that is also customized for the creation of distance learning resources. Their paper is a thoughtful and considered assessment of traditional ID models and their histories as well as a careful application of Emergence Theory to instructional design. For example, they believe a mistake traditional ID models make is to "assume that experts decide objectives, assessment criteria, outcomes, and learning activities" (p. 177). They disagree and use Emergence Theory to suggest that "design should proceed from the ground-up rather than from the top down" (p. 178) because learning environments are emergent systems rather than systems with top down control built in.

Another issue they raise has to do with the structure of the problem. They are critical of many traditional ID models because they assume the instructional design process is addressing a *well structured* problem ("the problem is clear and the solution is clearly specified" (p. 175). However, they disagree and argue that the problem is most often either *ill structured* ("it is not clear what the problem is or what the solution may be" (p. 175)) or *wicked* (an ill-structured problem that is also "compounded by no agreement about what the problem really is" (p. 175). From this redefinition of what ID is trying to do the authors propose a new way of thinking about ID.

> Within online environments, the complexity of interactions may even make the instructional design challenge a wicked problem, thus raising the question as to whether the process and systems thinking associated with many instructional design models are consistent with the ill-defined or wicked-problems that confront the online environment. (pp. 175–176)

The Three-Phase Design (3PD) Model asserts that there are both instructional implications of Emergence Theory ("provide a dynamic, emergent teaching and learning environment in which resources or strategies can be developed or modified during the actual delivery stage" (p. 179) and implications for the process of ID ("The 3PD model reinforces both the team-based approach to design and production of resources and the iterative and emergent processes of development and dynamic learning" (p. 179). It is perhaps the most fully developed ID model based on a chaos theory.

A similar model, the *Iterative Individual Instructional Development Model* (I3DM) was developed by Douglas Kranch (2008) after identifying both the strengths and potential weakness of the 3PD model. This model was created for individual teachers and instructors who are creating online lessons and courses without significant support from teams of experts. The emphasis in this model is on an iterative design and development process.

ID Models Based on Other Theories

Constructivist, Chaos and Emergence Theory are not the only foundations for new ID models. Li MingFen (2008) recently described an ID model for creating adult learning materials that is based on Critical Theory. Critical theories of education tend to focus on empowerment of learners and on a shift in the traditional power structure of the learning environment. As you might expect, MingFen's model proposes developing instruction that is empowering, but one innovation in her approach is the idea that design should be part of the learning environment. She proposes that teachers work with students to design instruction and that this participatory and collaborative way of doing "critical design inquiry" is a powerful way of facilitating the empowerment of adult learners.

ID Models for Other Purposes

An example of an ID model that was created to guide the development of specific types of learning environments is the work of Charlotte Gunawardena and her colleagues (2006) at the University of New Mexico.

The purpose of their WisCom model is to develop "online wisdom communities." This ID model is based on social constructivist and sociocultural philosophies and theories of learning. The paper explores in some detail the types of advanced learning that can occur in wisdom communities as well as the design procedures that facilitate the creation of such communities. The emphasis is on learning that is social and communicative, and on learning goals that are transformative. As the authors put it, "wisdom appears to be an integration of cognition, affect, and reflectivity. Reflective learning is a significant aspect of perspective transformations, and the instructional goal of the WisCom model" (p. 219).

Summary of the Design Process

While I have presented an "R2D2" process in this section I do not intend to propose it as The process that should be used in C-ID. As you have seen from the discussion of the processes used in other ID models, there are many different ways of thinking about and doing design. While I find the process I have proposed appealing, I also find the other ID processes discussed in this section appealing as well. My advice is to carefully evaluate the usefulness of different process models for the work you plan to do and then either use one of the many models of the ID process available, or create your own using what you have learned in your study of ID process and your own ID experience. In either case, you will probably make many process changes as you and your team do design work.

Before leaving this aspect of R2D2, the idea of scenarious should be discussed in a bit more detail. Scenario-based design (Carroll, 2000a) is one way of intimately involving a team in the design of a product. Essentially, the team envisions how the product would be used before it is design and then actually creates the product based on scenarios of use. There are many ways to begin design with scenarios. One involves getting your team to tell stories:

> I almost always begin design by talking with users. Initially, my goal is simply to collect people's stories. I believe that the stories people tell about what they do and how they do it contain information vital to designing good interfaces. Stories reveal what people like about their work, what they hate about it, what works well, what sorts of things are real problems. But although stories can contain a lot of valuable information, I believe that it is the process of collecting stories, rather than the content they contain, that is their most valuable contribution to design. Stories are a natural way of beginning dialog with users. (Erickson, n.d.)

There are also other methods of using scenarios. Most of them involve getting end-users or the design team to envision how a product will be used before it is designed. The visionary scenarios created then become a foundation for actually creating the product. A scenario may begin with a list of questions to ask those who will use the product. That list can be developed in a variety of ways, such as interviewing or observation. Then

> these questions are elaborated into scenarios: descriptions of the tasks people might undertake to address the questions. The set of scenarios is successively elaborated, ultimately incorporating device-specific details and other work contexts articulating exactly how particular tasks will be accomplished. (Caroll, Kellogg, & Rossan, 1991, p. 81)

This method takes the place of traditional design documents and, from the very beginning of the process, situates the work of the design team squarely in the context where the product will be used. Scenarios may be excellent tools for designing instructional materials. John Carroll's (2000a) book, papers (e.g., Carroll, 2000b; Go, Mitsuish, & Higuchi, (2005) and conference proceedings (e.g., Darses, Dieng, Simone, & Zacklad (2004) are all useful sources of information on the use of scenarios in design work.

Cooperative Inquiry

The section activity of Design and Development is cooperative inquiry. This aspect of the R2D2 model has been one of the most difficult to name and explain. The original R2D2 model used the term *formative evaluation* and contrasted it with summative evaluation. The essence of the concept of formative evaluation is important in ID work within the R2D2 framework. That essence involves gathering data that is used to make decisions about how to improve the material you are working on. However, formative evaluation, as it is used in many of the traditional design models, has overtones of objectivist empiricism. That is, many treatments of the concept of formative evaluation assume there is a "Truth" about the right way to do something and formative evaluation is a methodology the designer applies to find that Truth. R2D2 is not based on the assumption that a job of the design team is to find the one True design that is appropriate for a particular project. Rather, R2D2 is based on the assumption that a participatory team comes to some agreement or consensus on what its members consider to be local truth, and they base their work on that socially constructed truth. A different team, or the same team working at a different time or in a different context, might construct a different set of accepted truths and thus a different instructional package.

After considering several awkward terms such as "developing herme-neutic understanding," "participatory evaluation," "contextual inquiry," and "contextual formative evaluation," I selected the term *cooperative inquiry*. John Heron (1996), in his book on cooperative inquiry, defined the term this way:

> Cooperative inquiry involves two or more people researching a topic through their own experience of it, using a series of cycles in which they move between this experience and reflecting together on it. Each person is co-subject in the experience phases and co-researcher in the reflection phases. (p. 1)

Heron's idea of *cooperative inquiry* (sometimes referred to as collabora-tive inquiry) captures precisely what I have in mind in the R2D2 ID model. Members of the team cooperatively research and reflect with an eye toward improving the material being developed. This process is con-tinuous and begins when the idea of creating instructional resources is first discussed. It continues through the entire design and development process. It is not something that is done at particular points in a linear process; it is a continuous, integrated process that often cannot be sepa-rated from, or unraveled from, other aspects of design work. If you would like more information on the basic process of cooperative inquiry an excellent set of links to papers on the topic is available at http://www.human-inquiry.com/doculist.htm. Today John Heron and Peter Rea-son remain the two major proponents and Reason's books on action research (Reason & Bradbury, 2007, 2006) contain excellent treatments of cooperative inquiry. Reason's Web site for the Center for Action Research in Professional Practice at Bath University (http://www.bath.ac.uk/carpp/index.html) contains a number of introductory and advanced papers on cooperative inquiry as well as the full text of all the papers about coopera-tive inquiry in a special issue of the journal, *Systematic Practice and Action Research* (2002).

The concepts of cooperative inquiry have been applied to many differ-ent areas of practice and the exact way the theory and concepts are applied vary from context to context. In the field of instructional design, many of the methods of traditional ID models are well suited to coopera-tive inquiry. The type of data and the way it is collected is often the same, but the way it is interpreted and understood is different. For example, Smith and Ragan (2004) provide detailed guidelines for conducting for-mative evaluations in their book on instructional design. They begin for-mative evaluations with design reviews. Then they progress through a series of different types of formative evaluations, including expert reviews and learner validations. Within each type of evaluation they suggest dif-ferent forms of evaluation.

Smith and Ragan provide very useful suggestions about how inquiry or evaluation can be conducted in instructional design. Their methods will generally fit the R2D2 model. There are, however, some differences. In discussing design reviews, for example, they say: "These reviews may be conducted as each phase of design is completed. They serve to confirm the accuracy of the design process at each stage" (1993, p. 389). In R2D2, there is not so much an emphasis on "accuracy" as there is on helping the team cooperatively develop a vision of what this instructional material will be and why it is being created. The way Smith and Ragan suggest using some of the methods implies a required sequence of steps—one-on-one learner validation comes when you have a rough version of the instructional material for students to work through, for example. In R2D2, the one-on-one evaluation might happen when the team is thinking about developing materials. A student who has already completed the course for which material is being developed might be interviewed, for example, or asked to join the participatory team. In R2D2, these methods would not be used in sequence at particular points in the development process. Instead, they would be used iteratively throughout the process.

A related difference is the size of the iterations. Most ISD models do "big iterations" when they do iterations at all. That is, students and teachers may only look at the instructional material a few times after big changes have been made. In the R2D2 model, there is an emphasis on "small iterations." That is, as small changes are made, the designer is encouraged to get student input on them (as well as expert input). (Or, students and experts are a part of the design team from the beginning and thus have input throughout the design process.) This difference was made more broadly by Dick and Carey (1996):

> Formative evaluation was originally used as a process to improve instruction after the first draft of instruction was developed. Experienced designers, however, found that it was better to try out earlier components of the design process, thereby avoiding a lot of problems that would otherwise not be discovered until after the draft of the instruction was complete. (p. 256)

A final difference between R2D2 and the approach taken in many traditional ID models is that traditional ID models tend to put more emphasis on the importance of performance on objective tests. In the R2D2 model, student assessment in tryouts of the material tend to be more qualitative (e.g., interviews, observations, debriefing, think aloud protocols, teacher evaluations, portfolios, artifacts). The reason for the emphasis on qualitative assessment is that many of the higher level cognitive skills constructivist teaching and learning emphasizes are difficult to assess with traditional objective methods such as multiple-choice tests. That does not mean an objective test would never be used. Sometimes that would be one of the appropriate tools for evaluating the success and worthiness of the material being developed.

In addition to the resources on cooperative inquiry that were mentioned earlier, good sources of detailed guidelines for evaluation that can also be used for cooperative inquiry include Section II of the Smith and Ragan (2004) instructional design text book, Patton's (2002) book on qualitative research and evaluation, Jonassen, Tessmer, & Hannum's (1999) book on task analysis methods, and the 1998 *W. K. Fellog Foundation Program Evaluation* book which is available free on the Internet at http://www.wkkf.org/Pubs/Tools/Evaluation/Pub770.pdf. There are, of course, many other excellent sources on evaluation and assessment in the instructional design process, as well as on qualitative research methods. One caution that should be kept in mind when evaluating sources of information is the fact that much of the material on evaluating instruction emphasizes direct instruction while the R2D2 model may be comfortably used to create a broader range of instructional materials. Another concern is the tendency to emphasize pre-configured forms. The literature is filled, for example, with Likert scale forms you can ask students to complete after they have tried out a version of instructional material. In my opinion, most forms limit the range of responses a student can make so much that they limit the usefulness of student tryouts of material. A much more productive approach is often simply to observe users with the material. Gomoll (1990), who was part of the team that developed the Macintosh user interface, makes this point when she says,

> THE WORD IS OUT: Users should be involved in interface design. But how many people practice what they preach? Until I started observing users, I didn't know the excitement, the value, and the ease of involving users in design. Each time I set up an observation, I find myself discovering something new about the way people think and work. I've become such an advocate that I try to observe users at every stage of the design process, brainstorming, prototyping, building, and evaluating. (p. 85)

Gomoll (1990) goes on to discuss how she involved users in every iteration of the interface design work she and her colleagues were doing at Apple Computer. She concluded that "By observing users early and often we've been able to catch problems in the prototype stage, rather than waiting until just before the product ships" (p. 86).

The reliance on formal questionnaires reflects a broader trend. American social science and education have been dominated by a quantitative approach for almost a hundred years (Willis, 2007).

The challenges to qualitative research are many. Qualitative researchers are called journalists, or soft scientists. Their work is termed unscientific, or only exploratory, or entirely personal and full of bias. It is called criticism and not theory, or it is interpreted politically, as a disguised version of Marxism, or humanism. (Denzin & Lincoln, 1994, p. 4)

In the 15 years since that was written qualitative approaches to understanding our world better have become more acceptable and more widely used. They still, however, occupy a second class status in many settings. During the George W. Bush years in the White House, for example, they were systematically excluded from many types of federal funding. In spite of that, I am optimistic about the future of qualitative methods because I find that qualitative data provides me with more of what I want to know about.

In the case of cooperative inquiry in instructional design, it is often true that qualitative data – observations, interviews, discussions – provide a much richer and more useful source of guidance for revising and improving material. Gomoll's chapter in Brenda Laurel's (1990) book, *The Art of Human-Computer Interface Design,* is a very valuable source of information on how to observe users trying out your material. Gomoll emphasizes the idea that users should be asked to "think aloud" as they work with the material because

by listening to users think and plan, you'll be able to examine their expectations for your product, as well as their intentions and their problem-solving strategies. You'll find that listening to users as they work provides you with an enormous amount of useful information that you can get in no other way. (p. 88)

Observing people using drafts of your material is an important strategy in R2D2 that is not emphasized as much in some other ID models which tend to emphasize objective methods of formative evaluation and to de-emphasize or ignore qualitative ways of knowing, such as observing users, evaluating portfolios and artifacts from the learning process, reading journals kept by users, debriefing users, focus groups, expert/connoisseur studies, and the like. Fortunately, there is a growing body of literature on the use of qualitative methods of understanding our social and educational world better. Much of this material comes from fields other than instructional design, but it is as applicable to ID as it is software design, industrial design, and social science as well as educational research. One very good source of guidance on the use of ethnographic or observational methods in design is a paper by Werner Sperschneider and Kirsten Bagger (2003) at the User-Centered Design Group of the University of Southern Denmark. After introducing the use of ethnographic methods in participatory design methods, the authors present case studies of five innovative approaches to integrating ethnography into design projects. While the examples and the discussion focus on industrial design, there is much that can be transferred to instructional design. Another paper on the use of ethnographic field methods in design was published as a book chapter (Blomberg, Giacomi, Mosher, & Swenton-Wall, 1993. These authors, who were at Xerox Corporation, presented a theoretical framework for conducting qualitative research as part of the design process. "The ethnographic approach, with its emphasis on 'natives point of view,' holism, and natural settings, provides a unique perspective to bring to bear on understanding users' work activities" (p. 123). And,

> as practiced by most ethnographers, developing an understanding of human behavior requires a period of field work where the ethnographer becomes immersed in the activities of the people studied. Typically, field work involves some combination of observation, informal interviewing, and participation in the ongoing events of the community. (p. 124)

Blomberg and her colleagues provide some detailed suggestions for observing, interviewing, and analyzing ethnographic data gathered while engaged in design work. They then compare ethnographic approaches to traditional approaches to gathering data in design work and point out some of the significant differences. For example, in traditional approaches,

> users are not collaborators in the technology development. Traditional approaches provide little room for collaboration between designers and users over the evolving design, but instead rely on the users' ability to verbalize their needs or to expose inadequacies of the design in isolated tests which take place on single occasions. (p. 146)

The authors view the use of ethnographic methods as one way to develop a mutual understanding that will lead to product of higher quality.

Blomberg and her colleagues believe, however, that using ethnographic approaches in software design projects is not easy because they call for skills the designer may not have, and they require a reformulation of the design process. "Designers must develop skills in interviewing, observation, analysis, and interpretation, while development teams must be willing to shift their emphasis to support early and continued user involvement" (p. 147). These are real issues that must be addressed if participatory approaches to design are to succeed.

Observation is one major tool of the ethnographic researcher. Another crucial tool that is also helpful in design work is interviewing. In my own experience more open and flexible interviewing has been more helpful in design work than formal, structured interviewing methods. For information on open interviewing methods see Willis (2007), Hoffman (2007), Bossen (2002), and Knapp (1997). Holtzblatt and Jones (1995, 1993) have described in some detail the way to conduct what they refer to as a *contextual interview* because it is conducted in context. "The best way to understand the work is to talk to people in their actual work environment. Design information is present in its richest form when we speak with people during ongoing work or using work artifacts" (1993, p. 241). A contextual interview is a partnership:

> To design effective systems, we need to understand users' experience of work and systems. The information is invisible, we cannot access it by standing on the outside of the process, watching people's behavior and writing down what happens. We need to talk with users to understand their experience. To have an effective dialogue, we form partnerships with our users. The principle of partnership recognizes that only through dialogue can designers become aware of users' experience of work and tool use. Together designs and users create a shared understanding of work practice that reveals technological opportunities and problems that occur in work process and in system use. (p. 241)

That a contextual interview is not simply a way for the expert to mine information from users is illustrated by the four principles of partnership:

1. The user is the expert.
2. Share control during the inquiry.
3. Help create shared meaning.
4. Reflection and engagement are keys.

While many types of interviews can be used in collaborative inquiry, the contextual interview of Holtzblatt and Jones fits very well with the basic principles of the R2D2 model. Their papers provide an example of how contextual interviews are done, as well as detailed guidelines and suggestions. Another useful resources is a paper by Yuxin Ma (2006) that describes the use of contextual interviewing in a project to create and deploy an online library of teaching cases.

Product Design and Development

The last element of the Design and Development focus is, predictably, design and development. In the original paper on the R2D2 model

(Willis, 1995), the example used to illustrate how the model could be applied was a piece of learner-centered simulation software for adults who were learning to read. The organizing framework was a simple and well-known categorical scheme for thinking about designing the various components of the software:

1. *Surface Design*—screen layout, typography, language, graphics, illustrations, sound.
2. *Interface Design*—look and feel, user interaction, help support, navigation, metaphors.
3. *Scenario*—sequence of simulation options/choices, results.

This way of slicing up the product is not suited to the design of all types of instructional materials. (It is probably not even be the best way of thinking abut the design of all student-centered simulations.) Since 1994 the R2D2 model has been used to develop a wide range of educational materials, from videos to electronic books to graduate programs to multimedia training programs to web sites. We did not usually divide up work on the material according to list above, however. The particular way you think about the product will vary from project to project. The simple three part framework of surface, interface, and scenario may be satisfactory for some work, but other projects will call for very different ways of thinking.

There is, however, a more serious issue involved in the way we think about a product. It relates to how broadly we define the term *instructional material*. Many of the ID models in current use restrict that term to a few types of instructional materials that deliver or support what is generally termed *direct instruction*, or *teacher-centered instruction*. Tutorials, drills, and some types of simulations are well known forms of direct instruction. In my opinion, if we limit the use of ID to the creation of forms of direct instruction, we make a serious mistake. Direct instruction will probably play a much smaller role in the teaching and learning of this century than it did the twentieth century. That perspective has been expressed by many scholars including Charles Reigeluth (1997) who urged us to consider a very wide range of pedagogies when doing ID and to adopt broad theories of learning:

> not just in the cognitive domain, where we need theories for fostering understanding, building higher-order thinking skills, developing metacognitive skills, designing problem-based and interdisciplinary or thematic learning environments, and tailoring instructional guidance to specific content-area idiosyncrasies, but also in the affective domain, where we need guidance for developing what Daniel Goldman calls "emotional intelligence" and for what Thomas Lickona calls "character education," as well as

how to develop attitudes and values and so forth. Instructional theory has been construed much too narrowly in the past. (p. 1)

Reigeluth (1996) outlines a wide range of instructional approaches that include everything from drills and games to debates, cases, and role playing. I believe Reigeluth is right; those of us in the field have tended to restrict our thinking to a narrow range of instructional strategies. As we work on instructional design, one of the responsibilities of the designer may be to both learn about innovative pedagogies the other team members know and use, and to encourage to the team to consider a wide range of possibilities. This includes approaches mentioned by Reigeluth, such as problem-based learning as well as Perkins' (1992) *facets of learning environments:*

- Learning banks
- Symbol pads
- Construction kits
- Phenomenaria
- Task managers

All these facets are appropriate targets for ID.

There are, in fact, six families of pedagogies in use today in American schools. They are based on six families of learning or social theories:

1. Behaviorism (e.g. direct instruction
2. cognitivism (e.g., cognitive science),
3. cognitive constructivism and student-centered learning,
4. social constructivism,
5. humanism and humanistic education, and
6. critical theory or emancipatory learning or radical pedagogy

Instructional designers have tended to approach these five theories of learning, and the pedagogical families associated with them, as if they were fundamentalist religions that demand total devotion to their practices if the designer is to be considered a pure and ethical person. I find this approach unsatisfactory at the level of pedagogy. In my own teaching and learning I do use forms of direct instruction such as tutorials, lectures, just-in-time teaching and so on. But, I also use pedagogies based on cognitive science theories including memory enhancement strategies such as mnemonics, advance organizers, and data mining. From cognitive constructivism I use problem-based learning pedagogies and from social constructivism I use many forms of collaborative and cooperative learning pedagogies. From humanistic education I often use narrative, storytelling, and fiction to help students develop a holistic understanding of significant issues they will face as professionals and citizens. Finally, I find many of the emancipatory pedagogies of critical theory very useful in a range of learning environments.

Thus, while I work primarily from a social constructivist perspective I use pedagogies developed and nurtured within all six of the influential frameworks active in American education today. I believe it is important that the wide range of potential pedagogies be considered when doing ID, which means that members of the design team must either bring an understanding of those pedagogies to the ID project or be willing to learn about those that seem potentially useful as a part of the design process.

Expanding beyond direct instruction to include the creation of student-centered learning environments, for example, is certainly a step in the right direction. There is also something else that should also be considered as well. Rosa (1997) has pointed out that even the design of student-centered learning environments involves making many decisions that will, of necessity, limit some of the options available to students. An alternative would be to design electronic tools, information resources, and online worlds that students can use to create their own learning environments. The recent emergence of "social networking websites" such as Facebook, MySpace, and BuddyPress is one example of online communities that may have significant educational potential. Even more exciting from a learning environment perspective are online communities, synthetic worlds, and virtual worlds such as the wildly popular *SecondLife*. These online systems offer participants an opportunity to create their own identities and build their own vortia; worlds in an online environment where others are doing the same thing. Linden Labs, the owner and operator of SecondLife, even has a program *Campus: Second Life* to support teachers create and offer courses through environments on SecondLife. SecondLife maintains a Wiki on the educational uses of the online worldat http://www.simteach.com/wiki/index.php?title=Second_Life_Education_Wiki. As you consider what ID model you want to use one question to ask is "Would this model help my team develop innovative learning environments such as a SecondLife community?"

DISSEMINATION FOCUS

Traditional ID models generally include four activities in their dissemination stage: summative evaluation, final packaging, diffusion, and adoption. With the exception of summative evaluation, the R2D2 model adopts a similar approach. However, the constructivist emphasis on the importance of context suggests that diffusion and adoption should not emphasize using mateusprinussrial in the "right" way. Instead, the focus should be on helping teachers, instructors, and learners adapt the resource your team has created to the local context and use it in ways that are appropriate to that context. It may also involve some learning on your part - learning about the innovative and unexpected ways teachers use the material your team created that should be shared with other educators.

A second difference in the work associated with the Dissemination Focus has to do with the way material is distributed and shared. Traditional ID models often assume the best way to distribute new resources is via commercialization. While commercialization is always an option, it tends to be limited to low cost, high profit, high demand items such as basal reading programs for elementary schools or textbooks for high volume college courses such as *Introduction to Western Civilization* or *English Composition*. Over the past 25 years, commercial publishers have tried many ways to sell multimedia instructional resources at a profit. The result has generally been disappointing, both in terms of the number sold and the quality of the product – which is sometimes limited by the cost of disseminating complex, multimedia resources. Another option is Open Source or Open Access dissemination. Microsoft's Windows operating system is created and sold for a profit, but the many versions of the Linux operating system are distributed through a system of online sites that allow you to download and use the software for free (and through magazines that insert a DVD in each issue that contains the newest versions of different versions of Linux). There are also many types of Open Source software such as OpenOffice that competes with Microsoft Office.

Few groups make huge profits on the commercialization of instructional materials that are distributed commercially by for-profit publishers, and Open Source is an alternative that allows anyone interested in using the material to acquire it, adapt it, and put it to use. In fact, several types of educational software is now both revised and maintained by a worldwide community of volunteers and distributed to anyone who wants to use it without charge. A good example is Moodle, the Open Source course management system designed to support constructivist approaches to teaching and learning. This program began life as the dissertation project of Australian Martin Dougaimas. The main Web site, http://www.moodle.com, has a great deal of information about Moodle including many resources created and developed by teams from many countries. Another Open Source program that began as a dissertation is the Balch Internet Research and Analysis Toolkit, which is a system for conducting surveys over the internet. It was created by Charles Balch and is available at http://birat.net/br/.

Instructional designers may want to consider the open source method of distributing instructional materials. One of the most successful multimedia resources that is available online and without charge is *InTime* (Integrating New Technologies into the Methods of Education), a set of videocases developed at the University of Northern Iowa in collaboration with several other colleges and universities. The Web site for Insite is www.intime.uni.edu. For more information about Open Source and education the following resources are useful:

- *A Vision for free, global (online), education.* A YouTube video talk by Richard Baraniuk at Rice Univesity. He introduces TED (www.ted.com), an Open Source educational resoruce. Available at http://www.youtube.com/watch?v=RRymi-lFHpE&feature=related
- *How the Open Source Movement Has Changed Education.* Published in 2007, this paper is a good introduction and overview. Available: http://oedb.org/library/features/how-the-open-source-movement-has-changed-education-10-success-stories
- *MERLOT – Multimedia Educational Resources for Learning and Online Teaching* – is one of the most active and well established groups supporting open source dissemination of materials. http://www.merlot.com
- *www.Teachertube.com* is a web site where educational videos can be uploaded and shared with other educators without charge.
- The *Open Educational Resources* site is at http://www.oercommons.org/
- *Open Courseware* is a movement among higher education institutions to share courses with anyone. The address of the Open Courseware Web site at MIT is http://ocw.mit.edu/OcwWeb/web/home/home/index.htm
- www.schoolforge.net is a site for organizing participatory teams to create Open Source educational materials and to distribute them
- *Dg communities* is a collaboration for the creation and dissemination of Open Source educational materials. There are many links to other projects on the home page: http://openeducation.developmentgateway.org/
- Freeman Murray discusses one Open Source education dissemination project in a video on YouTube. This is part 1 of 3 parts. Available at http://www.youtube.com/watch?v=7Reb32nYMHQ

The other aspect of the Dissemination Focus, summative evaluation, also requires some additional comment. In many ID models, summative evaluation generally involves conducting a systematic study that entails collecting objective data on the effectiveness of the program. The typical purpose is to "prove" that the instruction developed actually does what it is supposed to do. Constructivists tend to question all efforts to *prove* something in a general and universal way. If context is important in education, then the success of instructional materials, pedagogies, and curricula will likely vary from one context to another. Thus, trying to prove in some universal way that an instructional package works is a waste of time. So too is any attempt to generalize the results to other, similar materials, and conclude that they "work" as well. Effectiveness, success, and similar concepts are better considered as local rather than universal characteristics of instruction. In the R2D2 model, there are too many local, context-based variables involved in the use of any instructional package to make valid generalizations to other settings. A summative evaluation in the R2D2 model is much less ambitious. It is, instead, the story of what happens when the material is used in a particular context in a particular way with a particular group of instructors and learners. Such summative evaluations can be very helpful to others who are considering adapting and using the material, but potential adopters must consider many variables before they decide that findings from a summative evaluation may also apply to their setting. And, because the context is so important, summative evaluations in the R2D2 model should collect rich, thick data on the way the package was used as well as a range of measures of how well the material worked. Those measures may include quantitative assessments such as test scores as well as many forms of qualitative assessment.

Another difference between summative evaluations in traditional ID and C-ID relates to the way "success" is measured. Traditional summative models tend to rely on universal, objective measures in part because a specific set of objectives guided the development of the instructional package. Thus, it makes sense to try and determine whether all the students achieved those objectives. However, constructivist approaches tend to encourage individual goal setting by students and to encourage diverse learning activities. Therefore, objective tests are not always an appropriate way of evaluating the success of instruction, since different students learn different things in different ways. Alternative methods of assessment, such as journals, projects, artifacts, portfolios, and activity logs, may be more appropriate.

SUMARY AND CONCLUSIONS

R2D2 is one of several ID models based on constructivist theory. It has been used primarily in academic and research contexts, but other, similar, models have been used widely in industrial design and software engineering for many years. In fact, R2D2 may seem most radical when it is compared to other ID models in the linear tradition. In other fields of design there are innumerable papers on iterative (recursive), user-centered, non-linear, and participatory design approaches.

Other fields, such as organizational change and technology diffusion have also adopted non-linear, participatory approaches. Fullan's (1991, 2007) models of change, for example, come to mind. For example, he argues that there must be buy-in and participation by teachers in any change effort, and he is empathic that the "vision" of a change effort must emerge across the process of change rather than be given or handed to participants at the beginning. Similarly, emerging concepts in education such as site-based management, teacher leaders, participatory learning, and participatory action research are all reflections of the same basic theoretical foundations used by C-ID.

As a field, ID have been slow to adopt approaches that are already well established in other design fields. ID seems somewhat isolated from trends and developments in related fields. However, there are examples of innovations that have crossed the moat and established themselves in the castle of ID. Rapid prototyping (Jones & Richey, 2000; Nixon & Lee, 2001; Tripp & Bichelmeyer, 1990) was nurtured in the field of design engineering, but it has been adopted by a number of instructional designers. Rapid prototyping is but one of many examples of innovations that have been developed and applied in other design fields. There is a rich body of literature we can mine for alternative and innovative ways of thinking about design. Over the next 15 years I expect the acceptance and use of C-ID models to accelerate and the scholarship of C-ID to expand and deepen.

That brings me to a final point about the selection of ID models: It is that the ID model we decide to use is based on our beliefs, our experiences, and our perspectives about what design is. They are not, however, selected on the basis of superior research or "the right theoretical foundations." There is no research that conclusively (or even inconclusively) demonstrates that one family of ID models is the better choice. Design work is probably so complex and so context-dependent that it would be impossible to empirically test the usability of one design model versus others. We must for the most part, fall back on Winn's (1990) concept of *first principles* to guide us. Selecting an ID model is a rational, not an empirical process. Your worldview, your preferred paradigm, your view of your profession and the role of design in it, are all factors that influence your choice of ID model. Those are, in turn, influenced by your experiences, your reading, your discussions, and your reflections on them. In other words, you construct your choice of an ID model within communities of scholars and practitioners you interact with. ID is thus a field with an abundance of theoretical and conceptual frameworks within which to make decisions. I believe the situation calls for considerable dialog and sharing both within and among proponents of different ID models and frameworks. When it is all said and done, none of us are very good at this, and we will more likely become better through broad sharing and exchange of ideas than by working in isolated enclaves surrounded by theories that serve as moats to keep out "foreign" ideas and practices.

REFERENCES

Baecker, K., Nastos, D., Posner, I., & Mawby, K. (1995). The user-centered interactive design of collaborative writing software. In R. Baecker, J. Grudin, W. Buxton, & S. Greenberg (Eds), *Readings in human-comuter interaction: Toward the year 2000* (pp. 775–782). San Francisco: Morgan Kaufmann.

Bannon, L. (1995). From human factors to human actors: The role of psychology and human-computer interaction studies in system design. In R. Baecker, J. Grudin, W. Buxton, & S. Greenberg (Eds.), *Readings in human-comuter interaction: Toward the year 2000* (pp. 205–214). San Francisco: Morgan Kaufmann.

Blomberg, J., Giacomi, J., Mosher, A., & Swenton-Wall, P. (1993). Ethnographic field methods and their relationship to design. In D. Schuler & A. Namoika (Eds), *Participatory design: Principles and practices* (pp. 123–156). Hillsdale, NJ: Erlbaum.

Bodker, S. Gronbaek, J., & Kyng, M. (1995). Cooperative design: Techniques and experiences from the Scandinavian scene. In R. Baecker, J. Grudin, W. Buxton, & S. Greenberg (Eds.), *Readings in human-comuter interaction: Toward the year 2000*. San Francisco: Morgan Kaufmann.

Boehm, B. (1995). A spiral model of software development and enhancement. In R. Baecker, J. Grudin, W. Buxton, & S. Greenberg (Eds), *Readings in human-comuter interaction: Toward the year 2000* (pp. 281–292). San Francisco: Morgan Kaufmann.

Borenstein, N. (1991). *Programming as if people mattered.* Princeton, NJ: Princeton University Press.

Braden, R. (March-April, 1996). The case for linear instructional design and development: A commentary on models, challenges and myths. *Educational Technology, 36*(2), 5–23.

Bossen, C. (2002). *Ethnography in design: Tool-kit or analytic science?* Retrieved from http://www.nwow.alexandra.dk/publikationer/ClausBossen.pdf

Cahill, C. (2007). Including excluded perspectives in participatory action research. *Design Studies, 28,* 323–340.

Carroll, J. (2000a). *Making use: Scenario-based design of human-computer interactions.* Boston: MITA Press.

Carroll, J. (2000b, September). Five reasons for scenario-based design. *Interacting with computers, 13*(1), 43–60.

Carroll, J., Kellogg, W., & Rossan, M. (1991). The task-artifact cycle. In J. Carroll (Ed), *Designing interaction: Psychology at the human-computer interface* (pp. 74–102). Cambridge, England: Cambridge University Press.

Cennamo, K. (2003). Design as knowledge construction: Constructing knowledge of design. *Computers in the Schools, 20*(4), 13–35.

Cennamo, K., Abell, S., & Chung, M. (1996). A "layers of negotiation" model for designing constructivist learning materials. *Educational Technology, 36*(4), 39–48.

Darses, F., Dieng, R., Simone, C., & Zacklad, M. (2004). *Cooperative systems design: Scenario-based design of collaborative systems.* Amsterdam: IOS Press.

Denzin, N., & Lincoln, Y. (1994). Introduction: Entering the field of qualitative research. In N. Denzin & Y. Lincoln, (Eds.), *Handbook of qualitative research.* Thousand Oaks, CA: SAGE.

Dick, W. (1996). The Dick and Carey model: Will it survive the decade? *Educational Technology Research and Development, 44*(3), 55–63.

Dick, W., & Carey, L. (1996). *The systematic design of instruction* (4th ed.). New York: HarperCollins.

Erickson, T. (n.d.). *Design as storytelling.* Retrieved from http://www.pliant.org /personal/Tom_Erickson/Storytelling.html

Fullan, M. (2007). *The new meaning of educational change* (4th ed.). New York: Teachers College Press.

Gibbs, P. (2007, November). Practical wisdom and the workplace researcher. *London Review of Education, 5*(3), 223–235.

Gomoll, K. (1990). Some techniques for observing users. In B. Laurel (Ed.), *The art of human-computer interface design* (pp. 85–90). Reading, MA: Addison-Wesley.

Good, M. (1995). Participatory design of a portable torque-feedback device. In R. Baecker, J. Grudin, W. Buxton, & S. Greenberg (Eds.), *Readings in human-comuter interaction: Toward the year 2000* (pp. 225–232). San Francisco: Morgan Kaufmann.

Gould, J. (1988). How to design usable systems. In M. Helander (Ed.), *Handbook of human-computer interaction* (pp. 757–790). Amsterdam: North-Holland.

Greenbaum, J., & Kyng, M. (Eds.). (1991). *Design at work: Cooperative design of computer systems.* Hillsdale, NJ: Erlbaum.

Greenbaum, J., & Madsen, K. (1993). Small changes: Starting a participatory design process by giving participants a voice. In D. Schuler & A. Namoika (Eds.), *Participatory design: Principles and practices* (pp. 289–298). Hillsdale, NJ: Erlbaum.

Gunawardena, C., Ortegano-Layne, L., Carabajal, K., Frechette, C., Lindemann, K., & Jennings, B. (2006). New model, new strategies: Instructional design for building online wisdom communities. *Distance Education, 27*(2), 217–232.

Heron, J. (1996). Cooperative inquiry: *Research into the human condition.* Thousand Oaks, CA: SAGE.

Hoffman, E. (June, 2007). Open ended interviews, power, and emotional labor. *Journal of Contemporary Ethnography, 36*(3), 318–346.

Holtzblatt, K., & Jones, S. (1993). Contextual inquiry: A participatory technique for systems design. In D. Schuler & A. Namoika (Eds.), *Participatory design: Principles and practices* (pp. 177–210). Hillsdale, NJ: Erlbaum.

Holtzblatt, K., & Jones, S. (1995). Conducting and analyzing a contextual interview. In R. Baecker, J. Grundin, W. Buxton, & S. Greenberg (Eds.), *Readings in human-computer interaction: Toward the year 2000* (pp. 240–253). San Francisco: Morgan Kaufmann.

Hunter, A., & Ellis, A. (2000). *The development process for courseware material: A computing methodology approach.* Paper presented at the ASCILITE 2000 Conference. Retrieved from http://www.ascilite.org.au/conferences/coffs00/papers/andrew_hunter.pdf

Irlbeck, S., Kays, E., Jones, D., & Sims, R. (2006, August). The phoenix rising: Emergent models of instructional design. *Distance Education, 27*(2), 171–185.

Iskold, A. (October 16, 2007). The future of software development. *ReadWriteWeb.* Retrieved from http://www.readwriteweb.com/archives/the_future_of_software_development.php

Israel, B., Kreiger, J., Vlahov, D., Ciske, S., Foley, M., Fortin, P., Guzman, J., Lichtenstein, R., McGranaghan, R., Palermo, A., & Tang, G. (2006, November). Challenges and facilitating factors in sustaining community-based participatory research partnerships: Lessons learned from the Detroit, New York City, and Seattle urban research centers. *Journal of Urban Health, 81*(6), 1022–1044.

Jarche, H. (2007, October 17). *Instructional design needs more agility.* Retrieved from http://www.jarche.com/2007/10/instructional-design-needs-more-agility/

Johnson, S. (2001). *Emergence: The connected lives of ants, brains, and software.* New York: Simon & Schuster.

Jonassen, D., Tessmer, M., & Hannum, W. (1999). *Task analysis methods for instructional design.* Hillsdale, NJ: Erlbaum.

Jones, T,. & Richey, R. (June, 2000). Rapid prototyping in action: A developmental study. *Educational Technology Research and Development, 48*(2), 63–80.

Kaner, S. (2007). *Facilitator's guide to participatory decision making.* San Francisco: Jossey-Bass.

Karat, J., & Bennett, J. (1995). Working within the design process: Supporting effective and efficient design. In R. Baecker, J. Grundin, W. Buxton, & S. Greenberg (Eds.), *Readings in human-computer interaction: Toward the year 2000* (pp. 269–285). San Francisco: Morgan Kaufmann.

Laurel, B. (Ed.). (1990). *The art of human-computer interface design.* Reading, MA: Addison-Wesley.

Kindon, S. (2008). *Participatory action research methods: Connecting people, participation, and place.* New York: Routledge.

Knapp, N. (1997). Interviewing Joshua: On the importance of leaving room for serendipity. *Qualitative Inquiry, 3*(3), 326–342.

Kranch, D. (2008). Getting it right gradually: An iterative method for online instructional development. *The Quarterly Review of Distance Education, 9*(1), 29–34.

Leffingwell, D. (2007). *Scaling software agility: Best practices for larger enterprises.* Boston: Addison-Wesley/Pearson.

Luck, R. (2007). Learning to talk to users in participatory design situations. *Design Studies, 28,* 217–242.

Ma, Y. (2006). Contextual Interview as a Data Gathering Method to Examine User Perceptions of the Conceptual Design of a Faculty Development Resource. In E. Pearson & P. Bohman (Eds.), *Proceedings of World Conference on Educational*

Multimedia, Hypermedia and Telecommunications 2006 (pp. 602–605). Chesapeake, VA: AACE.

Marschall, S. (1998). Architecture as empowerment: The participatory approach in contemporary architecture in South Africa. *Transformation, 35,* 103–123.

MingFen, L. (2008, October), Transforming adult learning through critical design inquiry. *Systemic Practice & Action Research, 21*(5), 339–358.

Nixon, E., & Lee, D. (2001). Rapid prototyping in the instructional design process. *Performance Improvement Quarterly, 14*(3), 95–116.

Oancea, A., & Furlong, J. (2007, June). Expressions of excellence and the assessment of applied and practice-based research. *Research Papers in Education, 22*(2), 119–137.

Patton, M. (2002). *Qualitative research and evaluation methods.* Thousand Oaks, CA: SAGE.

Perkins, D. (1992). Technology meets constructivism: Do they make a marriage? In T. Duffy & D. Jonassen (Eds.), *Constructivism and the technology of instruction: A conversation* (pp. 45–56). Hillsdale, NJ: Erbaum.

Pilemalm, S., & Timpka, T. (2007). Third generation participatory design in health informatics – Making user participation applicable to large-scale information system projects. *Journal of Biomedical Informatics, 41,* 327–339.

Reason, P., & Bradbury, H. (2007). *The Sage handbook of action research: Participative inquiry and practice.* Thousand Oaks, CA: SAGE.

Reason, P. & Bradbury, H. (2006). *Handbook of action research, brief edition.* Thousand Oaks, CA: SAGE.

Reigeluth, C. (1996). A new paradigm of ISD? *Educational Technology, 36*(3), 13–20.

Reigeluth, C. (1997). *What is the new paradigm of instructional theory? IT Forum Paper #17.* Retrieved from http://itech1.coe.uga.edu/ITFORUM/papger17.html

Rosa, K. (1997). Personal communication.

Riel, M. (2007) Understanding action research. *Center For Collaborative Action Research. Pepperdine Univerity.* Retrieved from September 2, 2008, from http://cadres.pepperdine.edu/ccar/define.htmlg

Roodt, M. J. (2002). "Participatory development": A jargon concept?. In J. Coetzee & J. Graaff (Eds.), *Reconstruction, development and people.* Johannesburg: Oxford University Press Southern Africa.

Salomon, G. (1995). A case study in interface design. In R. Baecker, J. Grundin, W. Buxton & S. Greenberg (Eds), *Readings in human-computer interaction: Toward the year 2000* (pp. 223–233). San Francisco: Morgan Kaufmann.

Sanginga, P., Kamugisha, R., Martin, A., Kakuru, A., & Stroud, A. (2004). Facilitating participatory process for policy change in natural resource management: Lessons from the highlands of southwestern Uganda. *Uganda Journal of Agricultural Sciences, 9,* 958–970.

Sanoff, H. (2007). Special issue on participatory design. *Design Science, 28,* 213–215.

Schuler, D., & Namoika, A. (1993). *Participatory design: Principles and practices.* Hillsdale, NJ: Erlbaum.

Seels, B., & Glasgow, Z. (1998). Making instructional design decisions (2nd ed.). Columbus, OH: Merrill.

Smith, P., & Regan, T. (2004). *Instructional design* (3rd ed.). Columbus, OH: Merrill.

Smith, P., & Regan, T. (1993). *Instructional design*. Columbus, OH: Merrill.

Sperschneider, W., & Bagger, K. (2003). Ethnographic fieldwork under industrial constraints: Toward design in context. *International Journal of Human-Computer Interaction, 15*(1), 41–50.

Thoresen, K. (1993). Principles in practice: Two cases of situated participatory design. In D. Schuler & A. Namoika (Eds.), *Participatory design: Principles and practices* (pp. 271–287). Hillsdale, NJ: Erlbaum.

Tripp, S., & Bichelmeyer, B. (1990). Rapid prototyping: An alternative instructional design strategy. *Educaitonal Technology Research and Development, 38*(1), 31–44.

White, S. (2000). *The art of facilitating participation*. Thousand Oaks, CA: SAGE.

Willis, J. (1995). A recursive, reflective instructional design model based on constructivist-interpretivist theory. *Educational Technology, 35*(6), 5–23.

Willis, J. (2007). *Foundations of qualitative research*. Thousand Oaks, CA: SAGE.

Winn, W. (1990). Toward a rational and theoretical basis for educational technology. *Educational Technology Research and Development, 37*(1), 35–46.

You, Y. (1994). What can we learn from chaos theory? An alternative approach to instructional design. *Educational Technology Research and Development, 41*(3), 17–32.

CHAPTER 15

DESIGN AS KNOWLEDGE CONSTRUCTION

Constructing Knowledge of Design

Katherine Cennamo

In this chapter, I present a model of instructional design that has evolved from analysis and reflection on the process of designing materials for constructivist teaming environments. I observed how we addressed the critical questions for instructional design, comparing the process to traditional instructional design models and to my emerging ideas of instructional design involving Layers of Negotiation. Observations of the design and development effort confirmed tentative assumptions that instructional design is at process of knowledge construction, involving reflection, examining information at multiple times for multiple purposes, and social negotiations of shared meanings. The design process evolved in a spiral fashion, progressing through stages of analysis, design and development, and evaluation. Through this project, the Layers Of Negotiation Model was confirmed, yet evolved to (a) identify critical areas around which negotiations should occur and (b) illuminate the nature of the recursive process

Previously published in *Computers in the Schools* (2003), *20*(4), 13–35.

Constructivist Instructional Design (C-ID): Foundations, Models, and Examples, pp. 357–378

Design is part of many jobs—architctural design, interior design, graphic design, instructional design, and so forth. Design implies a careful planning prior to development. As a planning process, instructional design is a process of sense-making, of problem-solving, of constructing knowledge that will inform the development of instruction (Rowland, 1993; Willis, 1995).

In this chapter, I present a model of instructional design that has evolved from analysis and reflection on the process of designing materials for constructivist learning environments. I believe in the power of anecdotes to generate discussions and ideas about substantive issues (Jaworski, 1989) so throughout the following pages, I offer my story as at means of illustrating my evolving ideas about the nature and process of designing instruction. I offer my experience and insights on the process of design as simply one experience in designing instruction for constructivist learning environments.

LAYERS OF NEGOTIATION, AN INITIAL MODEL

In a previous project, I worked with a science educator to develop a series of case-based interactive videodiscs that provided a context for pre-service elementary science teachers to engage in reflective thinking (Cennamo, Abell, George, & Chung, 1996). As the instructional designer on that project, I explored the application of instructional design to constructivist learning environments (Cennamo, Abell, & Chung, 1996). As constructivism implies that it is inappropriate to set learning objectives for the students, we assumed that we could not simply apply a traditional instructional design model to develop these materials.

We approached the design process with the belief that instructional design should be grounded in what we know about the process of knowledge construction. Just as learners come to an instructional experience with understandings, beliefs, and values that have been shaped by their prior experiences, members of the instructional design team also begin the process of designing instruction with their individual sets of understandings, values, and beliefs. Therefore, we attempted to design the materials guided by our assumptions about teaching and learning, and once the materials had been developed, to reflect on the process and compare our procedures with those prescribed by traditional models.

ASSUMPTIONS ABOUT TEACHING AND LEARNING

The design and development of our interactive video eases were guided by the assumptions that:

- Learners bring their individual set of beliefs, values, and experiences to an instructional event. Each individual's experience in an instructional setting is filtered by a unique combination of beliefs, values, and prior knowledge (Osborne & Freyberg, 1985). Thus, the meaning one learner derives from an instructional experience may be different from the meaning another learner derives from the same experience.

- Learning occurs through assimilating new ideas into existing schemes and adjusting existing schemes to accommodate new information. New ideas must be consistent with the learner's prior experience in order to be acceptable to the learner; modified ideas must account for all new and previous data. Learners must become dissatisfied with their existing knowledge and beliefs in order for assimilation and accommodation to occur, thus, disequilibrium is essential to knowledge acquisition (Posner, Strike, Hewson, & Gertzog, 1982).

- Learners make sense of the world through social interactions, as well as through action on objects. When learners compare their ideas with the ideas of others, they may become dissatisfied with their existing knowledge and find new ideas that are useful, plausible alternatives. Thus, social interactions with experts and peers are an important source of cognitive development (Solomon, 1989).

Design Process

These assumptions had implications for the learning materials and, we believed, for the instructional design process. As we transformed recommendations for constructivist learning environments (Driscoll, 1993, p. 360) to the process of designing materials from a constructivist perspective, we attempted to:

- Embrace the complexity of the design process.
- Provide for social negotiations as an integral part of designing the materials.
- Examine information relevant to the design of the instruction at multiple times from multiple perspectives.
- Nurture reflexivity in the design process.
- Emphasize participatory design.

As we embraced the complexity of the design process and emphasized participatory design, social negotiations became an integral part of the instructional design process. The process of designing instruction included negotiating a set of shared beliefs that guided the development of the materials. As we negotiated these shared perspectives, all parties were required to become reflexive and articulate their thought processes. Through this process, ideas and data were examined from multiple perspectives. Examining instructional decisions from the multiple perspectives and "cultural knowledge" of the individuals on the design team enhanced the possibilities that emerged from the design process. Solutions gradually emerged though a recursive, reflective process.

As we reflected on the design process, we found that it was similar to, yet different from, the procedures depicted in traditional models.

- Although we proceeded through stages of analysis, design, development, and evaluation, we did not proceed in a linear fashion where the output of one stage served as the input for the next stage. Instead, we proceeded in an iterative, spiral fashion, proceeding through analysis, design, development, and evaluation as we developed each component of the instruction. For example, we proceeded through analysis, design, development, and evaluation activities as we prepared our instructional materials, then proceeded through analysis, design, development, and evaluation activities as we developed our assessments.

- Although we did not proceed through the steps of a traditional model in a linear fashion, we answered most of the questions posed by traditional models. For example, we considered the characteristics of our learners as we designed the interface. We identified a set of possible learning outcomes after we investigated how students reacted to the instructional materials.

Throughout the design process, decisions were made based on the data that were available and relevant. As additional data became apparent or relevant, we spiraled back and revisited prior decisions. We did not revisit decisions in order to stall progress; on the contrary, we made decisions, with the knowledge that we could revisit them, to allow us to move on. As we revisited the same issues, at different times, for different purposes, we built on ideas generated previously in iterative, knowledge-building cycles (see Figure 15.1)

Spiral process, dependent upon:

- embracing the complexity of the design process
- social negotiations of shared means
- examining information numerous times from multiple perspectives

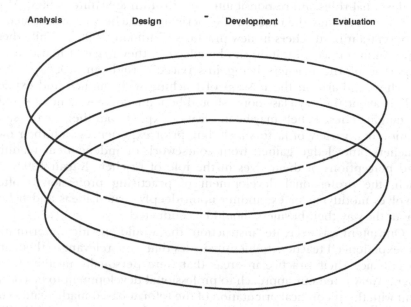

Analysis   Design   -   Development   Evaluation

Figure 15.1.   The Layers of Negotiation Model of instructional design.

RECREATING MATHEMATICS INSTRUCTION: A CONTRASTING CASE

A subsequent project provided an opportunity for me to test my emerging ideas about designing materials for constructivist learning environments in a novel situation. The purpose of the project was to develop a set of instructional materials for use in the professional development of elementary teachers who desired to recreate their practice of teaching mathematics in a manner consistent with reform initiatives (National Council of Teachers of Mathematics, 1989). As in the previous project, I served as the instructional designer and in this role, attempted to observe the manner in which we designed instruction, rather than direct it. I observed how we addressed the critical questions for instructional design, comparing the process to traditional instructional design models and to my emerging ideas of instructional design involving "layers of negotiation."

Context of Investigation

The professional development of individuals with experience in a given area of practice may require that they reconceptualize their professional roles and responsibilities (Goldsmith & Schifter, 1993). Most attempts to change the practice of experienced teachers have consisted of experts training teachers in new practices (Feldman, 1996). "While these types of in-service education can be effective, they make little use of the expertise of the teachers being in-serviced" (Feldman, 1996 p. 513). Teachers embark on the mission of teaching with implicit and explicit beliefs about the way classrooms should be managed and content should be taught. These beliefs may be implicit or explicit, but they are the summation of a variety of factors including prior experiences with their own teachers, knowledge gained from coursework or independent reading, and perceptions of themselves in the role of teacher (Kagan 1992). As such, the professional development of practicing professionals often involves modifying or expanding upon deeply seated ideas and beliefs about the way their business should be conducted.

Our intent was to create "instruction" that would provide opportunities for experienced teachers to identify issues that were relevant to them and to enhance their practice in areas that were personally meaningful to them. We wanted our approach to professional development to be consistent with the theoretical orientation of the reform-based mathematics curriculum that the teachers would use in their elementary classrooms. The curriculum was based on a Piagetian approach to learning. Consistent with this theoretical orientation, our professional development program was based on the assumption that conceptual change occurs through experiences that induce cognitive conflict and indicate inadequacies in current thinking (Piaget, 1985). Learners should be actively involved in trying things out to see what happens, posing their own questions and seeking their own answers, reconciling what was found at one time with what was found at another, and comparing their Endings with those of others (Driscoll, 1994). Through comparing their ideas with the ideas of others, the teachers may become dissatisfied with their existing knowledge and find new ideas that are useful alternatives to their current conceptions (Goldsmith & Schifter, 1993). Thus, we wanted to create opportunities for teachers to express and examine their beliefs, values, and practice of teaching, and to compare those personal beliefs, values, and practices to those of others.

Initial Conversations

The professional development program was targeted to elementary school teachers who desired to modify their teaching of mathematics to be consistent with reform initiatives (National Council of Teachers of Mathematics, 1989). We wanted to create "instruction" that would provide opportunities for experienced teachers to identify issues that were relevant to them and to enhance their practice of teaching mathematics in areas that were personally meaningful to them. Conversations among the instructional design team and the subject matter expert began with discussions of what kinds of "changes in thinking" we may see as teachers recreated their practice to be consistent with reform initiatives.

Informal observations of teachers who had attended 21 professional development workshops on teaching with a reform-based elementary math curriculum indicated there were differences in the way the teachers taught mathematics, even though all the teachers were using the same set of materials and all had attended the same workshop. However, we were unsure of the nature of these differences. As we were unable to identify specific differences among classrooms or "changes in thinking" that would lead to teaching in a manner consistent with reform initiatives, we moved to a discussion of how we would know that changes in thinking had occurred. We decided to analyze videotapes of classroom mathematics lessons to identify differences in the practice of teaching mathematics among elementary teachers. Using a data resource which consists of approximately 200 videotaped mathematics lessons collected over a period of one year in five second-grade classrooms, we developed a classroom coding scheme, coded, and interpreted the teachers' and students' actions (Wood, Turner-Vorbeck, & Walker, 1996b).

Analyzing teachers' behavior prior to developing instruction initially appeared to be similar to the task analysis prescribed by traditional instructional design models. I wondered if we would use the results of the analysis to identify goals for the learners. Would we move through the steps of analysis, design, development, and evaluation in a step-by-step, linear manner? In the science project, I had observed the team moving in a cyclical, recursive process through analysis, design, development, and evaluation. I observed us considering certain data (analysis), making tentative decisions (design), developing part of the product (development), evaluating that part (evaluation), then continuing additional analysis as the need for additional data became evident (see Cennamo, Abell, George, & Chung, 1996, for a description of that process). Analyzing the task in detail prior to development seemed more consistent with the discrete stages of traditional instructional design models than the recursive process that we previously used to develop the science materials. And it concerned me. It caused cognitive conflict. But then it hit me.

I recalled that the essence of systematically designed instruction is the alignment among the goals, activities, and assessments of instruction (see Figure 15.2). In the science project, we focused our initial discussions on the design of activities. We knew that we wanted to create interactive videodisc cases to facilitate reflective practice and so we proceeded to

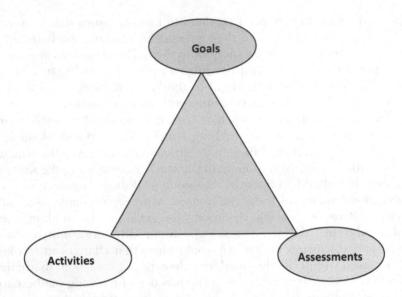

Figure 15.2.   The "essential triangle" of instructional design.

develop the activities through a reflective, recursive process. As I reflected on the development of both the math and the science materials, I became aware that, in the science project, we had cycled through several rounds of analysis, design, development, and evaluation as we created the activities, or materials, then began to cycle through similar stages in creating an assessment, or means for the students to confront their beliefs about teaching science. As the materials were initially implemented, we developed a means of assessing the learners' thinking. As we observed students using the materials and assessments, we identified several goals held by the learners. We were able to see what types of changes in thinking occurred through students' interactions with the materials and assessments.

As I reflected on the three elements of successful instruction, I realized we were creating our assessments first in this project. We were not simply developing a means of analyzing the task in order to derive goals and prescriptions for learning. The coding scheme could be used to identify the critical issues around which we would develop our instructional materials, but in addition, the coding scheme could provide teachers with a way of viewing their classroom actions in order to reflect upon issues they identified through self-assessment. After all, the coding scheme originally emerged out of discussions about the three essential elements of instruction:

1. What kinds of changes in thinking do we want to occur?
2. How will we know if these changes have happened?
3. How will we provide opportunities for learners to examine such issues?

As we attempted to develop an understanding of the dimensions around which the practice of teaching elementary mathematics varies, we chose to begin in the middle, creating assessments, creating a means for our teachers-as-learners, as well as ourselves, to identify differences in practice.

Development of the Coding Scheme

As we created the classroom coding scheme, the process proceeded through analysis, design, development, and evaluation in a reflective, recursive, participatory process characterized by complexity, social negotiations, examining data from multiple perspectives, and reflexivity. We began the process of developing our coding scheme by examining the work of Dillon (1990) and Ainley (1988). From these analyses, we developed tentative coding categories as a way of viewing the classroom videotapes. We viewed several videotapes and examined transcripts of the tapes to determine if the ideas we derived from our analysis were appropriate to the data. As the design and development team reviewed the tapes and transcripts, we concluded that Dillon's (1990) work did not provide adequate guidance for capturing the nature of the classroom interactions. We perceived that "statements" and "questions" seemed to hold different meanings in the classrooms we analyzed and in the classrooms analyzed by Dillon. We considered the work of Antaki (1994) and Branham (1991), and drafted a revised version of coding categories. Then we returned to an evaluation of our assessment. Individually, we attempted to apply the coding categories to transcripts of the videotaped lessons. Following our individual coding, we met as a group to discuss differences in our coding and to come to consensus on the meaning of particular coding categories and classroom actions. We initially categorized questions and statements into "teacher questions," "student questions," "teacher statements," and "student statements." We discussed the meaning of the questions and statements for both the listener and the speaker. Through our discussions, we altered our categories to refer to "listener" questions/statements and "speaker" questions/statements. We generated examples of questions and statements for each category. As we engaged in intense negotiations, we recognized that examples of statements and questions needed to capture the nature of the classroom interactions and norms. We began to invent our own coding categories. We became aware that our categories were becoming increasingly consistent with the work of scholars such as Pintrich, Marx, and Bowels (1993) and previous analysis by Wood, Cobb, Yaeklel, and Dillon (1993); thus, we examined works from these perspectives to gain additional insight.

Reflecting on the process of developing the coding categories, it is evident that the development occurred through a recursive process of (a) analyzing others' work, (b) drafting or modifying our coding categories, and (c) trying our categories on the data (see Table 15.1). This recursive pattern is characteristic of the "Layers of Negotiation" approach to designing instruction. Progress came in "iterative, knowledge building cycles" (Rowland, 1992, p. 66). Solutions gradually emerged through continually revisiting the same data (the transcripts) at different times, for different purposes. Reflexivity and social negotiations of shared meanings were central to the process.

Development of Instructional Materials

The development of the coding categories took approximately three months of meeting at least once a week. However, this period of intense negotiations eventually resulted in a simple yet effective set of coding categories that allowed us to identify differences in practice among the various classroom teachers. Once we had created this critical piece of our work, we were able to complete the development of the materials very quickly.

Through our analysis of videotapes of classroom practice, we identified three dimensions around which the teachers' practices varied (Wood, Turner-Vorbeck, & Walker, 1996a; Wood, Walker, & Turner-Vorbeck, 1996b):

1. Although all the teachers had their students vocalize their ideas about mathematics, the teachers varied in the way they had the students express their mathematical thinking. Some teachers had students report their solutions to the math problems, others required them to tell how they solved the problems, and others asked students to tell why they performed the steps they performed to solve the problems.

2. Teachers varied in the extent to which they allowed disagreement: Some set up exchanges that were primarily between teacher and student; others established exchanges among students and the rest of the class.

3. Finally, when children shared their ideas, teachers varied in their expectations for active participation by the "listener" and "explainer."

Having identified issues around which teachers' practices vary, our challenge was to develop activities that allowed teachers to explore these issues in their classrooms in personally meaningful ways. Using the work of Barbara Jaworski and the Open University (1989; Mathematical Association, 1991) as a guide, we developed a tentative "way of working" with teachers to promote reflection on the practice of mathematics teaching. We tested our way of working with seven teachers who participated in ten two-hour teacher development sessions. In these sessions, teachers were asked to videotape segments from their own classrooms to illustrate issues of importance to them. Although we had developed a tentative pattern for discussion, we continually reflected on and modified our way of working throughout the sessions to determine the sequence and questioning approach that was most effective in generating discussions of multiple solutions to problems and issues. As we conducted these teacher development sessions, we recorded our expected actions for the sessions prior to each session, videotaped the sessions, recorded field notes, and made reflective notes following the sessions. These videotapes were logged for analysis. The participants also recorded their reflections and assignments in a notebook. We analyzed these data to identify patterns of behaviors and responses, to determine if the issues teachers identified as important in their classrooms were similar to the issues we identified as important based on our coding of classroom videotapes, and to refine our way of working with the teachers.

As the problematic issues we identified from our coding of classroom videotapes were consistent with those identified by teachers who participated in the teacher development sessions, we felt confident in developing instructional materials to focus discussions on the issues we had previously identified. The materials consisted of a series of five videotaped segments of mathematical discussions as they occurred in elementary classrooms. The videotapes were supplemented with a "leader's guide" that outlined our way of working, and indicated possible issues that may arise from the videotaped segments or possible paths the discussion may take. The leader's role was to chair the teacher development sessions and move the discussions forward to the identification of issues and plans for action.

As I reflected on the design and development of these materials, it was apparent that we proceeded through a cyclical process of analyzing problems, designing and developing possible solutions, and seeking confirmation as to the appropriateness of these solutions (see Table 15.2). These

**Table 15.1.   Recursion at the Microlevel as
Applied to Developing the Coding Scheme**

Analysis:	Examined work of Dillon (1990) and Ainley (1988)
Solutions:	Developed tentative coding categories.
Checking:	Viewed several videotapes and examined transcripts to determine if the coding categories were appropriate to the data. Concluded that Dillon's (1990) work did not provide adequate guidance for capturing the nature of the classroom interactions: Statements and questions seemed to hold different meanings.
Analysis:	Considered the work of Antaki (1994) and Branham (1991).
Solutions:	Drafted a revised version of the coding categories.
Checking:	Individually applied the coding categories to transcripts of the lessons.
Analysis:	Met as a group to discuss individual differences in coding classroom actions and to come to consensus on the meaning of particular categories and actions. Discussed the meaning of "questions" and "statements" for both listeners and speakers.
Solutions:	Altered coding categories to refer to listener questions/statements and speaker questions/statements. Generated examples of questions and statements for each category. Recognized that additional statements and questions were needed, so began to invent our own categories.
Checking:	Categories were becoming increasingly consistent with the work of scholars such as Pinnich, Marx and Bowels (1993) and previous analysis by Wood et al. (1993); thus, examined works from three perspectives to confirm and provide additional insight to our categories.

stages parallel the common stages of analysis, design and development, and evaluation reflected in the majority of instructional design models. However, whereas classic models ascribe certain activities to these stages and most often progress through these stages in a linear manner, the recursive nature of design- as-knowledge construction also finds the designer cycling through these stages at a "micro" level as individual components of the instruction are developed.

Reflections and Implications

Designing this instruction, and reflecting on designing this instruction, provided for me a contrasting case, a means of trying out my newly constructed ideas in a novel situation that did not appear to follow the pattern of development identified in the previous project. In the science project, we developed materials and then examined the nature of the knowledge constructed in response to the materials. In the math project, we looked at differences in the way teachers taught mathematics to identify changes that may occur during the evolution of teaching practice. And from these analyses, we developed activities that provided opportunities for learners to examine such issues. My observations of the design and development effort confirmed my tentative assumptions that instructional design is a process of constructing knowledge, involving reflection, examining information at multiple times for multiple purposes, and socially negotiating shared meanings. The design process evolved in a spiral fashion, progressing through stages of analysis, design and development, and evaluation. Through this project, the Layers of Negotiation Model was confirmed, yet evolved to (a) identify critical areas around which negotiations should occur and (b) illuminate the nature of the recursive process.

CRITICAL AREAS OF NEGOTIATION

As an instructional designer, I believe I have a responsibility to ensure certain things. The goals, activities, and assessments of instruction should be in alignment. Instructional decisions should be made based on what we know about the learners' needs, characteristics, beliefs, and values. In addition, the instructional materials should be tested with a group of learners prior to distribution and modified based on the results of a pilot test.

Of course, the way we approach these critical issues is influenced by the theoretical perspective from which the materials are designed; thus, instructional design requires that we first negotiate our assumptions about the nature of teaching and learning. In the projects described in

Table 15.2. Recursion at the Macrolevel

Analysis	Analyzed videotapes of classroom practice to determine ways that practice varied.
Solutions	Developed a "way of working" with the teachers.
Checking	Tried the "way of working" in teacher development sessions.
Analysis	Analyzed videotapes of teacher development sessions.
Solutions	Refined "way of working" and developed videotapes and leader's guide.
Checking	Tested revised way of working and leader's guide with teachers.

this article, we believed that knowledge is constructed through conflict between our existing knowledge and new ideas, and that existing ideas are modified through interactions with others. Within the context of this and similar theoretical frameworks, conversations around these critical issues may involve the following considerations:

1. *What are the information needs; prior experiences, beliefs, and values of the learners relative to the topic of instruction?*

Learners come to an instructional setting with their individual set of prior experiences, values, and beliefs, and there is no way we can ensure that instruction will affect individual learners in similar ways. However, instruction can be enhanced by considering the possible range of beliefs, values, and knowledge relative to the topic. "Learner profiles" can be developed historically (based on past experience with similar groups), or through examining theoretical and research literature (common misperceptions in science for example) to piece together a "picture" of the target audience. An awareness of "typical" beliefs, values, and knowledge of the learners allows us to become advocates for the learners.

2. *What kinds of changes in thinking do we want to occur? How will we know if these changes have occurred? How will we provide opportunities for learners to examine issues that may stimulate changes in thinking? In other words, how can we ensure a match among the goals, activities, and assessment of instruction?*

Although learner's individual goals for change may emerge through interactions with the instructional materials and with others, all instructional activities are created with some goal in mind. The goal may be broad and open-ended in nature, directed to providing opportunities for the learners to determine what is important for them as individuals. Goals may be metacognitive and interpersonal in nature (Cognition and Technology Group at Vanderbilt, 1992), such as developing a reflective practice or engaging in constant improvement of practice.

Activities that support such goals are similarly open-ended in nature. They provide opportunities for learners to identity issues that are important to them, and through promoting disequilibrium and conflict, provide opportunities for the learners to reflect on the issues to seek resolution of conflict. Often, the activities provide opportunities for the learners to compare their ideas with others, in order to construct shared meanings. Situated learning, cognitive apprenticeship, case studies, and anchored instruction are examples of activities that provide learners with the opportunity to construct their own knowledge through social interactions.

Within constructivist learning environments, assessments are designed to be an integral part of the instruction (Choi & Hannafin, 1995). They are developed to provide learners with a picture of where they are at a particular point in time, rather than to provide teachers with a way of determining whether students have obtained a desired learning outcome at the end of instruction. Portfolios, case analyses, and reflective journals are examples of assessments that are consistent with this view of teaching and learning. Assessments such as these focus on the process of learning rather than the product. More importantly, assessments are at part of the instructional activities, rather than a "test" at the end of instruction (Young, 1993).

3. How do learners respond to the materials?

Testing the instruction with a group of learners in order to determine how they respond to the materials provides designers with the opportunity to verify the usefulness of a product prior to distribution. Evaluation, within this context, does not involve pretest to posttest improvement scores; rather, evaluations should focus on how learners make sense of the information, the nature of their discussions, and the ability of the activities to generate thinking about issues of importance to the learners.

These three questions parallel the stages of analysis, design and development, and evaluation that are common in instructional design. However, design-as-knowledge construction may not involve addressing the stages of instructional design in a linear, step-by-step manner as suggested by traditional instructional design models.

THE NATURE OF RECURSION

Over and over again, we see models that include analysis, design and development, and evaluation activities (Gustafson & Branch, 1997). In traditional instructional design models, these stages typically include specific tasks:

- The analysis stage includes activities such as needs assessment, learner analysis, task analysis, context or job analysis, and content or instructional analysis.
- The design stage includes activities such as writing performance objectives, developing assessment items, specifying learning activities and instructional strategies, and media selection.

- The development stage includes creating the instructional materials.
- The evaluation stage includes formative evaluation to identify revisions necessary prior to distribution, revisions, and summative evaluation.

But the analysis, design and development, and evaluation cycle is more pervasive than reflected in these broad stages of instructional design. In addition, this cycle represents a basic problem-solving approach that can be applied across the design tasks.

According to Rowland (1992), design involves recalling a template or mental model for the type of information needed to solve the design problem, and seeking to fill slots in the mental model. It involves pausing at certain points in the process to determine what is known (in terms of "slots filled") and what still needs to be known. The designer attempts to understand or "analyze" the problem, gathering information to fill slots in his or her mental model. At various times in the process, the designer may reach a point where no additional information can be determined, either because the client, learners, or ether contact people are not able to provide any more information, or because time constraints require that the designer move on. At that point, the designer must make decisions about possible answers to the questions of design (to fill the slots) and choose tentative solutions, constantly subjecting those decisions to confirmation as the design process proceeds. These "hypotheses" may take the form of hypothesized solutions to the problem in the form of instructional materials or may involve nothing more complex than summarizing the discussion in progress. Throughout the process, tentative decisions, design documents, or products are presented to the clients or learners for confirmation or "checking." Although evaluation, as the term is commonly used in instructional design, occurs at certain paints in the design process, checking does not always involve the evaluation of some "thing"; instead, it is a dynamic process of confirming or disconfirming "hypotheses." Tentative decisions are modified as more information becomes available or relevant.

Throughout this process, the designer proceeds through a cycle of (a) seeking information that will illuminate the nature of the problem or provide insight as to solutions (analysis), (b) generating tentative solutions (design/development), and (e) checking these hypothesized solutions. This iterative cycle of analysis, generating solutions, and checking solutions continues over and over again, until the process yields a suitable solution to the initial design problem.

CONCLUSIONS

As I have attempted to design instruction for constructivist learning environments, the ideas that evolved through reflection on practice are remarkably similar to those proposed by Willis (1995, 1998, 2000; Willis & Wright, 2000). Nearly 10 years ago, we were working on instructional design projects simultaneously; his reflections on the process resulted in the R2D2 model (Willis, 1995), and mime in the Layers of Negotiation Model (Cennamo, Abell, & Chung, 1996). Unfortunately, I did not have access to Willis' thinking as I worked on the project leading to the Layers of Negotiation Model. The two of us, as independent researchers, have concluded similar things. Willis (1995, 1998, 2000; Willis & Wright, 2000) bases much of his rationale on the literature backed up by practice; most of my rationale was based on practice, backed up by literature. The fact that we have both come to similar conclusions, taking different routes to that knowledge, makes the similarities more powerful.

We both propose an iterative model that involves reflection and participation by a wide variety of stakeholders. Activities that Willis and Wright (2000) propose such as "progressive problem solutions" and "cooperative inquiry" are key to both of our approaches to instructional design. We agree that analysis is an ongoing process rather than a preliminary phase in a design and development project. "The same issues or questions may be addressed many times throughout the process. Solutions, decisions, and alternatives gradually emerge over the course of the project" (Willis, 1995, p. 14), We both acknowledge that design is a process of solving multiple problems in context, where " 'solutions,' such as the set of objectives for a project, progressively emerge across the entire design and development process instead of being completed early and then used to guide design and development work" (Willis & Wright, 2000, p. 7). And we both agree that the majority of conversations and negotiations among members of a collaborative team should occur in the context of the development of the instructional materials.

Where we differ is in the role of the instructional designer. Willis does not believe that instruction can be developed by an outside "expert" designer (Willis, 1995; Willis & Wright, 2000). Willis and Wright state that "there is no such thing as a 'general ID expert' who can take his or her skills from one setting to another, and design well without understanding the context" (2000, p. 8). While I agree that it is critical for the designer to understand the context of the instruction, and to have individuals who live and work in the context in question as part of the participatory design team, my experience leads me to believe that a "general ID expert" does indeed have skills that transfer from one setting to another, regardless of context.

The role of the instructional designer is to serve as a facilitator of the design and development process. The instructional designer participates in conversations with other members of the design and development team. Through these conversations, the designer seeks to extract information relative to the design of instruction. The designer listens, asks questions, and organizes the ideas presented by the team, then reflects that information back to the team for confirmation or modification. Engaging in a constant process of sense-making, the designer works with the materials, the ideas of others, to convert them to an instructional plan. Through the process of analysis, proposing tentative solutions, and checking the reasonableness of those solutions, the design process moves toward a concrete plan in iterative knowledge-building cycles. Throughout the process, the designer constantly seeks answers to the critical questions of design, and works to ensure that all aspects of the instruction are in alignment. Three beliefs are essential to this approach to instructional design:

1. Design is a *collaborative* activity among individuals offering different perspectives and expertise.
2. Design involves conversations around critical *questions*.
3. Design is an *iterative* knowledge-building cycle.

Design Is Collaborative

When designing materials consistent with a philosophy of design-as-knowledge construction, we need to create an environment conducive to knowledge construction, reflection, and social interactions. Paraphrasing Driscoll's "five conditions for constructivist learning environments" (Driscoll, 1994, p. 360), the designer should:

1. Embrace the complexity of the design process.
2. Provide for social negotiations as an integral part of designing the materials.
3. Examine information relevant to the design of the instruction at multiple times from multiple perspectives.
4. Nurture reflexivity in the design process.
5. Emphasize participatory design.

Critical Questions

Instructional designers should ask certain questions, of themselves and others, to ensure certain things.

- Instructional decisions should be made based on what we know about the learners' needs, characteristics, beliefs, and values.
- The goals, activities, and assessments of instruction should be in alignment.
- Instructional materials should be tested with a group of learners prior to distribution, and modified based on the results of the evaluation.

Conversations should center around questions such as:

- What are the information needs, prior experiences, beliefs, and values of the learners relative to the topic of instruction?
- What changes in thinking or performance do we want to occur?
- How will we know if these changes have occurred'? How will we provide opportunities for learners to examine issues that may stimulate changes in thinking or performance?
- How do learners respond to the materials?

Iterative Process

At both a macro and micro level, the instructional design process involves a recursive cycle of analysis, solutions, and checking.

- Analysis involves asking questions, reviewing literature, observing, examining artifacts, listening to conversations, and similar techniques that provide answers to the "essential questions" of instructional design.
- Generating solutions may involve activities as simple as summarizing conversations or as complex as developing prototypes. At this stage, the designer takes the information provided during the analysis process and organizes it in a way that makes sense relative to the design process.
- He or she then reviews this information with the rest of the team for confirmation or modification, through the process of "checking."

Like many instructional designers, I was trained to use classic instructional design models such as Dick and Carey (1996). But I've found that in practice, instructional design is not just about following a model. It's about a systematic, but collaborative, process of constructing knowledge to inform the development of instructional materials.

You start with conversations around the critical issues. You embrace the complexity of the design process. You emphasize participatory design. As you come together with the design team, you nurture reflexivity in the design process. You examine information relevant to the design of the instruction at multiple times from multiple perspectives.

You engage in social negotiations as you develop shared knowledge that leads to the development of tentative solutions. Throughout the process, you, as the designer, seek answers to the critical questions of design. You proceed through iterative, knowledge-building cycles of analysis, solutions, and checking, again and again until you have found a workable alternative. That's the nature of design to me: collecting information, making tentative decisions, trying them out, then evaluating the results, in order to make new decisions and repeat the process all over again.

REFERENCES

Ainley, J. (1988). Perceptions of teachers' questioning styles. In *Proceedings of the Twelfth International Conference on the Psychology of Mathematics Education* (pp. 92–99). Vesprim, Hungary: Psychology of Mathematics Education.

Antaki, C. (1994). *Explaining and arguing.* London: SAGE.

Branham, R. L. (1991). *Debate and critical analysis: The harmony of conflict.* Hillsdale, NJ: Erlbaum.

Cennamo, K. S., Abell, S. K., & Chung, M. (1996, July–August). Designing constructivist materials: A Layers of Negotiation Model. *Educational Technology, 36,* 39–48.

Cennamo, K. S., Abell, S. K., George, E. J., & Chung, M. (1996). The development of integrated media cases for use in science teacher education. *Journal of Technology and Teacher Education, 4*(1), 19–36.

Choi, J., & Hannafin, M. (1995). Situated cognition learning environments: Rules, structures, and implications for design. *Educational Technology Research and Development, 43*(2), 53–69.

Cognition and Technology Group at Vanderbilt. (1992). The Jasper experiment: An exploration of issues in learning and instructional design. *Educational Technology Research and Development, 40*(1), 65–80.

Dick, W., & Carey, L. (1996). *The systematic design of instruction.* Glenview, IL: Scott Foresman.

Dillon, J. T. (1990). *The practice of questioning.* London: Routledge.

Driscoll, M. P. (1993). *Psychology of learning for instruction.* Boston: Allyn & Bacon.

Feldman, A. (1996). Enhancing the practice of physics teachers: Mechanism for the generation and sharing of knowledge and understanding in collaborative action research. *Journal of Research in Science Teaching, 33,* 513–540.

Goldsmith, L. T., & Schifter, D. (1993). *Characteristics of a model for the development of mathematics teaching.* Newton, MA: Center for the Development of Teaching, Education Development Center, Inc.

Gustafson, K. L., & Branch, R. (1997). Revisioning models of instructional development. *Educational Technology Research and Development, 45*(3), 73–89.

Jaworski, B. (1989). *Using classroom videotape to develop your teaching.* Milton Keynes, England: The Open University

Kagan. D. M. (1992). Professional growth among preservice and beginning teachers. *Review of Educational Research, 62*(2), 129–169.

Mathematical Association. (1991). *Develop your teaching.* Cheltenham, UK: Stanley Thornes.

National Council of Teachers of Mathematics. (1991). *Professional standards for teaching mathematics.* Reston, VA: Author

National Council of Teachers of Mathematics. (1989). *Curriculum and curriculum standards for school mathematics.* Reston, VA: Author.

Osborne, R., & Freyberg, P. (1985). *Learning in science: The implications of children's science.* Portsmough, NH: Heinemann.

Piaget, J. (1985). *The equilibration of cognitive structures.* Chicago: University of Chicago Press.

Pintrich, P. R., Marx, R. W., & Boyle, R. A. (1993). Beyond cold conceptual change: The role of motivational beliefs and classroom contextual factors in the process of conceptual change. *Review of Educational Research, 63,* 167–195.

Posner, G., Strike, K., Hewson, P., & Gertzog, W. (1982). Accomidation of a scientific conception: Toward a theory of conceptual change. *Science Education, 66,* 211–227.

Rowland, G. (1992). What do instructional designers actually do? An initial investigation of expert practice. *Performance Improvement Quarterly, 5*(2), 65–86.

Rowland, G. (1993). Designing and instructional design. *Educational Technology Research and Development, 41*(3), 79–91.

Solomon, J. (1989).The social construction of school science. In R. Millar (Ed.), *Doing science: Images of science in science education* (pp. 126–136). Philadelphia: Falmer Press.

Willis, J. (1995, November/December). A recursive, reflective instructional design model based on constructivist-interpretivist theory. *Educational Technology, 35,* 5–23.

Willis, J. (1998, May–June). Alternative instructional design paradigms: What's worth discussion and what isn't. *Educational Technology, 38,* 5–16.

Willis, J. (2000, January/February). The maturing of constructivist instructional design: Some basic principles that can guide practice. *Educational Technology, 40,* 5–15.

Willis, J., & Wright, K. E. (2000, March-April). A general set of procedures for constructivist instructional design: The new R2D2 model. *Educational Technology, 40,* 5–20.

Wood, T., Cobb, P., Yackel, E., & Dillon, D. (1993). *Rethinking elementary school mathematics: Insights and Issues. Journal of Research in Mathematics Education, Monograph No. 6.* Reston, VA: National Council of Teacher of Mathematics.

Wood, T., Turner-Vorbeck, T., & Walker, W. (1996a, April). *Mathematical discussion or argumentation? Differences among teachers in the constitution of norms for interaction.* Paper presented at the annual meeting of the American Educational Research Association, New York.

Wood, T., Turner-Vorbeck, T., & Walker, W. (1996b, April). *Recreating mathematics teaching: A theoretical and methological framework for analysis.* Paper presented at the annual meeting of the American Educational Research Association, New York.

Young, M. (1993). Instructional design for situated learning. *Educational Technology Research and Development, 41,* 43–58.

CHAPTER 16

FROM THREE-PHASE TO PROACTIVE LEARNING DESIGN

Creating Effective Online Teaching and Learning Environments

Roderick Sims

For many tertiary educational institutions, enterprise learning management systems have become a significant component of the teaching and learning environment. At the same time, pedagogies applied to these systems have emphasized socio-constructivist frameworks to align with the collaborative and interactive affordances of online learning. However, these same institutions are faced with a double dilemma; first, many academic and support staff have limited skill-sets in both the technology and the emerging pedagogy and second, the design models applied to the creation of teaching and learning resources are often founded on transmission models of education.

The original articulation of the Three-Phase Design (3PD) model was presented in Sims, R. & Jones, D. (2003). Where practice informs theory: Reshaping instructional design for academic communities of practice in online teaching and learning. *Information Technology, Education and Society, 4*(1), 3–20.

Consequently it is not unusual to observe online learning classes exhibiting traditional face-to-face strategies which minimise opportunities for teachers and learners to engage in interactive and meaningful dialogue. To address this situation, the majority of Australian universities have established teaching and learning centres to support curriculum design and effective strategies for online teaching and learning.

Within this context, this chapter presents an argument that it is the processes and resources applied to the development of online teaching and learning resources which must be consistent with the institutional framework, the teaching and learning environment, the technological infrastructure and, most importantly, an appropriate online pedagogy. However this must be achieved in an environment still "characterized by demands from students for quality face-to-face and distance education, staff concern over workloads, institutional budgeting constraints and an imperative to use learning management systems" (Sims & Jones, 2003, p. 4).

This chapter begins by revisiting Sims and Jones' (2003) Three-Phase Design (3PD) model and the way it has been successfully applied to the implementation of two online learning development projects. Using 3PD as a framework, the discussion elaborates on other key elements of the design process such as proactive evaluation (Sims, Dobbs, & Hand, 2002), interactivity (Sims, 2006), emergence (Irlbeck, Kays, Jones, & Sims, 2006), competencies (Sims & Koszalka, 2008) and authentic, contextual learning (Sims & Stork, 2007) to create a fully-operational model for the implementation of engaging, collaborative and constructive online teaching and learning environments. Known as Proactive Design for Learning (PD4L), the model encapsulates all elements required to be successful in online teaching and learning—from the competencies of the stakeholders to the most effective design strategies to the interactions between course participants. Underpinning PD4L is an ethos of collaborative design and development teams who can, over time, establish *communities of practice* (Wenger, 1998) where the shared learning and interest of its members maintain the functionality and integrity of the course.

THE THREE-PHASE DESIGN MODEL

According to Sims and Jones (2003), the value of Three-Phase Design (3PD) would be released through a three-step process of *develop functionality, evaluate/elaborate/enhance and maintain* rather than the more traditional sequence of *design, develop, implement, evaluate*. While not attempting to replace accepted instructional design models, the intent of 3PD was to provide a new lens for the overall development process, especially with respect to online teaching and learning. A key component of 3PD was that online course creation could not be viewed as a short-term development process, but rather as a long-term collaborative process which would "generate and evolve into focused communities of practice with shared understanding and a philosophy of continuous improvement" (Sims & Jones, 2003, p. 18).

Three-Phase Design (3PD) also integrates the three essential competency sets for unit or course development (design, subject matter, production) in a cohesive rather than disparate manner. Rather than process driving development, it is the context of the educational components which determine the members of development teams in a targeted and effective manner. Ideally, these teams would remain for the duration of the project, potentially over a number of semesters. Finally, as shown in Figure 16.1, 3PD specifies *baselines* that align with implementation iterations—the first focusing on building functional and essential course components, the second on enhancement or interactivity and the third to ongoing maintenance. These three phases of development integrate methodological approaches to unit development, scaffolding of participants and quality control and assurance.

Phase 1: Building Functional Components

The aim of the initial phase is to create a functional teaching and learning online environment that will achieve learning outcomes as well

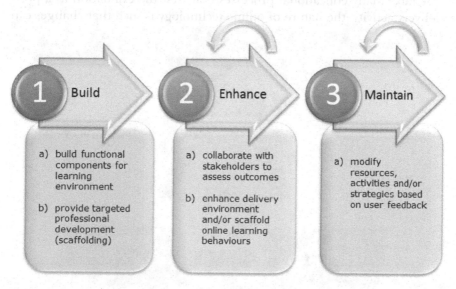

Figure 16.1.   Three-Phase Design & Scaffolding (adapted from Sims & Jones, 2003).

as meet organization standards for display. Production is neither focused on completion of a final course of study nor does it encourage 'bells and whistles' that can cause interference with learning (Sims, 2006). This phase involves specifying key course components including resources (e.g., study guides, readings), their mode of access (e.g., print, online), assessment-based outcomes, preferred teaching strategies (e.g. experiential, problem-based) and learning/learner activities designed to achieve outcomes. In this way the teacher with minimal experience with online teaching and learning environments has access to functional learning structures as well as ongoing support will enable the generational development of that environment. With each phase of 3PD, it is anticipated that there will be a dynamic interplay between three key stakeholders in the process – teachers, learners and designers (who include technical specialists). As illustrated in Figure 16.2, the communication between each of these three roles is critical to the success of the teaching and learning environment, and also leads to potential role variation—*teacher* as designer and learner, *designer* as teacher and learner and *learner* as teacher and designer.

Phase 2: Evaluate, Elaborate, and Enhance

Unlike many educational processes that view the evaluation as a post-delivery activity, the nature of online technology is such that changes can

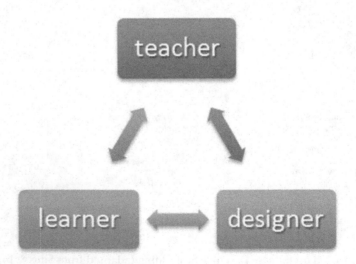

Figure 16.2.   Dynamic collaboration.

be made immediately, as long as those changes do not compromise the integrity of the course outcomes. Therefore the second phase can be conceptualized to take place during course delivery, with feedback from both teachers and learners being used to modify and/or enhance delivery.

This phase continues the dynamic collaboration (as shown in Figure 16.2) during delivery. While the more technically-oriented roles will assess the efficacy of the environment, those involved in the pedagogy (teachers and learners) can address the integration of additional content, interactive learning objects or collaborative activities. At the same time, targeted professional development or scaffolding for effective online teaching and learning can be delivered *in situ*. This second phase enables generational changes in the course structure, with emphasis on the production (completion) of resources, and where learners can take a role of research and evaluation assistants. By developing and building effective communication paths between each of these three roles, a shared understanding of the course goals and learning outcomes can be established, thereby minimizing any compromise in educational quality and effectiveness. As illustrated in Figure 16.1, the recursive arrow indicates that the enhancement process would be repeated until the course of study is considered complete.

Phase 3: Maintain

Ultimately the course will reach a state where the teaching strategies and learning activities are working well, allowing the course to shift into a maintenance phase until it is targeted for formal review. The implications of applying the 3PD model is that the original functional system will always be subject to change, and that development environments need to schedule resources for the life-time of that course. The continual process of gathering and incorporating evaluation data caters to the goal of sustainability of the course.

Academic Professional Development

Within the context of these three phases, and for academics new to online learning, maximum exploitation of the online environment can mean reassessing teachers' overall approach to the content, how it should be presented or accessed and the ongoing relationship between teacher and learner. The options for course content must therefore be considered in terms of their interaction with design strategies and their relevance to the learning community (Sims, Dobbs, & Hand, 2002). Competencies in online pedagogy manifested by many academics is such that they do not have the necessary background to work online and therefore require support to assist in the transfer of their good classroom practice to good online practice (Sims, 2006).

Therefore there is an ongoing requirement for staff in support units to provide targeted professional development to academic staff such that it functions as a scaffold for their teaching and learning practice. If this can be implemented into the day-to-day work of academics and their relationship with course participants, all the better. This approach emerges from the work of Vygotsky (1978), in that scaffolding can be understood to relate to supporting academic staff who may not be able to accomplish all aspects of online teaching and learning independently. While this support is likely to diminish over time, as academic staff gain in competency, it is important that the relationships established be maintained.

INFORMING FACTORS

While the 3PD model provides a strategic context from which to build and maintain online teaching and learning environments, its original design did not focus on more detailed aspects of the design process. In this section six key informing factors are presented which, it is argued, are essential to achieving engaging, interactive and memorable learning experiences.

As illustrated in Figure 16.3, three-phase design provides a core for these factors, establishing the build-enhance-maintain process as core to successful project implementation. Integral to the success of this model, labelled Proactive Design for Learning (PD4L), are the six factors which imply that design for effective online teaching and learning is:

- *theory-based*, ensuring that decisions are based on contemporary approaches to teaching and learning.
- *innovative* and relevant (incorporating elements of proactive evaluation documented (by Sims, Dobbs, & Hand, 2002).
- *team-based*, with team members having the relevant and appropriate competencies to engage with and complete the design tasks (Sims & Koszlaka, 2008).
- *emergent*, allowing (where appropriate) the interactions between course participants to establish and introduce course content (Irlbeck, Kays, Sims, & Jones, 2006).

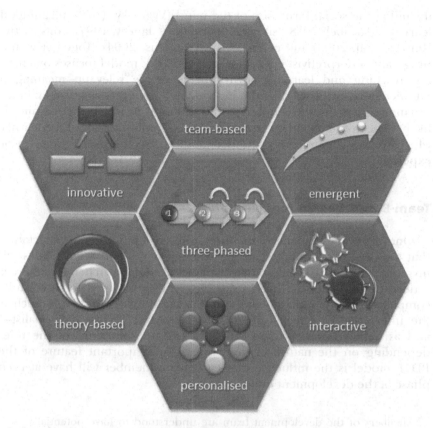

Figure 16.3.   Components of proactive design for learning (PD4L).

- *interactive,* enabling participants to actively explore the relevance and application of the course content (Allen, 2003; Sims, 2006).
- *personalized,* such that participants are able to apply their own context and situation to the learning outcomes (Sims & Stork, 2007).

These six factors are discussed in more detail in the following sections.

Theory-Based Design for Learning

The first key element of the model relates to the range of theories and approaches which underpin the success of online teaching and learning. In terms of learning theories, the PD4L model is founded on theories including the social formation of the mind (Vygotsky, 1978), meaningful learning (Ausubel, 1968), situated cognition (Clancey, 1997), constructivism (Driscoll, 2005) and connectivism (Siemens, 2004). Together with a pragmatic, interpretivist epistemology, the PD4L model focuses on creating teaching and learning environments where relevant, meaningful knowledge is constructed by the individual. These theories also embrace learning as activities which mirror the situation in which the learning is likely to be applied, and as such propose an empowerment of the learner which values their current understanding of the world and prior experience.

Team-Based Design for Learning

One of the key elements of the three-phased "engine" is an expectation that the creation of online teaching strategies and learning activities will involve a long-term, collaborative team of stakeholders. Therefore there is a need to articulate the likely roles that will be required to successfully complete the development. Sims and Jones (2003) identified roles such as the Interactive Architect, Content Specialist and Technical Specialist— and as detailed earlier there can be a certain "blurring" of the roles depending on the nature of the project. An important feature of the PD4L model is the influence that each team-member will have at each phase of the development effort:

> members of the development team are understood to have potential levels of influence at any stage of the development and delivery process. For example the Interactive Architect, who has the main responsibility (influence) for creating the design specifications, may also be active in the quality review of the project as it nears completion. (Sims & Jones, 2003, p. 15)

And within this context there is the importance for team members to embrace new sets of competencies to ensure a strong alignment with the affordances on the online environment (Sims & Koszalka, 2008).

The development of a team that is also focused on continuous improvement will also help develop a community of practice (Wenger, 1998), where all participants on the online development environment actively contribute and participate, integrating the three dimensions identified by Wenger (1998): as a joint enterprise understood and continually renegotiated by its members, as a mutual engagement that binds members together into a social entity and as a shared repertoire of communal resources.

Design for Personalized Learning

The notion of personalized or individualized learning is by no means new, and much was written about adaptive learning systems to meet the specific needs of the learner. More recently, these needs have been addressed through concepts such as Multiple Intelligences (Gardner, 1983), Learning Styles (e.g., Kolb, 1984) and Cognitive Load (Sweller, 1988). The perspective taken with PD4L is, however, somewhat different, especially when considering the online learning context which theoretically may consist of diverse learners spread across the globe. As documented in Sims and Stork (2007) the role of the designer shifts from assessing the characteristics of the learner (such as through the traditional target audience analysis) to focusing on creating an environment where the learner is able to apply individual traits such as their prior knowledge, specific context and situation, preferred learning style, cultural framework, and media preferences.

Design for Interactive Learning

What makes a learning experience interactive? As a designer it is essential to create conditions for learning whereby at any point within the course structure the learner is engaged in an activity that is directly related to the learning activity—and that completion of that activity is critical to moving towards finishing the learning activity. A number of self-paced courses have been designed so that, literally, a learner could gain mastery by closing their eyes and pressing the enter key repeatedly. This is not interaction and it is not interactive learning. Without risk, or challenge, or a means to engage (Allen, 2003) the design effort will have failed.

The majority of research on interactivity identifies the conditions for interaction—such as learner-to-learner or learner-with-content—but the value of that interaction is what is important when applying PD4L principles. As detailed by Sims (2006) the most important of these is to ensure the learner is engaged in some cognitive activity, such as adjusting variables (experimenting), testing assumptions (hypothesizing), introducing content (modifying), constructing solutions (manipulating) or situating the learning within their own experience and environment (contextualizing).

A second area that can enhance the interactive dynamic is where the learner has a clear understanding of their role in the learning process; it is more than interacting with other learners or the content, it is about the learning having an individual purpose. In addition, interaction without feedback (whether immediate, conditional or delayed) will detract from the learning process. For the designer, the key issues is to be able to respond by articulating, for any part of the course, what the learner will be doing in terms of cognitive engagement.

Innovative Design Practice

Integrating the concepts proposed for proactive evaluation of online teaching and learning (Sims, Dobbs, & Hand, 2002), this component of PD4L addresses a set of factors which, if applied and considered, will ensure the integrity of the online teaching and learning environment. The first of these relates to ensuring there is a clear definition of the *strategic intent* of the course—why it is required, who it is for and what the desired outcomes are. Often academic staff are familiar with a course that is content-rich but outcome poor—and establishing a clear link between assessment, learning outcomes and delivery strategy is an essential part of effective educational practice. By addressing this, a clear focus can be placed on optimal teaching strategies, associated learning activities (including assessment) and the context in which those activities will take place.

The second factor concerns *content*. Traditionally, many courses have been designed around the delivery of content, but with the advent of extensive and reliable Web-resources, it is argued that it is the learners who can take more responsibility for accessing and identifying relevant content, and the teachers who need to focus on the strategies by which learners will make sense of that content.

A third factor relates to the actual environment and *interface* in which the teachers and learners will be working. While there are well-accepted standards for interfaces, and the learning management systems in use have clear but unexciting designs, when considering the interface as representing a learning environment, it becomes incumbent on the designer to consider the options. Within the PD4L context it is recommended that an interface provide the course participants with a clear context or metaphor within which to work. For example, creating the sense for participants that they are part of an overarching narrative where each task they complete is related to working through that narrative sequence. Importantly this does not mean focusing on extensive audio-visual effects and artifacts, but a clear context and purpose for the various learning activities.

The fourth factor focuses on assessment, and the importance of ensuring that all learning activities are related to the assessment outcomes—to the extent that learning activities and the resulting products become the assessment item. In working towards this goal, there are innovative options other than all assessment being teacher-managed; strategies such as self or peer assessment can also be used effectively to develop a sense of reflective practice within the course participants.

Finally, the *delivery* elements of the system must be considered. This includes the support provided to students, especially as there are many off-campus and distance students making use of the affordances of online learning. In addition delivery must include an assessment of the value the content will have to the student through copyright and accessibility provisions. Finally, in packaging a course for delivery, the reliability, reusability, flexibility and scalability of resources must be taken into consideration.

Design for Emergent Learning

Emergence is a concept that Steven Johnson (in Sims & Dornfest, 2002) suggests is "what happens when the whole is smarter than the sum of its parts" and occurs from the bottom-up without any explicit control. If emergence is applied to the educational setting, then we have the notion of knowledge generation through the interaction of participants, and without the presence of a "teacher." The design for an emergent learning component does not advocate anarchy in education, but rather emphasizes the potential to allow knowledge emergence to occur instead of defining a curriculum based on fixed content. In the online environment where collaboration and communities of practice are considered desirable because of the social learning outcomes, emergence has particular relevance because it allows the designer to focus on the students being able to take advantage of the dynamic nature of the environment through contributing content and knowledge based on prior-experience, rather than relying on teacher-prescribed materials (Irlbeck et al., 2006).

Rather than assuming in our respective fields that knowledge is fixed and that domain knowledge is something students should learn, what if environments could be created where the potential for enhancing that knowledge base was enabled? Empowering students to challenge and question existing theories and principles is an essential component of the model presented in this chapter and, it is argued, is critical for the successful implementation of online teaching and learning environments. The implications for the "traditional" instructional designer is that control must be relinquished by not designing for someone, but assigning that someone design privileges. The online environment is complex, flexible, organic and dynamic, and creating the space from which new perspectives on existing knowledge can emerge is recommended, even more so when there is the potential for informal participants in course structures.

CONCLUSION

As higher education is changing, and online learning becoming more embedded, new models are required to meet these changes and challenges, especially those that support academic staff with minimal experience in online design and development. The Proactive Design for Learning (PD4L) model proposed in this chapter articulates a re-shaping of the traditional instructional design process where specific aspects of development and delivery are viewed in parallel and interconnected rather than in sequence. Integrating these six components within the three-phase framework creates a model for designers and developers that not only supports the team-based and continuous improvement requirements but also provides theory-based prescriptions for sound learning design.

REFERENCES

Allen, M. W. (2003). *Michael Allen's Guide to e-Learning*. Hoboken, NJ: Wiley.

Ausubel, D. P. (1968). *Educational psychology: A cognitive* view. New York: Holt, Rinehart & Winston.

Clancey, W. J. (1997). *Situated cognition: On human knowledge and computer representations*. New York: Cambridge University Press.

Driscoll, M. P. (2005). *Psychology of learning for instruction*. Boston: Pearson.

Gardner, H. (1983). *Frames of mind: The theory of multiple intelligences*. New York: Basic Books.

Irlbeck, S., Kays, E., Jones, D., & Sims, R. (2006). The phoenix rising: Emergent models of instructional design. *Distance Education, 27*(2), 171–185.

Kolb, D. A. (1984) *Experiential Learning: experience as the source of learning and development* Englewood Cliffs, NJ: Prentice-Hall.

Siemens, G. (2004). *Connectivism: A learning theory for the digital age*. Retrieved from http://www.elearnspace.org/Articles/connectivism.htm

Sims, D., & Dornfest, R. (2002). *Steven Johnson on "Emergence."* Retrieved from http://www.oreillynet.com/pub/a/network/2002/02/22/johnson.html

Sims, R. (2006). Beyond instructional design: Making learning design a reality. *Journal of Learning Design, 1*(2), 1–8. (Keynote Paper) Retrieved from http://www.jld.qut.edu.au/

Sims, R., Dobbs, G., & Hand, T. (2002). Enhancing quality in online learning: Scaffolding design and planning through proactive evaluation. *Distance Education, 23*(2), 135–148.

Sims, R., & Jones, D. (2003). Where practice informs theory: Reshaping instructional design for academic communities of practice in online teaching and learning. *Information Technology, Education and Society, 4*(1), 3–20.

Sims, R., & Stork, E. (2007). Design for contextual learning: Web-based environments that engage diverse learners. In J. Richardson & A. Ellis (Eds.),

Proceedings of AusWeb07. Lismore, NSW: Southern Cross University. Retrieved from http://ausweb.scu.edu.au/aw07/papers/refereed/sims/index.html

Sims, R., & Koszalka, T.A. (2008). Competencies for the new-age instructional designer. In J. M. Spector, M.D. Merrill, J. van Merriënboer, & M. P. Driscoll (Eds.), *Handbook of research on educational communications and technology* (3rd ed., pp. 569–575). New York: Taylor & Francis.

Sweller, J. (1988). Cognitive load during problem solving: Effects on learning. *Cognitive Science, 12*, 257–285.

Vygotsky, L. S. (1978). *Mind and society: The development of higher mental processes*. Cambridge, MA: Harvard University Press.

Wenger, E. (1998). *Communities of Practice: Learning as a social system*. Retrieved July 29, 2002, from http://www.co-i-l.com/coil/knowledge-garden/cop/lss.shtml

CHAPTER 17

DESIGN-BASED RESEARCH AND TECHNOLOGY-ENHANCED LEARNING ENVIRONMENTS

Feng Wang and Michael Hannafin

During the past decade, design-based research has demonstrated its potential as a methodology suitable to both research and design of technology-enhanced learning environments (TELEs). In this chapter, we define and identify characteristics of design-based research, describe the importance of design-based research for the development of TELEs, propose principles for implementing design-based research with TELEs, and discuss future challenges of using this methodology.

During the past decade, literature on the design of technology-enhanced learning environments (TELEs) has flourished. Multiple TELE theoretical frameworks, especially those based on constructivist epistemology (Cognition and Technology Group at Vanderbilt [CTGV], 1992a, 1992b;

Previously published in *Educational Technology Research and Development* (2005), *54*(4), 5–23.

Hannafin, Land, & Oliver, 1999; Savery & Duffy, 1996), have been proposed. TELEs are technology-based learning and instructional systems through which students acquire skills or knowledge, usually with the help of teachers or facilitators, learning support tools, and technological resources (Aleven, Stahl, Schworm, Fischer, & Wallace, 2003; Land, 2000; Shapiro & Roskos, 1995). In recent years, with the rapid development of new technologies (e.g., computers, wide-area Internet, and PDAs), TELEs have generated considerable enthusiasm within the design community. However, as with previous teaching-learning innovations, design and research have evolved in a largely sequential manner, with little direct influence on practice. As a result, TELEs have not been widely used by either students or teachers (Cuban, 1986, 2001; Kent & McNergney, 1999).

The design-based research paradigm, one that advances design, research and practice concurrently, has demonstrated considerable potential. Advanced initially by Brown (1992) and Collins (1992) as design experiments, design-based research posits synergistic relationships among researching, designing, and engineering. Design experiments manifest both scientific and educational values through the active involvement of researchers in learning and teaching procedures and through "scientific processes of discovery, exploration, confirmation, and dissemination" (Kelly, 2003, p. 3). Design-based research challenges the assumption that research is contaminated by the external influence of the researcher (Barab & Kirshner, 2001). Instead, researchers manage research processes in collaboration with participants, design and implement interventions systematically to refine and improve initial designs, and ultimately seek to advance both pragmatic and theoretical aims affecting practice.

In many ways, design-based research is intrinsically linked to, and its development nourished by, multiple design and research methodologies. Researchers assume the functions of both designers and researchers, drawing on procedures and methods from both fields, in the form of a hybrid methodology. For example, design-based research requires significant literature review and theory generation, uses formative evaluation as a research method, and utilizes many data collection and analysis methods widely used in quantitative or qualitative research (Orrill, Hannafin, & Glazer, 2003; Reigeluth & Frick, 1999). In these regards, design-based research does not replace other methodologies, but rather provides an alternative approach that emphasizes direct, scalable, and concurrent improvements in research, theory, and practice.

In other ways, however, the convergence of design research, theory, and practice extends current methodologies. For example, participatory action research-a qualitative approach akin to design-based research-involves collaboration between researchers and participants, local practices that support systematic theorizing, and improvement in both theory and practice. However, local improvements in participatory action research typically derive from participants' own research that is facilitated by researchers rather than interventions designed and progressively refined jointly with researchers (see Kemmis & McTaggart, 2000; Patton, 2002; Stringer, 1999). Likewise, intervention design-sometimes equated with formative evaluation-is often undertaken to generate evidence used to guide possible revisions in an ongoing design (Reeves & Hedberg, 2003). Design-based research is both based on, and conducted in order to generate, theory; the simultaneous pursuit of theoretical goals differentiates design-based research from formative evaluation (Barab & Squire, 2004).

In TELE practice, design-based research methods have been utilized widely, including technology-supported inquiry learning (TSIL; Edelson, Gordin & Pea, 1999), Web-based inquiry science environment (WISE) and its forerunner knowledge integration environment (KIE; Bell & Linn, 2000; Linn, Clark, & Slotta, 2003; Linn, Davis, & Bell, 2004), the Jasper Woodbury Series (CTGV, 1992a, 1992b, 1997), biology guided inquiry learning environment (BGuILE; Reiser et al., 2001; Sandoval & Reiser, 1998, 2004), and computer-supported intentional learning environments (CSILE; Hewitt & Scardamalia, 1998; Scardamalia & Bereiter, 1994). The purposes of this paper are to define and identify characteristics of design-based research, describe its importance for the development of TELEs, propose principles for implementing design-based research in TELEs, and identify future prospects for design-based research in TELE and instructional design.

AN INTRODUCTION TO DESIGN-BASED RESEARCH

We use the term design-based research (Design-Based Research Collective [DBRC], 2003) to encompass a paradigm described using different terms in the literature, including design experiments (Brown, 1992; Collins, 1992), design research (Cobb, 2001; Collins, Joseph, & Bielaczyc, 2004; Edelson, 2002), development research (van den Akker, 1999), developmental research (Richey, Klein, & Nelson, 2003; Richey & Nelson, 1996), and formative research (Reigeluth & Frick, 1999; Walker, 1992). As summarized in Table 17.1, each has a slightly different focus, but the underlying goals and approaches are similar.

To underscore both the similarities among and distinctions between design-based research and related methods, we define design-based research as a systematic but flexible methodology aimed to improve educational practices through iterative analysis, design, development, and implementation, based on collaboration among researchers and practitioners in real-world settings, and leading to contextually-sensitive design principles and theories. The five basic characteristics: (a) pragmatic; (b) grounded; (c) interactive, iterative, and flexible; (d) integrative; and (e) contextual, are summarized in Table 17.2 and illustrated in the following sections, and represent a synthesis of related approaches shown in Table 17.1. As noted previously, many characteristics are not unique to design-based research, but rather the nature of their use varies and the approaches are often extended in design-based research.

Pragmatic

Researchers address practical issues to promote fundamental understanding about design, learning, and teaching (Orrill et al., 2003). The Jasper Series (CTGV, 1997), for example, was developed and improved through its applications, progressively refining the theory of anchored instruction that has widely informed TELE design and practices. Similarly, the fostering communities of learners (FCL) project (Brown & Campione, 1996), conducted in inner city elementary schools for more than a decade, typifies this synergy as researchers collaborate with teachers and students. Following iterative design, development, and implementation, learning principles useful for both conceptual understanding and practical dissemination are generated, based on the research procedures and settings.

From a design-based research perspective, theory development is inextricably linked to practice (Brown & Campione, 1996); research should refine both theory and practice (Collins et al., 2004) as well as provide new possibilities. Ultimately, the value of theory is appraised by the extent to which principles and concepts of the theory inform and improve practice (Cobb, Confrey, diSessa, Lehrer, & Schauble, 2003; DBRC, 2003; Greeno, Collins, & Resnick, 1996). In addition to asking whether a theory works, researchers further question how well the theory works; that is, whether a given theory is better (i.e., more effective in achieving the design goals, cost efficient, and appealing to stakeholders) than known alternatives to attaining a desired outcome, and how research might refine the theory (Reigeluth & Frick, 1999). The pragmatic goal of design-based research is continually reified through disciplined application of its methodologies and research processes.

Table 17.1. Design-Based Research Variants and Methods

Variant and Reference	Method
Design-based research (Design-Based Research Collective (2003)	• Often conducted within a single setting over a long time. • Iterative cycles of design, enactment, analysis, and redesign. • Contextually dependent interventions. • Document and connect outcomes with development process and the authentic setting. • Collaboration between practitioners and researchers. • Lead to the development of knowledge that can be used in practice and can inform practitioners and other designers.
Design experiments (Collins, 1992, 1999)	• Comparison of multiple innovations. • Characterizing the messy situation. • Multiple expertise in design. • Social interaction during design. • Flexible design revision and objective evaluation. • Developing a profile as findings.
Design research (Edelson, 2002)	• Designs both directly propel the development of practice and improve researchers' understanding. • Four characteristics: research driven, systematic documentation, formative evaluation, generalization. • Design generates three types of theories: domain theories, design frameworks, design methodologies; these theories go beyond the specific design context.
Developmental research (van den Akker, 1999) Development research (van den Akker, 1999)	• Begin with literature review, expert consultation, analysis of examples, and case studies of current practice. • Interaction and collaboration with research participants to approximate interventions. • Systematic documentation, analysis, and reflection on research process and outcomes. • Using multiple research methods; formative evaluation as the key activity. • Empirical testing of interventions. • Principles as generated knowledge in the format of heuristic statements.
Developmental research (Richey, Klein, & Nelson, 2003)	• Type 1 (emphasizing specific product or program) and • Type 2 (focusing on the research process). • Begin with defining research problem and reviewing related literature. • Different participating populations in Type 1 and Type 2 developmental research during different phases. • Various forms of data collection depending on the research focus. • Employ multiple research methods, such as evaluation, field observation, document analysis, in-depth interview, expert review, case study, survey, etc. • Data analysis and synthesis includes descriptive data representations, quantitative and qualitative data analysis. • Reports of development research are long and can be published in variou

Table continues on next page.

Table 17.1.   Continued

Variant and Reference	Method
Formative research (Reigeluth & Frick, 1999)	• Drawn from case-study research and formative evaluation. • Used to improve instructional systems and to develop and test design theory in education. • Preferability (i.e., effectiveness, efficiency, and appeal) over validity. • Two types: (a) designed case studies and (b) naturalistic case studies.

Table 17.2.   Characteristics of Design-Based Research

Characteristics	Explanations
Pragmatic	• Design-based research refines both theory and practice. • The value of theory is appraised by the extent to which principles inform and improve practice.
Grounded	• Design is theory-driven and grounded in relevant research, theory and practice. • Design is conducted iin real-world settings and the design process is embedded in, and studied through, design-based research.
Interactive, iterative, and flexible	• Designers are involved in the design process and work together with participants. • Processes are iterative cycle of analysis, design, implementation, and redesign. • Initial plan is usually insufficiently detailed so that designers can make deliverate changes when necessary.
Integrative	• Mixed research methods are used to maximize the credibility of ongoing research. • Methods vary during different phases as new needs and issues emerge and the focus of the research evolves. • Rigor is purposefully maintained and discipline applied appropriate to the development phase.
Contextual	• The research process, research findings, and changes from the initial plan are documented. • Research results are connected with the design process and the setting. • The content and depth of generated design principles varies. • Guidance for applying generated principles is needed.

Grounded

Before conducting design-based research, researchers select a theory about learning and instruction. They examine literature and available design cases, and identify gaps to ensure the value of the research (Edelson, 2002) and to identify existing problems or issues (e.g., specific student learning abilities; Cobb et al., 2003). In subsequent efforts, they seek to revise and refine that theory-an "anchor" that determines which interventions should (or should not) be introduced and which should be eliminated. BGuILE (Reiser et al., 2001), for instance, is derived from an analysis of literature related to scientific inquiry. This analysis led to its two major theoretical goals-(a) observational investigations and (b) theory articulation-which are supported by all design efforts, ranging from determining the characteristics of inquiry products, through selecting investigation strategies, to designing tools and artifacts (Reiser et al., 2001). The theory-driven nature of design-based research is important in that its approaches are considered more a research paradigm than an evaluation method. Without underlying theory support for both the framework and design procedures employed, results often fail to inform theory development for design innovation in education (Collins, 1992). Thus, the methods need to be grounded in relevant research, theory, and practice to develop future innovations and designs.

Design-based research is also grounded in real-world contexts where participants interact socially with one another, and within design settings rather than in laboratory settings isolated from everyday practice (Brown & Campione, 1996; Collins, 1999). Thus, design-based researchers address simultaneously the multitude of variables evident in real-world settings (Collins, 1992, 1999). Researchers observe different aspects of the design using both quantitative and qualitative methods, address associated problems and needs, and document why and how adjustments are made (Collins et al., 2004). Furthermore, by embedding research within practical activities, the design processes themselves are studied. The resulting principles are perceived as having greater external validity than those developed in laboratory settings (Greeno et al., 1996) and as better informing long-term and systemic issues in education (Bell, Hoadley, & Linn, 2004). Thus, the design process is embedded in, and studied through, design based research.

Interactive, Iterative, and Flexible

Design-based research stresses collaboration among participants and researchers throughout the processes (Cobb et al., 2003). Because of dynamic and complex relationships between theory and practice, direct theory application without practitioner interaction is often not feasible; thus, researchers and practitioners work together to identify approaches and develop principles to address these problems (Schwartz, Lin, Brophy, & Bransford, 1999; van den Akker, 1999). Although distinctions among designers, researchers, and participants are blurred in design-based research processes (Bannan-Ritland, 2003), researchers manage the design process, cultivate the relationship with practitioners, and most importantly, develop their understanding of the research context (Cobb et al., 2003). In WISE, Linn et al. (2003) provided an initial flexible framework that incorporated both general technology features and instructional resources, and teaching-learning strategies. However, many features ultimately emerged from or were adapted based on input from participants. Project partner-participants (i.e. design teams, including classroom teachers, pedagogy researchers, curriculum designers, technologists, and discipline experts) developed, tested, and refined their individual inquiry projects based on the framework, helping to refine WISE content, user interface, and affordances. Hence, design researchers seek to shape the local learning environment by applying their expertise to improve educational practice (Barab & Kirshner, 2001; van den Akker, 1999); likewise, research and theory evolve in concert with advances in practice, ensuring that complementary expertise and different perspectives contribute to the design (DBRC, 2003). With the involvement of both researchers and participants, emerging local issues can also be addressed in an efficient and timely manner. Consequently, the design may be better optimized given the constraints of the local setting and addressing participant concerns.

Design-based research is also characterized by an iterative cycle of design, enactment or implementation, analysis, and redesign (DBRC, 2003). Outcomes from previously conducted designs provide explanatory frameworks "that specif[y] expectations that become the focus of investigation during the next cycle of inquiry" (Cobb et al., 2003, p.10). For example, research conducted prior to the emergence of the Jasper Series revealed inert-knowledge problems (i.e., knowledge recallable but not applied to solving problems). To address this problem, CTGV researchers initiated and subsequently focused their research on anchored instruction through which instruction was situated in meaningful, problem-rich learning environments (CTGV, 1992a, 1997).

Design-based research processes are also flexible, as collaborators seek to improve an initial design plan through implementation. Schwartz et al. (1999) suggested that designs should be flexibly adaptive but "consistent with important principles of learning" (p. 189). During implementation, a theory emerges based on the accumulated data collected during successive iterations as well as the implementation experiences of the designers (Edelson, 2002). The theoretical framework upon which the design is based may be extended and developed; in some cases, a new framework may emerge. Initial design plans may be insufficiently detailed to account for emerging patterns, so changes are anticipated and implemented when necessary (Cobb et al., 2003; Collins, 1999; Edelson, 2002).

In addition to improving the ongoing design, researchers also consider the influence of en route changes on the integrity of the design. During FCL development, Brown and Campione (1996) found that adaptive reciprocal teaching (RT) reading strategies were less useful than research seminar (RS) for older students. However, the researchers chose not to simply replace RT with RS for older students because the RT functions were also linked to other key activities in the project (e.g., information sharing, student writing or publishing). In other words, "any changes to one aspect of the design" need to be compatible "with other aspects of the design" (Collins et al., 2004, p. 19). Thus, researchers need to balance their roles as designer and researcher to ensure that practical constraints are considered, alternative perspectives are provided, and discipline in the inquiry is ensured (van den Akker, 1999).

Integrative

Design-based research draws from a variety of widely used approaches, such as survey, expert review, evaluation, case study, interview, inquiry methods, and comparative analysis (see, e.g., McCandliss, Kalchman, & Bryant, 2003; Richey et al., 2003). By using a combination of methods, data from multiple sources increase the objectivity, validity, and applicability of the ongoing research. Sandoval and Reiser (2004) conducted design-based research on Explanation-Constructor, a tool "designed to support students' construction and evaluation of explanations through their inquiry" (p. 348). In order to understand the role of their tool in students' epistemic practices, researchers videotaped groups of student activities and analyzed them using interaction analysis. Likewise, in order to "understand students' practices of explanation evaluation" (p. 363), researchers collected self-assessments and peer critiques from student activities and subsequently conducted documentation analyses.

Methods may also vary as new needs and issues emerge and the focus of the research evolves. Researchers may initially conduct observations to document changes in the classroom environment while using surveys or tests to collect data on student performance. During development, the emphasis on quality shifts from validity to practicality and effectiveness, and design researchers may employ expert appraisals, tryouts, microevaluations, or field tests (van den Akker, 1999) as warranted by the changing research focus. For example, during early stages of a TELE design, researchers may focus on the robustness of its theoretical anchors and the consistency between the planned interventions and the theoretical goals of the research. When developing and implementing the design, however, they may put more emphasis on the feasibility of the design in the classroom, and assess whether the theoretical goals can be achieved through the interventions. Rigor is purposefully maintained, ensuring adherence to discipline and scientific research standards and conventions (Shavelson, Phillips, Towne, & Feuer, 2003).

Retrospective analysis and formative evaluation are employed by some design researchers. Through retrospective analysis of collected data and design events, evidence-based claims and results are examined in concert with the underlying design theory; implicit design elements become explicit to further guide subsequent analysis and research activities (Battista & Clements, 2000; Cobb et al., 2003; Edelson, 2002). During diSessa and Cobb's (2004) design-based research on the teaching of physics, several issues emerged beyond the planned focus of their study. The researchers expected that students would neither intensively engage in designing graphs about motion nor continue to discuss possibilities for improving their graphs after class. Retrospective analysis of newly emerged issues enabled the researchers to identify a phenomenon known as meta-representational competence—students' prior knowledge that supported their abilities to create, critique, and adapt scientific representations. This competence was not initially anticipated from existing literature, but became the focus of their subsequent inquiry.

Formative evaluation typically focuses on the local design, exposes issues to be addressed through design research, and enables researchers to identify problems and gaps (Edelson, 2002; Reigeluth & Frick, 1999; van den Akker, 1999). In the Jasper project (CTGV, 1997), for instance, formative evaluation revealed that the Jasper challenge series was especially effective when students have opportunities "to engage in problem-based curricula" (p. 108). Consequently, CTGV researchers recommended that "teachers provide students with multiple opportunities to 'Identify problems to be solved, Develop plans, Act on them, Receive feedback and Revise as necessary.' " (p. 104).

Contextual

According to design-based research advocates (e.g., Brown & Campione, 1996; DBRC, 2003; van den Akker, 1999), research results need to be connected with both the design process through which results are generated and the setting where research is conducted. The findings generated from design-based research take many forms. They may be comparative profiles akin to a consumer report (Collins et al., 2004), principles in the form of heuristics, case studies, or longitudinal studies. The findings are more than prescribed activities to be followed by other designers; they transcend the immediate problem setting and context to guide designers in both evolving relevant theory and generating new findings. According to van den Akker (1999), the generalizability of findings increases when they are validated in "successful design of more interventions in more contexts" (p. 9).

Two studies on CSILE underscore the importance of context. Consistent with literature indicating that student online discussion promotes equality, Scardamalia et al. (1992) and Hewitt (1996) found that students at different ability levels perform equally when using computers as their discourse medium. However, compared to face-to-face communications, students' online communication also results in less immediate feedback from others on their individual work. To address these issues in classroom practices, teachers encourage collaboration between students to review their peers' work during their CSILE sessions. These studies and implementations led CSILE researchers to identify a design principle- "support educationally effective peer interactions." This principle is "particularly effective in fostering educationally-beneficial distributed practices" (Hewitt & Scardamalia, 1998, p. 56) when used together with other related CSILE principles (e.g., integrating different forms of discourse, emphasizing the work of the community).

The research process, the research findings, and changes from the initial research plan have been documented; warrants, claims, and guidance on the use of resulting principles have been provided (Shavelson et al., 2003). Thus, interested researchers or designers can trace the emergence of an innovation or combinations of innovations according to their interests, examining closely contextual factors or conditions that led to particular effects (Baumgartner & Bell, 2002). The content and depth of design principles vary. Principles may be generic and based on the findings of multiple research results, or, content specific to assist direct action (Bell et al., 2004). A series of design studies on the Computer as Learning Partner curriculum, focusing on science learning and instruction, generated four generic, but cornerstone, principles of the scaffolded knowledge framework: (a) making science accessible, (b) making thinking visible, (c) helping students learn from others, and (d) promoting autonomy and lifelong learning (Linn & Hsi, 2000). In contrast, Edelson et al.'s (1999) research on TSIL, featuring scientific visualization technologies in the geosciences, generated two content-specific principles: (a) "the design of investigation tools could ... [address] the challenges of motivation, accessibility, and practical constraints" (p. 442); (b) "knowledge resources and record-keeping tools" (p. 444) are necessary process supports for inquiry-based learning. These principles were particularly helpful for the specific inquiry-based learning case under study, but may not apply across domains.

Given the assumption that comparable performance is most likely in similar settings, contextually relevant design principles and knowledge are important for design-based researchers. Because of the complex and dynamic nature of education, a myriad of context-specific and context-dependent variables influences any given innovation (Brown & Campione, 1996; Collins, 1999; van den Akker, 1999). Therefore, the results from broadly contextualized research methods may prove too global and abstract to be useful in many settings (Baumgartner & Bell, 2002; Cobb, 2001; Cobb et al., 2003).

In contrast, the principles derived from typical design-based research are relevant to designs and development tasks where parallel contextual conditions exist (van den Akker, 1999). Brophy (1998) designed the Questioning Environment to Support Thinking (QUEST) project that "structures media resources to help students sustain their own inquiry during problem solving" (p. 6). QUEST employs a four-stage problem-solving model: (a) problem presentation, (b) information exploration, (c) discovery, and (d) reflection on solutions. QUEST'S designers referenced anchored instruction research and theory (CTGV, 1997) extensively, providing relevant references on design and implementation of meaningful, problem-oriented activities to facilitate learning. More importantly, these principles are systematically aligned with the research context. They may prove ineffective when used alone but they can be modified, replaced, or adapted by others provided the system itself remains unaffected. Thus, researchers attempt to analyze the relationship between principles (e.g., the order of implementing them, the interdependencies between them) so that the design procedures they employ in the original setting will likely prove effective in new settings (Brown & Campione, 1996). Guidance for applying generated principles is needed to increase the adaptability, and ultimately the generalizability, of the research.

IMPORTANCE OF DESIGN-BASED RESEARCH FOR TELES

Design-based methodologies are especially important considering that TELEs have often been developed using incompatible or contradictory theoretical and epistemological foundations (Hannafin, Hannafin, Land, & Oliver, 1997). Consequently, gaps are evident between what a TELE is and how it should be used in theory compared with what it is and how it is used in practice. Alternative approaches are needed to align learning environments with their fundamental assumptions (Hannafin et al., 1997; Jonassen & Rohrer-Murphy, 1999) and "encourage flexibility as well" (Schwartz et al., 1999, p. 189). Design-based research emphasizes closely linked strategies for developing and refining theories rather than testing intact theories using traditional methodologies (Edelson, 2002). Design-based research guides theory development, improves instructional design, extends the application of results, and identifies new design possibilities (Cobb et al., 2003; Edelson, 2002; Gustafson, 2002; Reigeluth & Frick, 1999). Design-based research can "help create and extend knowledge about developing, enacting, and sustaining innovative learning environments" (DBRC, 2003, p. 5).

Several aspects of design-based research are consistent with TELE design theories (e.g., iterative design process, collaboration with participants), which in turn are helpful to the development of design-based research methods. In the following sections, we highlight three implications of design-based research for TELEs: (a) encouraging continuous synergy, (b) refining TELE theory, and (c) encouraging socially responsible and responsive inquiry and practice.

Encouraging Continuous Synergy

In traditional instructional design (ID) and instructional systems design (ISD) approaches, design and research are typically related, but separate, activities. Research is usually conducted after ID/ISD processes have been completed, to test the design's effectiveness rather than to address issues of educational practice (Cobb et al., 2003). The emergence of grounded design practice and design-based research addresses a core TELE problem: the lack of clearly defined and enacted theoretical frameworks applicable to practice. In ID/ISD practice, the need for an integral relationship between design and research is underscored in Hannafin et al.'s (1997) criteria for grounded design practice: (a) designs must be based on a defensible or widely acknowledged theoretical framework; (b) methods must be consistent with the outcomes of research conducted to test, validate, or extend the theories on which they are based; (c) designs are generalizable; and (d) designs and their frameworks are validated iteratively through successive implementation.

Design-based research posits synergy between practice and research in everyday settings. This synergy engenders simultaneous refinements of theory and practice as theory is generated and refined through its application; in effect, educational approaches and theory emerge reciprocally (Bell et al., 2004). Synergy helps to generate principles that inform the design itself as well as the thinking and actions of researchers, designers, and practitioners. Design-based research can extend and develop both grounded design practices generally and TELE design theories specifically.

Accordingly, TELE design and research activities can become more reciprocal: the design of learning environments and development of learning theories can be intertwined (DBRC, 2003). Strong theoretical anchors support design work, forming "the most immediate foundation for the discipline in which the original problem arose" (Winn, 1997, p. 38). Theories generated from designs are often supported by examining learning in naturalistic contexts and through developing innovations, technological tools, and theories (Barab & Squire, 2004). Inconsistencies between theory and practice can be revealed through "the practical process of applying a theory to construct a design" (Edelson, 2002, p. 118). The theories are of practical use to resolve problems and cannot be generated by "either isolated analysis or traditional empirical approaches" (p.118).

Research supports design reciprocally in design-based research, providing frequent and often subtle refinement guided by detailed data (Cobb, 2001). Designs are evidence based, that is, they engender tangible changes in TELE practice, ranging from the impact to the ongoing design resulting from a specific innovation or a combination of innovations, to the influences of theories generated from other TELE research. TELE designers use evidence to refine the design, to address new or emerging issues, to support new theory or approaches to deepen understandings of TELE research and practice, and to guide further research and theory construction. Moreover, design-based research enables the creation and study of learning conditions that are presumed effective but are not well understood in practice, and the generation of findings often overlooked or obscured when focusing exclusively on the summative effects of an intervention (DBRC, 2003). In effect, design is embodied in research, and research is embodied in design.

The synergy between design and research is typified in Sandoval and Reiser's (2004) refinements of Explanation Constructor. Initial classroom research provided detailed confirmatory evidence that their tool guided student inquiry as predicted from its conceptual framework, which focused on "the influence of epistemological commitments on strategies for pursuing inquiry" (p. 347). However, their data indicated that the tool was unsuccessful in supporting student ideas or interpreting the data they needed to explain. The tool was revised accordingly, enabling students to cite data directly in their explanations and incorporating a review feature to support assessment. Studies on the revised Explanation Constructor resulted in principles related to the design of tools for supporting students' scientific practice, and structuring explanations, and subsequent evaluation, as well as limitations in supporting scientific argumentation.

Refining TELE Theory

Richey et al. (2003) proposed two types of developmental research. Type 1 research is context specific; conclusions typically take the form of lessons learned from the development of a specific product and conditions that improve the effectiveness of that product. Type 2 research, in contrast, yields generalizable design procedures or principles. Likewise, Edelson (2002) identified three types of theories of potential relevance to TELE: (a) domain theories, (b) design frameworks, and (c) design methodologies. Domain theories are descriptive in nature and concern the nature of the problem or issue under study, such as the challenges and opportunities in a middle-school science course, and findings associated with students' using an online learning environment of scientific investigation. Design frameworks are systemic guidelines and generalized solutions to achieve an array of goals in a specific context, such as open-ended learning environments (Hannafin et al., 1999) and goal-based scenarios (Schank, Fano, Bell, & Jona, 1994). Design methodologies are generic procedures that guide the process, such as how to achieve a design goal and develop the needed expertise. Both design frameworks and design methodologies are prescriptive in nature.

In TELE designs, new ideas can emerge from decision-making processes in the form of context-based knowledge and meta-design knowledge. Both Type 1 and Type 2 research identified by Richey et al. (2003) are emphasized in design-based research. Context-based knowledge focuses on problems and issues specific to a given TELE design, including relevant domain theories and knowledge generated from Type 1 research. Meta-design knowledge emphasizes principles, procedures, and frameworks that provide more generally useful design guidance, including design frameworks, design procedures, and knowledge generated through Type 2 research. Context-based knowledge and meta-design knowledge are interwoven in design-based research iterative design, development, and implementation processes.

Both types of knowledge were generated during the development of the Jasper Series. Context-based knowledge resides in many domains, including curriculum design, instruction and assessment, formative assessment, and teacher learning and learning communities. Meta-design knowledge includes new design frameworks, such as anchored instruction, and the looking at technology in context framework (CTGV, 1992a, 1992b, 1997). Likewise, Brophy (1998) compared problem-solving processes of students using the simulated QUEST environment or wet-lab equipments. The context-based knowledge that emerged focused on the effectiveness of treatments used in this study: "problem solving contexts encourage qualitative thinking" (p. 25), and instruction started with problems could result in more self-directed learning. The meta-design knowledge that emerged-establishing a learning context for student knowledge building-focused on clarifying the goals of, and evaluating success in, classroom technology integration.

With design-based TELE research, meta-design knowledge and context-based knowledge transcend specific designs for theory development purposes (Edelson, 2002). Meta-design knowledge becomes more credible and applicable because it is based on research results from not only the current design but from related studies as well. Additionally, both context-based knowledge and meta-design knowledge are fully specified. Multiple aspects of practice are reflected through iterative research and continuous refinement (Greeno et al., 1996; van den Akker, 1999). In complex designs, TELE designers can identify the relevance of context-based knowledge derived from other TELEs and reliably anticipate the effectiveness and efficiency of new tools, models, and principles (Richey et al., 2003). They can also avoid mistakes, assimilate valuable experiences from both results and processes of the designs, and decide whether to use or adapt proven approaches in their designs.

Encouraging Socially Responsible and Responsive Inquiry and Practice

According to Fullan (2001), educational researchers must strive harder to improve the circumstances of individuals, as well as policies and resources in both local and remote settings. In recent years, researchers have questioned why educational research has failed to influence practice, the trends and directions of research and development, and the strength of the link between research and practice (Berliner, 2002; Burkhardt & Schoenfeld, 2003). In Collins's (1992) criticism of traditional experimental studies, he concluded that only significant effects are typically tested in a single design; designs are too variable for any valid class conclusions to be drawn; and underlying theories are rarely provided to support the design. In addition, van den Akker (1999) criticized that complex and ambitious reform policies in educational practice are often ill specified; the effectiveness of proposed interventions is unknown, and the implementation process in various contexts is uncertain. Many researchers now seek pragmatic methodologies that invest more genuinely in the practitioners who implement innovation in everyday settings, encouraging the refinement of goal-oriented theories that support practice (Peterson, 1998; Reigeluth, 1997; Robinson, 1998).

Design-based research has the potential to generate theories that both meet teachers' needs and support educational reforms (Reigeluth & Frick, 1999). For example, Fishman, Marx, Blumenfeld, and Krajcik (2004) described their research on integrating technology-enhanced, inquiry-based science curricula as part of the Detroit Public Schools educational reform initiative. They initially designed and implemented technology innovations with design-based research approaches, then attempted to expand the innovations to other settings in the city. During the process, they encountered unforeseen challenges to their implementation, identified teacher preparation and organizational gaps "between the capacity of the district and demands of the innovations" (p. 56), and proposed research to narrow the gaps. Moreover, design reflected teacher perspectives and helped teachers to better understand the implications of TELEs in student learning. Likewise, WISE teachers contributed to its inquiry focus using both their knowledge and awareness of their classroom contexts, while becoming increasingly skilled in guiding student inquiry processes (Linn et al., 2003). Improvements in both local and remote settings, with teachers' concerns addressed and their expertise utilized, may help to promote wider classroom application of TELEs.

PRINCIPLES OF DESIGN-BASED RESEARCH

As with all disciplined inquiry, design-based research implementations need to be both purposeful and systemic. In these regards, design-based research parallels instructional design in many ways. Traditional ID activities are applied to address local design needs and requirements-a goal shared by design-based research. To generate practical, credible, and contextual design theories, however, rigorous, disciplined, and iterative inquiry is needed. Design-based research extends the immediate local goal shared by traditional ID designers to generate pragmatic and generalizable design principles. Therefore, design activities and research activities usually cannot be conducted separately; systematic ID processes can be referred to design-based research procedures. As described in the following sections, we identify nine principles central to planning and implementing TELE design-based research.

Principle One: Support Design With Research From the Outset

Prior to proceeding, designers need to identify resources relevant to their project needs using available literature and design cases from multiple sources, such as journal publications, research reports, conference proceedings, and technical reports. In instances where topic- or issue-specific research cannot be identified, consider literature indirectly linked to the theoretical foundation of the design or extrapolate guidance from related research (Richey et al., 2003).

By analyzing available literature and the design setting critically, designers may also gain different insights as to underpinnings and focus. For example, after reviewing literature on inquiry-based science learning and related design cases, Edelson et al. (1999) initially identified the purpose of the TSIL research as "to understand the opportunities and obstacles presented by scientific visualization as technology to support inquiry-based learning" (p. 392) and focused on technological issues. After analyzing the design setting, however, they identified the need to account for both technological and curricular strategies (e.g., students' management skills, motivation, background knowledge), ultimately developing visualization environments and curriculum to pursue this objective.

Designers can adapt a mature theoretical framework or initiate a new one according to the purpose of the design and features of the setting. For example, in KIE debates, Bell and Linn (2000) utilized the scaffolded knowledge integration framework, which was established through a series of prior investigations. Brown and Campione (1996), in contrast, initiated a new framework in the fostering communities of learners (FCL) project by adapting situated learning theory to support its design purpose-to promote critical thinking and reflection skills.

Principle Two: Set Practical Goals for Theory Development and Develop an Initial Plan

After the purpose has been clarified, designers set specific goals that can be pursued and attained through principled design. Researchers cannot study everything; setting reasonable goals helps to enhance rigor and enforce discipline of the effort (diSessa & Cobb, 2004). The goals are pragmatic in that they aim to address problems in educational practice. For instance, the goals of the Jasper Series (CTGV, 1992b), to improve student ability to solve complex problems, were achieved through "support [for] teachers as they learn to teach with the Jasper materials" (p. 300). Because time and effort may be wasted unnecessarily when significant changes are made late in the process, several design factors (e.g., design setting, available resources) need to be considered early in the process and prior to setting design goals.

Once theory goals have been defined, designers formalize their initial plan. The plan, viewed as an outline strategy designed to achieve the theory goals, will be supported by all design activities. For example, a central innovation in the FCL project is the research-share-perform cycles. All

FCL activity structures support these cycles, including guided viewing and writing, consulting experts, and peer teaching (Brown & Campione, 1996). The plan usually contains descriptions or arrangements of the anticipated research phases and steps, the design setting, design team members, research participants, research methods, and other factors considered initially in design. In addition, the plan is flexible to accommodate inevitable refinements necessary in the design processes.

Principle Three: Conduct Research in Representative Real-World Settings

The research problems associated with a given design arise from needs evident in educational practice. The innovations are derived from both the available literature and the analysis of the prospective real-world design settings. The innovations chosen by TSIL, for example, are used to address problems in classroom experiences, such as students' failure to engage in inquiry within available time and resources (Edelson et al., 1999). Thus, contexts in design-based research need to represent rather than oversimplify typical (but complex) settings to the extent possible. Designers need to account for the influence of social factors and dynamics that affect both design participants and the design processes (e.g., school culture, physical characteristics of classrooms). Brown and Campione's (1996) FCL efforts, for example, are situated in elementary schools as students engage in group and independent activities and share their expertise with other participants. The learning environment, as a consequence, is a natural classroom replete with the flow of potentially competing activities and influences typical in everyday schools. At the end of a design cycle, newly generated design principles are connected with the real-world design setting and related literature to ensure their practicality and usability.

Principle Four: Collaborate Closely With Participants

In design-based research, all participants are immersed in the setting and work as collaborators or coconstructors of the design. To ensure the feasibility of the initial plan and improve the design en route, designers consult with teachers and students, remaining mindful of their theory-generating goals as they balance the theoretical and practical. Thus, they neither adopt their clients' values nor impose their own, acting instead as facilitators and adapting to their clients' perspectives, beliefs, and strategies while aligning and extending the design processes (Hannafin, Hill, & Glazer, in press).

To collaborate successfully, coordination of the considerable range of resources and effort is often necessary (Collins et al., 2004). Consequently, designers need to become familiar with the people, resources, and constraints in the learning environment. Familiarization can help to lessen the obtrusiveness of the designer's presence in the learning environment. Moreover, designers need to ensure that their contributions transcend their immediate influence: They are intimately involved in the process, but cannot, themselves, cause research findings, nor can their continued presence become integral to the success of the effort. To the extent the process is managed ineffectively, the extra effort may inadvertently hamper the sustainability and scalability of a design (Fishman et al., 2004).

Consider the challenges involved in studying the cognitive factors affecting sixth graders' use of a Web-based learning environment on geography (WBLE-G). Mr. Stokes, the teacher, thinks that training should be provided to improve his students' map-reading abilities. As the head of the design team, Dr. Carter, however, does not just simply adopt Mr. Stokes's suggestions. Instead, Dr. Carter negotiates with Mr. Stokes to determine whether providing the training suggested is appropriate and consistent with the goals and values of the effort. Dr. Carter may ask design team members to document their influences when they help students to read maps; alternatively, he may determine that the research findings will be confounded by the training provided and fundamentally bias the assessment of knowledge and expertise.

Principle Five: Implement Research Methods Systematically and Purposefully

Researchers use multiple methods, including observations, interviews, surveys, and document analysis (e.g., school policies, student records, and district documents). In addition, needs assessment and evaluation-formative and summative-are often employed in design-based research (Richey et al., 2003). Qualitative documentation methods are often especially useful in design-based research. Hutchinson (1990), for example, noted that both tape recordings and written field notes are widely used to collect original data; sometimes, because of their obtrusiveness, researchers employ them after rather than during implementations. Designers document closely their research procedures, anomalies, and interpretations and understandings using research journals and field notes: The more relevant the available documentation, the greater the decision-altering potential and the more persuasive the descriptions of interventions and findings.

Research methods are also aligned with data analysis and refinement needs of the design. For example, continuous documentation is needed from the outset for retrospective analysis and to generate contextual design principles (Shavelson et al., 2003). Formative evaluation methods are often used when examining intermediate design goals; survey, interview, and observation are helpful to address the theoretical and practical needs of the design.

Principle Six: Analyze Data Immediately, Continuously, and Retrospectively

Analysis is conducted simultaneously with data collection and coding to improve the design and to address theory-generation goals. Generally, two levels of coded data emerge. Level I Data describe the exact research setting and the research processes, such as notes from observations in classrooms, and specific revisions made in the design; Level II Data represent a distillation of Level I Data and are used to explain the design and to construct design principles. Comparative analysis and retrospective analysis are utilized to generate Level II Data by comparing Level I Data with the design context, earlier events, previously collected data, and knowledge in the available literature. Complementary expertise among team members contributes to the retrospective analysis because different interpretations can minimize the bias of a single designer (Cobb et al., 2003).

In the aforementioned WBLE-G design, Dr. Carter may find that 54% of the fourth-grade and 60% of the sixth-grade students use a notebook tool provided in the system (Level I Data). Through retrospective analysis, he compares it with student and teacher evidence gleaned from previous data, revealing that teacher facilitation is important for students to use WBLE-G tools (Level II Data). Based on this analysis, design refinements can be made accordingly.

Principle Seven: Refine Designs Continually

A flexible initial plan is refined iteratively until completion of corresponding design cycles. Refinements, based on Level II Data and constant comparative data analysis, deepen a researcher's understanding of the study context. FCL designers, for example, set age-appropriate goals for children in the design process based on their understanding of children's developmental thinking (Brown & Campione, 1996). Because the design's theoretical framework is valued more than differentiating whether or not a given activity is implemented, refinements are contingent upon the designer's theory-generating goals. Designers refine continually to reach intermediate design goals that collectively address ultimate design goals.

Designers may also reexamine available literature to refine design activities or even intermediate and ultimate theory goals. A new innovation may be introduced en route if proved necessary and feasible. In unexpected situations, designers may refine the design to deal with external or unanticipated influences, such as time constraints or pressure from school principals.

Principle Eight: Document Contextual Influences With Design Principles

Design principles should be context sensitive and of practical importance to other designers. Designers "must be able to specify ... principles ... in such a way that they can inform practice" (Brown & Campione, 1996, p. 291) and provide principles that are reciprocal and mutually reinforcing; otherwise, they may be perceived to be of limited value for classroom practices. For example, one TSIL design principle is to "identify a motivating context for inquiry early in the design process." This principle is accompanied by descriptions of designers' experiences and strategies in using the principle, and design examples, such as "the selection of global warming as a motivating context" (Edelson et al., 1999, p. 440).

Design-based research reports generally include purpose and goals, framework, setting and processes, outcomes, and principles. The purpose and goals section introduces relevant literature related to the design, states the design purpose, and explicates the goals and innovations of the design. The design framework section provides an in-depth description of the framework, its origin and source (i.e., adapted, adopted, or created), and how researchers can achieve their goals through it. The design setting and processes section details both the classroom where design research is conducted and larger system influences (e.g., the environment and culture of the school, student backgrounds), as well as the design phases, processes, intermediate goals, refinements and rationale for refinements, and data collection and analysis methods. Findings are described in the outcomes of the design section, supported by observed results, and linked to the research processes. In the design principles section, principles that transcend the local setting are presented with relevant contextual information; warnings and guidance for appropriate application of these principles may also be provided.

Principle Nine: Validate the Generalizability of the Design

Whereas traditional ID/ISD tends to emphasize the effectiveness of particular approaches to address a local need, design-based research strives to balance local effectiveness with design principle and theory development. Generalizability-the methods used, refinements made, and innovations introduced to support the purpose and theory-generating goals of the design-must be verified according to the theory goals of the design and discipline requirements of the research. Researchers need to optimize a local design without decreasing its generalizability, because effectiveness is a function of both success in addressing local needs and the applicability of design principles to other settings. For instance, through collaboration with teachers, researchers may recognize teacher concerns and enact refinements consistent with the immediate and ultimate research goals. These refinements, in turn, may improve the immediate effort of the local design and subsequent collaborations, but the idiosyncratic nature of the concerns and refinements may pose problems in different settings where the design might be implemented.

CHALLENGES OF DESIGN-BASED RESEARCH FOR TELE DESIGN

As an emerging methodology, design-based research has both advantages and limitations. Four issues are particularly challenging: (a) immature methodology, (b) applicability and feasibility, (c) paradigm shift, and (d) data utilization.

Immature Methodology

Methodological development is needed to both enhance rigor and account for the importance of local context (DBRC, 2003). For example, it is difficult to determine whether to continue or abandon an iterative design, because standards do not exist to judge its effectiveness (Dede, 2004). Moreover, even where the design is proved effective in a local context, it may prove difficult to determine if valid design principles can be generated. An otherwise effective design, capable of generating useful principles, could be discarded because it was ineffective in a specific local context.

Next, design-based research comprises a collection of multiple research frameworks that are internally consistent but assume many forms and reflect varying levels of discipline and rigor. Many differences exist between and among these frameworks. For example, during developmental research, researchers may or may not be involved in different aspects of the research processes (Richey et al., 2003); in design experiments, researchers are involved throughout (Cobb et al., 2003). Thus, while certain conceptual similarities exist, the methods themselves may differ in fundamental ways, making it difficult to identify a specific methodology to guide research and design.

Applicability and Feasibility in Current Education System

The accountability culture of present-day research and practice emphasizes methodologies that are deemed scientifically valid, that is, they demonstrate particular discipline and provide particular kinds of evidence. Design-based research may not satisfy the policymaker's requirement for scientifically based research (Cobb, 2001). The premium on compliance with accepted methods and measures may limit preemptively funding prospects for design-based research and development, discouraging its use and limiting its potential in otherwise ideal circumstances. In addition, the presence of researchers in the classroom throughout the process may be perceived as a distraction or intrusion rather than a contribution to local efforts. Teachers and administrators may prefer to use already developed products and approaches rather than to become deeply involved in their creation. Thus, pragmatic and political constraints may hamper or preclude design-based research approaches in many settings.

Paradigm Shift

Design-based research methods both share and extend conventional evaluation approaches. In some cases, the extensions are significant and represent fundamental changes in goals, scope, and methodology. For example, TELE designers are generally very familiar and comfortable with formative evaluation methods, but less familiar and comfortable with generating new theories and generalizable design models. Additionally, because designers work intimately with participants, unanticipated influences such as Hawthorne effects may result from their pervasive presence. The designer's influence-undocumented in the research process-may inadvertently affect research outcomes. This paradigm shift requires changes in both how designers plan and implement system approaches and how they interact with participant-collaborators.

Data Utilization

Design-based research has been characterized as over-methodologized-only a small percent of the data collected are used to report findings (Dede, 2004). Design-based research requires documenting the whole design process and using multiple research methods in real-world learning environments. The data are typically extensive and comprehensive, requiring both extended time and resources to collect and analyze (Collins et al., 2004). However, because time and resources are often limited, large amounts of data are routinely discarded, and research quality may be influenced negatively. If made accessible to other TELE researchers, however, "lost" data could both save time and improve quality. The gap between the methodology used to collect data and its meaningful utilization needs to be decreased.

CONCLUSIONS

Design-based research and TELE designs are reciprocal and, thus, need to be interdependent. In order to stimulate contextually-sensitive practices of learning and instruction in the design and implementation of TELEs, practical, detailed, and contextual advice is necessary. Design-based research, as a pragmatic methodology, can guide TELE designers while generating practical knowledge to be shared among a broad design community. Conversely, TELE design theories, models, and procedures need to ensure that design-based research methodologies can be made operational, formalized, and systematized.

Design-based research may not be applicable for TELE designs valuing local efficiency and economy over validity, theory refinement, and generalizable design principles. Nor are design-based approaches likely to fit all the varied needs and requirements of clients, policymakers, and designers. Design-based research advances instructional design research, theory, and practice as iterative, participative, and situated rather than processes "owned and operated" by instructional designers. They are neither easy nor intuitive to implement; indeed, they require a shift in perspective of the traditional ID/ISD enterprise and a sustained commitment to advancing theory and practice. TELE designers need systemic guidance to identify suitable interventions, integrate diverse research methods with design processes, implement designs appropriately, and document their effectiveness and impact. Future research should help to document both the effectiveness of local designs and the generalizability of research results, and ultimately improve applications of design-based research in TELEs.

ACKNOWLEDGMENTS

The authors gratefully acknowledge the assistance of Chandra Orrill for her in-depth review comments, and Bryon Hand and Craig Shepherd for their editorial suggestions on drafts of this paper. Thanis also to the anonymous ETR&D reviewers for their thoughtful revision suggestions.

REFERENCES

Aleven, V., Stahl, E., Schworm, S., Fischer, F., & Wallace, R. (2003). Help seeking and help design in interactive learning environments. *Review of Educational Research, 73*(3), 277–320.

Bannan-Ritland, B. (2003). The role of design in research: The integrative learning design framework. *Educational Researcher, 32*(1), 21–24.

Barab, S. A., & Kirshner, D. E. (2001). Guest Editors' introduction: Rethinking methodology in the learning sciences. *Journal of the Learning Sciences, 10*(1&2), 5–15.

Barab, S., & Squire, K. (2004). Design-based research: Putting a stake in the ground. *Journal of the Learning Sciences, 13*(1), 1–14.

Battista, M. T., & Clements, D. H. (2000). Mathematics curriculum development as a scientific endeavor. In R. A. Lesh & A. E. Kelly (Eds.), *Research on design in mathematics and science education* (pp. 737–760). Hillsdale, NJ: Erlbaum.

Baumgartner, E., & Bell, P. (2002). *What will we do with design principles? Design principles and principled design practice.* Paper presented at the Annual Conference of the American Educational Research Association, New Orleans, LA.

Bell, P., & Linn, M. C. (2000). Scientific arguments as learning artifacts: Designing for learning from the Web with KIE. *International Journal of Science Education, Special Issue (22)*, 797–817.

Bell, P., Hoadley, C. M., & Linn, M. C. (2004). Design-based research in education. In M. C. Linn, E. A. Davis, & P. Bell (Eds.), *Internet environments for science education* (pp. 73–84). Mahwah, NJ: Erlbaum.

Berliner, D. C. (2002). Educational research: The hardest science of all. *Educational Researcher, 31*(8), 18–20.

Brophy, S. P. (1998). *Sequencing problem solving and hands on activities: Does it matter?* Paper presented at the Annual Meeting of the American Educational Research Association, San Diego, CA.

Brown, A. L. (1992). Design experiments: Theoretical and methodological challenges in creating complex interventions in classroom settings. *Journal of the Learning Sciences, 2*(2), 141–178.

Brown, A., & Campione, J. (1996). Psychological theory and the design of innovative learning environments: On procedures, principles, and systems. In L. Schauble & R. Glaser (Eds.), *Innovations in learning: New environments for education* (pp. 289–325). Mahwah, NJ: Erlbaum.

Burkhardt, H., & Schoenfeld, A. H. (2003). Improving educational research: Toward a more useful, more influential, and better-funded enterprise. *Educational Researcher, 32*(9), 3–14.

Cobb, P. (2001). Supporting the improvement of learning and teaching in social and institutional context. In S. Carver & D. Klahr (Eds.), *Cognition and instruction: Twenty-five years of progress* (pp. 455–478). Mahwah, NJ: Erlbaum.

Cobb, P., Confrey, J., diSessa, A., Lehrer, R., & Schauble, L. (2003). Design experiments in educational research. *Educational Researcher, 32*(1), 9–13.

Cognition and Technology Group at Vanderbilt. (1992a). The Jasper experiment: An exploration of issues in learning and instructional design. *Educational Technology Research and Development, 40*(1), 65–80.

Cognition and Technology Group at Vanderbilt. (1992b). The Jasper Series as an example of anchored instruction: Theory, program description, and assessment data. *Educational Psychologist, 27*(3), 291–315.

Cognition and Technology Group at Vanderbilt. (1997). *The Jasper project: Lessons in curriculum, instruction, assessment, and professional development.* Mahwah, NJ: Erlbaum.

Collins, A. (1992). Towards a design science of education. In E. Scanlon & T. O'Shea (Eds.), *New directions in educational technology* (pp. 15–22). Berlin: Springer.

Collins, A. (1999). The changing infrastructure of education research. In E. Lagemann & L. Shulman (Eds.), *Issues in education research* (pp. 289–298). San Francisco: Jossey-Bass.

Collins, A., Joseph, D., & Bielaczyc, K. (2004). Design research: Theoretical and methodological issues. *Journal of the Learning Sciences, 13*(1), 15–42.

Cuban, L. (1986). *Teachers and machines: The classroom use of technology since 1920.* New York: Teachers College Press.

Dede, C. (2004). If design-based research is the answer, what is the question? A commentary on Collins, Joseph, and Bielaczyc; diSessa and Cobb; and Fishman, Marx, Blumenthal, Krajcik, and Soloway in the JLS special issue on design-based research. *Journal of the Learning Sciences, 13*(1), 105–114.

Design-Based Research Collective. (2003). Designbased research: An emerging paradigm for educational inquiry. *Educational Researcher, 32*(1), 5–8.

diSessa, A. A., & Cobb, P. (2004). Ontological innovation and the role of theory in design experiments. *Journal of the Learning Sciences, 13*(1), 77–103.

Edelson, D. C. (2002). Design research: What we learn when we engage in design. *Journal of the Learning Sciences, 11*(1), 105–121.

Edelson, D. C., Gordin, D. N., & Pea, R. D. (1999). Addressing the challenges of inquiry-based learning through technology and curriculum design. *Journal of the Learning Sciences, 8*(3&4), 391–450.

Fishman, B., Marx, R., Blumenfeld, P., & Krajcik, J. (2004). Creating a framework for research on systemic technology innovations. *Journal of the Learning Sciences, 13*(1), 43–76.

Fullan, M. (2001). *The new meaning of educational change* (3rd ed.). New York: Teachers College Press.

Greeno, J. G., Collins, A., & Resnick, L. (1996). Cognition and learning. In D. C. Berliner & R. C. Calfee (Eds.), *Handbook of educational psychology* (pp. 15–46). New York: Macmillan.

Gustafson, K. L. (2002). The future of instructional design. In R. A. Reiser & J. V. Dempsey (Eds.), *Trends and issues in instructional design and technology* (pp. 333–343). Upper Saddle River, NJ: Merrill/ Prentice-Hall.

Hannafin, M. J., Hannafin, K. M., Land, S. M., & Oliver, K. (1997). Grounded practice and the design of constructivist learning environment. *Educational Technology Research and Development, 45*(3), 101–117.

Hannafin, M. J., Hill, J. R., & Glazer, E. M. (in press). Designing grounded learning environments: The value of multiple perspectives in design practice. In G. Anglin (Ed.), *Critical issues in instructional technology*: Englewood, CO: Libraries Unlimited.

Hannafin, M. J., Land, S., & Oliver, K. (1999). Studentcentered learning environments. In C. M. Reigeluth (Ed.), *Instructional-design theories and models: Vol. 2. A new paradigm of instructional theory* (pp. 115–140). Mahway, NJ: Erlbaum.

Hewitt, J. (1996). *Progress toward a knowledge-building community*. Unpublished dissertation. University of Toronto, Toronto, Canada.

Hewitt, J., & Scardamalia, M. (1998). Design principles for distributed knowledge building processes. *Educational Psychology Review, 10*(1), 75–96.

Hutchinson, S. A. (1990). Education and grounded theory. In R. Sherman & R. Webb (Eds.), *Qualitative research in education: Focus and methods*. London: Falmer.

Jonassen, D. H., & Rohrer-Murphy, L. (1999). Activity theory as a framework for designing constructivist learning environments. *Educational Technology Research and Development, 47*(1), 61–79.

Kelly, A. E. (2003). Research as design. *Educational Researcher, 32*(1), 3–4.

Kemmis, S., & McTaggart, R. (2000). Participatory action research. In N. K. Denzin & Y. S. Lincoln (Eds.), *Handbook of qualitative research*, 2nd ed. (pp. 567–605). London: SAGE.

Kent, T. W., & McNergney, R. F. (1999). *Will technology really change education: From blackboard to Web*. Thousand Oaks, CA: Corwin Press.

Land, S. M. (2000). Cognitive requirements for learning with open-ended learning environment. *Educational Technology Research and Development, 48*(3), 61–78.

Linn, M. C, Clark, D., & Slotta, J. D. (2003). WISE design for knowledge integration. *Science Education, 87*(4), 517–538.

Linn, M. C., Davis, E. A., & Bell, P. (2004). *Internet environments for science education*. Mahwah, NJ: Erlbaum.

Linn, M. C., & Hsi, S. (2000). *Computers, teachers, peers: Science learning partners*. Mahwah, N.J.: Erlbaum.

McCandliss, B. D., Kalchman, M., & Bryant, P. (2003). Design experiments and laboratory approaches to learning: Steps toward collaborative exchange. *Educational Researcher, 32*(1), 14–16.

Orrill, C. H., Hannafin, M. J., & Glazer, E. M. (2003). Disciplined inquiry and the study of emerging technology. In D. H. Jonassen (Ed.), *Handbook of research for educational communications and technology* (2nd ed., pp. 335–353). Mahwah, NJ: Erlbaum.

Patton, M. Q. (2002). *Qualitative research & evaluation methods* (3rd ed.). Thousand Oaks, CA: SAGE.

Peterson, P. (1998). Why do educational research? Rethinking our roles and identities, our texts and contexts. *Educational Researcher, 27*(3), 4–10.

Reeves, T. C., & Hedberg, J. G. (2003). *Interactive learning systems evaluation.* Englewood Cliffs, NJ: Educational Technology.

Reigeluth, C. M. (January/February, 1997). Instructional theory, practitioner needs, and new directions: Some reflections. *Educational Technology,* 42–47.

Reigeluth, C. M., & Frick, T. W. (1999). Formative research: A methodology for creating and improving design theories. In C. M. Reigeluth (Ed.), *Instructional-design theories and models,* Vol. II, (pp. 633–651). Mahwah, NJ: Erlbaum.

Reiser, B. J., Tabak, I., Sandoval, W. A., Smith, B. K., Steinmuller, F., & Leone, A. J. (2001). BGuILE: Strategic and conceptual scaffolds for scientific inquiry in biology classrooms. In S. M. Carver & D. Klahr (Eds.), *Cognition and instruction: Twenty-five years of progress* (pp. 263–305). Mahwah, NJ: Erlbaum.

Richey, R. C., Klein, J. D., & Nelson, W. A. (2003). Development research: Studies of instructional design and development. In D. H. Jonassen (Ed.), *Handbook of research for educational communications and technology* (2nd ed., pp. 1099–1130). Mahwah, NJ: Erlbaum.

Richey, R. C., & Nelson, W. A. (1996). Developmental research. In D. Jonassen (Ed.), *Handbook of research for educational communications and technology* (pp. 1213–1245). London: Macmillan.

Robinson, V. M. J. (1998). Methodology and the research-practice gap. *Educational Researcher, 27*(1), 17–27.

Sandoval, W. A., & Reiser, B. J. (1998). *Iterative design of a technology-supported biological inquiry curriculum.* Paper presented at the Annual Meeting of the American Educational Research Association, San Diego, CA.

Sandoval, W. A., & Reiser, B. J. (2004). Explanation-driven inquiry: Integrating conceptual and epistemic scaffolds for scientific inquiry. *Science Education, 88*(3), 345–372.

Savery, J. R., & Duffy, T. M. (1996). Problem based learning: An instructional model and its Constructivist framework. In B. G. Wilson (Ed.), *Constructivist learning environments: Case studies in instructional design* (pp. 135–148). Englewood Cliffs, NJ: Educational Technology.

Scardamalia, M., & Bereiter, C. (1994). Computer support for knowledge-building communities. *Journal of the Learning Sciences, 3*(3), 265–283.

Scardamalia, M., & Bereiter, C., Brett, C., Burtis, P., Calhoun, C., & Smith Lea, N. (1992). Educational applications of a networked communal database. *Interactive Learning Environments, 2*(1), 45–71.

Schank, R. C., Fano, A., Bell, B., & Jona, M. (1994). The design of goal-based scenarios. *Journal of the Learning Sciences, 3*(4), 305–346.

Schwartz, D. L., Lin, X., Brophy, S., & Bransford, J. D. (1999). Toward the development of flexibility adaptive instructional designs. In C. M. Reigeluth (Ed.), *Instructional-design theories and models,* Vol. II (pp. 183–213). Mahwah, NJ: Erlbaum.

Shapiro, W. L., & Roskos, K. (1995). Technology-enhanced learning environments. *Change, 27*(6), 67–69.

Shavelson, R. J., Phillips, D. C., Towne, L., & Feuer, M. J. (2003). On the science of education design studies. *Educational Researcher, 32*(1), 25–28.

Stringer, E. (1999). *Action research* (2nd ed.). Thousand Oaks, CA: SAGE.

van den Akker, J. (1999). Principles and methods of development research. In J. van den Akker, N. Nieveen, R. M. Branch, K. L. Gustafson & T. Plomp (Eds.), *Design methodology and developmental research in education and training* (pp. 1–14). The Netherlands: Kluwer Academic.

Walker, D. F. (1992). Methodological issues in curriculum research. In P. Jackson (Ed.), *Handbook of research on curriculum* (pp. 98–118). New York: Macmillan.

Winn, W. (1997, January/February). Advantages of a theory-based curriculum in instructional technology. *Educational Technology,* 34–41.

CHAPTER 18

APPRECIATIVE INSTRUCTIONAL DESIGN (AiD)

A New Model

Karen E. Norum

In recent years, there has been a call for new instructional design models; models that meet the speed of change most organizations must now perform at (Carr, 1997; Gordon & Zemke, 2000; Gustafson & Branch, 1997). Gordon and Zemke (2000) suggest that the instructional systematic design (ISD) models that dominate the field have outlived their usefulness. Although instructional design models based on constructivist learning theory are emerging (see Reigeluth, 1996; Willis, 1995, 2000; Winn, 1992), the ISD models are still dominated by the behaviorist paradigm (Gordon & Zemke, 2000; Gustafson & Branch, 1997; Seels & Glasgow, 1998; Willis, 2000). Thus, the idea that "behavior can be observed, measured, planned for, and evaluated in reasonably valid and reliable ways" (Gustafson & Branch, 1997, p. 73) is reflected in current models of instructional systematic design. The word "systematic" in instructional design suggests that

Originally paper presented at the 2000 Association for Educational Communication and Technology (AECT) National Conference, Denver, Colorado, October 25–28, 2000 instructional design is a step-by-step, orderly, sequential, logical, linear process (Banathy, 1996; Gordon & Zemke, 2000). This is reflective of the "machine mentality," which creates the illusion that certain conditions will lead to certain outcomes, thus by following a "lock-step, engineering like" (Gordon & Zemke, 2000, p. 48) model, the instructional designer can create the right conditions for the desired outcomes.

This paradigm has lead to "fiendishly complex" (Gordon & Zemke, 2000, p. 43) models that emphasize efficiency in human learning, instruction, and performance (Carr, 1997). It also can lead to low expectations: Morrison, Ross, and Kemp (2001, p. 3) describe the role of instructional design as planning, developing, evaluating, and managing the instructional design process effectively "so that it will ensure *competent* performance by learners." Today's organizations find themselves in quickly changing environments and need to develop the capacity to change and adapt quickly. They need to be constantly learning, able to re-create themselves at will and with skill (Fitzgerald, 1995; Senge, 1990; Wheatley, 1999). This does not call for "competent" employees, it calls for quantum (Zohar, 1997) employees: people who are "mindful" (Daft & Lengel, 1998; Langer, 1997) and can think holistically, evoking and co-creating reality(s). This calls for an instructional design model that is *systemic* rather than *systematic* in nature.

FROM DEFICIT-BASED TO VALUE-BASED

Current models of instructional design are deficit-based: the gap between current performance and desired performance is systematically analyzed and as appropriate, instruction is designed to fill that gap. It is a problem-solving process: What is the problem and how do we solve it? often begins the analysis (e.g., Seels & Glasgow, 1998). The existence of a problem is often heralded by a gap in learning, evidenced by poor job performance or unacceptable error rates. During the analysis phase of instructional design (ID), a needs assessment that focuses on this gap is often conducted. By focussing on this gap, the unstated message becomes that there is an acceptable level of error. For example, it is OK for the airlines to lose luggage, they just need to lose less of it.

This deficit-based approach leads to unintended consequences. By focussing on the gap, we tend to focus on fragments: there is a danger we will analyze each puzzle piece instead of considering the place of piece within the puzzle (Capra, 1982; Cooperrider & Whitney, 1999b). The systematic approach to instructional design can set us up to design and develop excellent instruction and training that is not suited for the organization or context in which it needs to live (Tessmer, 1990). A deficit-based approach is slow and past-oriented: it has us looking at yesterday's causes. The assumption is that if we correctly identify the problem, we can then select the solution that corresponds to it (Banathy, 1996). But because of the interdependent nature of systems, we may never find "the" cause of the problem (Capra, 1996; Senge, 1990; Wheatley, 1999) and there is a good chance we will not actually solve the problem (Cooperrider & Whitney, 1999b). Instead, what will most likely happen is that a new problem that demands more attention will come along and the "problem" we were working on will fade into the background. As our attention jumps from problem to problem never actually solving any of them, a negativity is engendered. We become progressively enfeebled, resign ourselves to live with diminished expectations, and become visionless (Cooperrider & Whitney, 1999b). We come to believe that rather than designing instruction or training for the "best of" level of performance, we need to design it for an "acceptable" or "competent" level of performance. We design the training or instruction to eliminate what we *do not* want rather than to give us more of what we *do* want.

The Appreciative instructional Design model offers a new alternative. Appreciative instructional Design (AiD) takes its theoretical foundation from Appreciative Inquiry (AI). AI was developed by David Cooperrider, professor of organizational behavior at Case Western Reserve University. As opposed to problem solving, AI begins with a search for the best of "what is" rather than looking for exactly what is wrong or what needs to be "fixed." According to Cooperrider, every system has good and bad in it. We are trained to look for the "bad" and "fix" it. But what if we paid just as much attention to the "good" in the system? AI gives us a structure for searching out the "goodness" in the system, allowing us to appreciate "what is" and use that as inspiration for what "could be." It is a valued-based rather than deficit-based approach (Norum, 2000a).

Applying AI to Instructional Design

When AI is applied to the instructional design process, the goal is to discover the factors present when the system is operating at its "best of" level. Instruction or training is then designed around those generative factors. Thus, the gap in performance or learning becomes immaterial. There is no need to analyze the gap between current and desired levels of performance: what we want to know is what is the "best of" level of performance? AiD assumes there is something working in the current level of performance (after all, it cannot *all* be poor) and that what is "working" can be found and amplified (Bushe, 2000a). The instruction or training then will be designed to nurture, develop, and amplify the competencies needed to perform at the "best of" level and give the organization "more" of what it wants.

To contrast the ISD model with the AiD model, imagine you are charged with designing customer service training: the organization has been receiving what it considers to be too many complaints. Using a traditional ISD model, the gap between the current (unacceptable) level of performance and the acceptable level of performance is analyzed. One measure for this might be the number of complaints received: how many are received now and how many is an "acceptable" number? The training would then be designed to lower the number of complaints received into the "acceptable" range. Using the AiD model, instead of determining the gap between the current and acceptable levels of performance, a search to discover what the organization defines as the "best of" level of customer service performance is engaged in. When customer service is operating at its "best of" level, what does it look like? Training would then be designed to nurture, develop, and amplify the competencies needed to perform at that "best of" level. Thus, it is quite possible that the training developed will take the employees well beyond the "acceptable" level of performance! By focussing on the potential to create the best of what "could be," this model goes beyond filling gaps in performance.

The AiD model is systemic, advocating "a global conception of the problem and an understanding of the interrelationships and interconnections" (Carr, 1996, p. 17). Another distinctive characteristic of AiD is its focus on "inquiry." While conducting a needs assessment is important in traditional ISD models, it is critical in AiD. From Appreciative Inquiry, we learn that the questions we ask determine what we find and the data we gather determines what we design (Cooperrider & Whitney, 2000). Thus, in the AiD model, a fair amount of time is devoted to constructing questions: the inquiry is at the heart of the process. The AiD model is future-oriented, looking at generating "more of" what the organization wants rather than minimizing what it does not want. This is reflected in the difference in one of the first questions asked: rather than beginning the process by asking, "What is the problem and how do we solve it?" AiD begins with a question designed to evoke stories of what the "best of" level of performance looks like.

The Appreciative instructional Design Model: An Overview

The AiD model is based on the "4-D" cycle used in Appreciative Inquiry: Discovery, Dream, Design, Destiny. Cooperrider and Whitney describe Appreciative Inquiry as "the cooperative search for the best in people, their organizations, and the world around them" (1999a, p. 10). This involves a search into what gives "life" to the system. A hallmark of AI is the kinds of questions asked during the inquiry: questions that are unconditionally positive, designed to strengthen the positive potential in the organization. The inquiry is based on the assumption that there are untapped, rich, inspiring stories about the organization (Cooperrider & Whitney, 1999a) and that what is "working" can be amplified and fanned throughout the system (Bushe, 2000b). A four-phase model (Figure 18.1) is employed to conduct this appreciative inquiry. Cooperrider and Whitney (1999a) refer to it as the "4-D Cycle":

The AI cycle begins by crafting positive questions that are designed to uncover the "life-giving" forces of the system. The inquiry begins through an interviewing process. During this process, the task is to discover the best of what already is—to appreciate the good things about the system. The "best of what is" is used to inspire "what might be": the Dream phase. Possible (positive) futures are envisioned. The next step is to create the policies, procedures, infrastructures, governance systems, and so forth. that are needed to support "what should be." This is the Design phase. The new system is co-constructed. As the new system is implemented, the question turns to how to sustain and maintain this new system. The focus of the Destiny phase is how to continue to learn, improvise and adjust so that the system can continuously strengthen its affirmative capacity. This often leads back to the first phase in the cycle: Discovery. A new inquiry begins into the "best of" what has just been re-created.

This same process is reflected in AiD. As in an Appreciative Inquiry, the process begins with Discovery: questions are crafted to discover the "best of" performance level in the organization. Stakeholders (those who will be the audience for the training and/or those who need to support it) are interviewed to elicit stories about what the "best of" performance level looks like as well as their "best" instructional or training experiences. Questions are also asked about the "ideal" system and "ideal" instruction or training. Drawing inspiration from what is work-

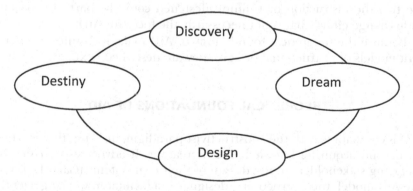

Figure 18.1.   The 4-D cycle.

ing, people are encouraged to dream about what could be. They are dared to expand the realm of the possible. The information gathered in the interviews is relevant to the Discovery and Dream phases of AiD. The goal is to discover what is already working well in the system's performance and to understand why it is working well. What life-giving factors are present at this "best of" level of performance and what does the system want "more of?"

As the Discovery and Dream data is analyzed to find themes, patterns, and refrains (Lawrence-Lightfoot & Davis, 1997), the Design Phase is entered. Competencies that need to be nurtured and developed, what needs to be learned, how to learn it, effective design elements, what needs to be amplified are all considered. Clues to these questions are contained in the stories gathered during the Discovery and Dream phases. Those clues will be used to create instruction or training designed to give the system "more" of what it wants. The goal is to create instruction or training that will nurture, develop, and amplify competencies that will in turn amplify the life-giving factors identified in the Discovery and Dream phases. The design takes place around generative factors.

The Destiny Phase is entered when the instruction or training is implemented. This phase reflects the plan to sustain, maintain, improve, or adjust the instruction or training. It also reflects how the system will know if it is getting "more" of what it wanted. The plan outlines what will be assessed and how. The appraisal of what is working about the instruction or training brings us back to the beginning of the cycle: Discovery.

AiD is a specific application of Appreciative Inquiry. It shares the same theoretical foundations and the "4-D" cycle. Where it differs is in its focus. While Appreciative Inquiry is a system-wide intervention with a focus on organizational change, Appreciative instructional Design is specifically concerned with designing instruction or training. While it is quite possible that the instruction or training designed could be part of a system-wide change effort, that is not necessarily the focus for AiD.

Because the theoretical foundations of AiD make it distinct from current models of instructional design, they are described next.

THEORETICAL FOUNDATIONS OF AiD

AiD is compatible with the constructivist paradigm, insisting that learners go beyond acquiring knowledge and create it. It advocates user-design: engaging stakeholders in the design of their own systems (Carr, 1997). In the AiD model, the instructional designer is a facilitator, working *with* the system to help it get more of what it wants. The AiD model also draws from action research methodology particularly in the data analysis process, which takes place in the Design phase. It recognizes that learning and performance take place *in a context* and that context can "facilitate or inhibit human enterprises" (Tessmer & Richey, 1997, p. 88). AiD understands that organizations and the people in them are living, dynamic systems, embracing the "new science" paradigm (Wheatley, 1999; Zohar, 1997) and ecological thinking (Capra, 1996). It evokes "idealized" design: design that is future-focused, based on what we want, yet grounded in current reality (Banathy, 1996).

Five principles are central to Appreciative Inquiry and thus are foundational to AiD. These five principles (Cooperrider & Whitney, 2000) are the:

- Constructionist Principle
- Principle of Simultaneity
- Poetic Principle
- Anticipatory Principle
- Positive Principle

Several propositions are related to these Principles. What follows is a description of each Principle and its related propositions.

Constructionist Principle

This Principle asserts that organizations are living, human constructions. They are constructed based on what we think we know, thus what we know and how we know it becomes fateful (Cooperrider & Whitney, 2000). "[T]he truth about an organization is what those involved agree the truth is" (Zemke, 1999, p. 30). This principle is strengthened by the proposition that stakeholders in the organization carry in their minds some sort of shared idea of what the organization is, how it should function, and what it might become (Cooperrider, 2000). The Constructionist Principle calls us to unearth and examine the mental models (Senge, 1990) that we hold about an organization and consider how those mental models have effected the fate of the current system.

Principle of Simultaneity

Change begins the minute we ask a question. The questions posed set the stage for what is found. What is found becomes the data we use to reconstruct the future. "Even the most innocent question evokes change" (Cooperrider & Whitney, 2000, p. 18). Thus, change is not something that happens after an analysis is conducted; change begins with the analysis. A corresponding proposition encourages us to create the conditions for organization-wide appreciation to "ensure the conscious evolution of a valued and positive future" (Cooperrider, 2000, p. 52).

The Poetic Principle

If organizations are constructed, they can be re-constructed. Just as a poem can be interpreted and re-interpreted as we bring new meaning to every reading of it, so can organizations be re-interpreted as the system they are embedded in changes. This is "The Poetic Principle": as the stories of the people in and attached to the organization change, the organization changes. "There is no such thing as an inevitable organization" (Cooperrider, 2000, p. 47). This principle teaches us that we can choose what to study in an organization: the good or the bad, the joy or the alienation, the creativity or mediocrity (Cooperrider & Whitney, 2000). A related proposition tells us that no matter what the previous history, every system can be altered and re-invented (Cooperrider, 2000).

The Anticipatory Principle

From this Principle we learn that the image of the future guides the current behavior and actions of the system (Cooperrider & Whitney, 2000). Positive images of the future lead to positive actions; negative images lead to negative actions. This image becomes the "referential core" of the system and determines its essential characteristics (Capra, 1996; Wheatley, 1999). It is possible for this image to be incoherent or unclear or even for it to be pathetic. Many organizations are better at articulating what they *do not* want than at being clear about what it is they *do* want. An image that is based on what we *do not* want is likely to engender negative behavior and actions. Malaise, mediocrity, angst, and dysfunction are likely to be present in such an organization. This principle is supported by the proposition that systems are limited only by their imaginations (Cooperrider, 2000). Paradoxically, even the best future images can hold the system back if those positive images become so cherished, they cannot be given up for even better images (Cooperrider, 2000). This proposition reminds us of the Constructionist Principle: our organizations are constructions.

The Positive Principle

The more positive the question asked is, the more longer lasting and successful the change effort will be (Cooperrider & Whitney, 2000). Problem solving is a null-sum game, directing the focus to what is wrong (Zemke, 1999). Building and sustaining momentum for change requires large doses of hope, inspiration, caring, excitement and commitment. Seeking out positive experiences and past successes and using those to build the future engenders positive affect and social bonding. The "heliotropic" proposition tells us that systems, like plants, move in the direction of light or positive imagery (Cooperrider, 2000). Thus, organizations move in the direction of what they study. To move in a positive direction, the system has to be studying the positive, not the negative. A related proposition asserts that the more an organization experiments with conscious evolution of positive imagery, the better it will become as its heliotropic and affirmative competencies strengthen (Cooperrider, 2000). This heliotropic tendency needs to be appreciated and understood—this proposition directs us to appreciate rather than fix our organizations (Cooperrider, 2000; Zemke, 1999).

THE IMPORTANCE OF LANGUAGE

There is great power in how we talk about things. "The words we use and the way we use them are powerful indicators of how we see, of our particular vision of reality" (Daloz, 1986, p. 233). We are constantly telling each other stories. As we share our stories, we also share what we believe about the way the world works. The stories we tell have the potential to expand our imaginations and enlarge our vision of what could be (Feige, 1999). When we tell stories about our organization in the break room, over the water cooler, in the hallway, we socially construct what we want our organization to be (Abma, 1999; Bushe, 2000b). These stories then end up guiding the organization's practices and policies (Abma, 1999; Cooperrider, 2000; Cooperrider & Whitney, 1999a). This is the Constructionist Principle in action: a system can be reconstructed by changing the stories that are being told (Norum, 2000b).

The Anticipatory Principle reminds us that the questions we ask will determine what we find. What we find will determine what we design. Thus, the questions we ask become powerfully consequential. This is illustrated in the movie *Apollo 13*. When Jim Lovell announces, "Houston, we have a problem," imagine how different the outcome might have been if flight director Gene Kranz had asked for a rundown of everything that was wrong or had failed vs. a status report on what was still working on the space craft. By keeping the focus positive, Kranz engendered creative energy and hope rather than failure and despair.

By how we talk about it, "we largely create the world we later discover" (Cooperrider & Srivastva, 2000, p. 92). Our current social vocabulary is largely deficit based, challenging us to consciously move to hopeful and appreciative language (Ludema, 2000). How we think and talk about things largely influences our destiny. If we think and talk like something is impossible to accomplish, it probably will be. By how we think and talk about it, we create the world we later discover.

AiD PRINCIPLES

The five Principles and the importance of language are reflected in the AiD model. AiD assumes that we are only limited by our imaginations and collective will. Thus, the instruction or training that can be designed is limited only by our own thinking. This is the role the Constructionist Principle plays in AiD.

The heliotropic hypothesis has an important place as well: the image we hold of what the instruction or training should accomplish will determine the direction the system will grow in. For example, it makes a difference if the image held is to minimize error versus perform at the "best of" level: the training developed will differ according to the image held. This is related to the importance of language as the questions asked in developing the training will be different according to the image held. AiD offers the potential to create images that can release a system caught in paradox (Bushe, 2000b).

User-design is a central premise. AiD engages the potential learners in creating their own instruction or training. This process garners support for the training and minimizes the potential of sabotage. Because they have been part of designing it, they are assured of its applicability and implementation of the instruction or training generally goes faster. Working together, the instructional designer and learners are creating and recreating the organization.

The design created through the AiD process is ultimately meant to expand the realm of the possible while still being realistic (Cooperrider & Srivastva, 2000). It is idealized design in that the training or instruction has to live in the current environment yet simultaneously be connected to the future (ideal) image. It is training or instruction that "enlarges [people's] conception of what can be implemented" (Banathy, 1996, p. 193) by providing a systemic rather than systematic perspective.

A NEW MODEL

Organizations are living, dynamic systems. They are not dissociated collections of parts but rather are holistic, interconnected and interdependent systems (Capra, 1996). Traditional instructional design models treat organizations as if they are machines to be engineered and re-engineered (Wheatley, 1999; Zohar, 1998). These models are systematic rather than systemic in nature, deficit rather than value based, tend to be deductive and reductionistic rather than inductive and additive, and expert rather than user driven. These models have come under attack for being cumbersome and not flexible and fluid enough to keep pace with the internet speed organizations operate at today (Carr, 1997; Gordon & Zemke, 2000).

Ackoff (in Banathy, 1996) describes three properties of an "idealized" design model: technological feasibility, operational viability, capable of rapid learning and adaptation. The AiD model meets these criteria. It is technologically feasible in that it uses technologies that are known and usable (interviewing, user-design, action research). It is operationally viable because what is designed through the AiD process can function and sustain itself once implemented. This is assured through the participation of stakeholders in the design process. The capability of rapid learning and adaptation is seen in the stakeholder's ability to modify what has been created through the introduction of "positive dissatisfaction" (Carr-Chellman, 2000) which leads into another round of the "4-D" cycle.

Taking its theoretical foundations from Appreciative Inquiry, AiD honors organizations as living dynamic systems. Every living system has three basic components:

- identity or pattern of organization
- information or life process
- relationships or structure

The identity or pattern of organization is found in the system's referential core: the qualitative features that determine the system's essential characteristics (Capra, 1996; Wheatley, 1999). Information or life process keeps the organization alive. As information is processed, interpreting that information through the identity of the organization helps the system to know in what direction to move (Capra, 1996; Wheatley, 1999). As information flows and is interpreted through the identity of the organization, the various components of the system determine how they need to be in relationship to each other to accomplish the work of the system. This becomes the structure of the system (Capra, 1996; Wheatley, 1999).

The three components of any living system are reflected in the AiD process. Determining what the organization wants more of becomes the referential core of the instruction or training that is being developed. Information begins to flow through the AiD interviews, conducted in the Discovery and Dream phases. When analyzing the data collected through the interview process, knowing what the organization wants more of helps to determine the data most relevant to the design of the instruction or training. The design of the training is like putting puzzle pieces together: what components are necessary and how will they work together to give the organization more of what it wants through the instruction or training? The relationship or structure component of living systems is reflected in the AiD Design phase.

The notion that organizations are not machines that can be controlled through engineering and re-engineering, but rather are living dynamic systems that at best can be disturbed (Wheatley, 1999) is reflected in the Destiny Phase of AiD. Rather than creating an analytic evaluation plan, a valuation plan and a plan for tinkering is outlined.

Organizations today live with uncertainty, rapid change, unpredictability and need to be responsive, flexible and fluid to adjust to their changing environment (Senge et al., 1999; Wheatley, 1999; Zohar, 1998). Instruction and training developed for such organizations needs to be the same. The models we use to design instruction and training need to be flexible, fluid, adaptable, and responsive which would suggest that they provide guidance but not directives for design. "The best we can do is work from flexible guidelines or principles that are subject to change and being overruled" (Willis, 2000, p. 9). Such models need to be simple, based on a few clear core assumptions; adaptable; easy to use with a knowledge base that is available or readily attainable; and capable of providing new insights (Banathy, 1996). This simplicity gives rise to complexity (Wheatley, 1999; Wheatley & Kellner-Rogers, 1996).

The Appreciative instructional Design model offers an alternative to traditional instructional systematic design, honoring organizations as living, dynamic systems. As the AiD model is employed, a holistic picture of the organization emerges. The instruction or training that is developed expands the realm of the possible, giving the organization more of what it wants, nurturing the capacity to evoke and co-create new realities.

REFERENCES

Abma, T. A. (1999). Powerful stories: The role of stories in sustaining and transforming professional practice within a mental hospital. In R. Josselson & A.

Lieblich (Eds.), *Making meaning of narratives: Vol. 6* (pp. 169–195). Thousand Oaks, CA: SAGE.

Banathy, B. H. (1996). *Designing social systems in a changing world*. New York: Plenum Press.

Bushe, G. R. (2000a). Advances in appreciative inquiry as an organizational development intervention. In D. L. Cooperrider, P. F. Sorenson, Jr., D. Whitney, T. F. Yaeger (Eds.), *Appreciative inquiry: Rethinking human organization toward a positive theory of change* (pp. 113–121). Champaign, IL: Stipes.

Bushe, G. R. (2000b). Five theories of change embedded in appreciative inquiry. In D. L. Cooperrider, P. F. Sorenson, Jr., D. Whitney, T. F. Yaeger (Eds.), *Appreciative inquiry: Rethinking human organization toward a positive theory of change* (pp. 99–109). Champaign, IL: Stipes.

Capra, F. (1982). *The turning point: Science, society, and the rising culture*. New York: Bantam Books.

Capra, F. (1996). *The web of life: A new scientific understanding of living systems*. New York: Anchor Books/Doubleday.

Carr, A. A. (1996). Distinguishing systemic from systematic. *TechTrends for Leaders in Education and Training, 41*(1), 16–20.

Carr. A. A. (1997). User-Design in the creation of human learning systems. *Educational Technology Research and Development, 45*(3), 5–22.

Carr-Chellman, A. A. (2000). The new sciences and systemic change in education. *Educational Technology, 40*(2), 28–37.

Cooperrider, D. L. (2000). Positive image, positive action: The affirmative basis of Organizing. In D. L. Cooperrider, P. F. Sorensen, Jr., D. Whitney, T. F. Yaeger (Eds.), *Appreciative inquiry: Rethinking human organization toward a positive theory of change* (pp. 29–53). Champaign, IL: Stipes.

Cooperrider, D. L., & Srivastva, D. (2000). Appreciative inquiry in organizational life. In D. L. Cooperrider, P. F. Sorensen, Jr., D. Whitney, T. F. Yaeger (Eds.), *Appreciative inquiry: Rethinking human organization toward a positive theory of change* (pp. 55–97). Champaign, IL: Stipes.

Cooperrider, D. L., & Whitney, D. (1999a). *Appreciative inquiry*. San Francisco: Berrett Koehler Communications.

Cooperrider, D. L., & Whitney, D. (1999b). *Appreciative inquiry: A constructive approach to organization development and social change (A workshop)*. Taos, NM: Corporation for positive change.

Cooperrider, D. L., & Whitney, D. (2000). A positive revolution in change: Appreciative inquiry. In D. L. Cooperrider, P. F. Sorensen, Jr., D. Whitney, T. F. Yaeger (Eds.), *Appreciative inquiry: Rethinking human organization toward a positive theory of change* (pp. 3–27). Champaign, IL: Stipes.

Daft, R. L., & Lengel, R. H. (1998). *Fusion leadership: Unlocking the subtle forces that change people and organizations*. San Francisco: Berrett-Koehler.

Daloz, L. A. (1986). *Effective teaching and mentoring: Realizing the transformational power of adult learning experiences*. San Francisco: Jossey-Bass.

Feige, D. M. (1999). The legacy of Gregory Bateson: Envisioning aesthetic epistemologies and praxis. In J. Kane (Ed.), *Education, information, and transformation: Essays on learning and thinking* (pp. 77–109). Upper Saddle River, NJ: Prentice-Hall.

·

Fitzgerald, L. A. (1995). *Building a learning organization: The series.* Workshop sponsored by Mountain States Employers Council, Denver, CO.

Gordon, J., & Zemke, R. (2000, April). The attack on ISD. *Training, 36*(6), 42–53.

Gustafson, K. L., & Branch, R. M. (1997). Revisioning models of instructional development. *Educational Technology Research and Development, 45*(3), 73–89.

Langer, E. J. (1997). *The power of mindful learning.* Reading, MA: Addison-Wesley.

Lawrence-Lightfoot, S., & Davis, J. H. (1997). *The art and science of portraiture.* San Francisco: Jossey-Bass.

Ludema, J. D. (2000). From deficit discourse to vocabularies of hope: The power of appreciation. In D. L. Cooperrider, P. F. Sorensen, Jr., D. Whitney, T. F. Yaeger (Eds.), *Appreciative inquiry: Rethinking human organization toward a positive theory of change* (pp. 265–287). Champaign, IL: Stipes

Morrison, G. R., Ross, S. M., & Kemp, J. E. (2001). *Designing effective instruction* (3rd ed.). New York: Wiley.

Norum, K. E. (2000a, July). *Appreciative Design.* Paper presented at the International Society for Systems Sciences Annual Meeting, Toronto, Canada.

Norum, K. E. (2000b, September). *Storying change: The power of the tales we tell.* Paper presented at the ALARPM/PAR World Congress, University of Ballarat, Victoria, Australia.

Reigeluth, C. (1996). A new paradigm of ISD? *Educational Technology, 36*(3), 13–20.

Seels, B., & Glasgow, Z. (1998). *Making instructional design decisions* (2nd ed.). Upper Saddle River, NJ: Prentice-Hall.

Senge, P. M. (1990). *The fifth discipline: The art and practice of the learning organization.* New York: Doubleday/Currency.

Senge, P. M., Kleiner, A., Roberts, C., Ross, R., Roth, G., & Smith, B. (1999). *The dance of change: The challenges to sustaining momentum in learning organizations.* New York: Doubleday/Currency.

Tessmer, M. (1990). Environmental analysis: A neglected stage of instructional design. *Educational Technology Research and Development, 38*(1), 55–64.

Tessmer, M., & Richey, R. C. (1997). The role of context in learning and instructional design. *Educational Technology Research and Development, 45*(2), 85–115.

Wheatley, M. J. (1999). *Leadership and the new science: Discovering order in a chaotic world* (2nd ed.). San Francisco: Berrett-Koehler.

Wheatley, M. J., & Kellner-Rogers, M. (1996). *A simpler way.* San Francisco: Berrett-Koehler.

Willis, J. (1995). A recursive, reflective instructional design model based on constructivist-interpretivist theory. *Educational Technology, 35*(6), 5–23.

Willis, J. (2000). The maturing of constructivist instructional design: Some basic principles that can guide practice. *Educational Technology, 40*(1), 5–16.

Zemki, R. (1999). Don't fix that. *Training Magazine, 36*(6), 26–33.

Zohar, D. (1997). *Rewiring the corporate brain: Using the new science to rethink how we structure and lead organizations.* San Francisco: Barrett-Koehler Publishers.

SECTION IV

C-ID in Practice: Examples From the Field

Jerry Willis

The fourth and final section of this book includes three examples of how C-ID was actually used to develop instructional resources. As you read the papers in this section I think there are several things that may stand out. The first is that when you are actually practicing ID instead of just theorizing about it, things quickly get fuzzy, muddled, and unclear. Few, if any, of the four examples in this section are "perfect" illustrations of the particular model of C-ID that was used by the developers. The clarity that can be achieved in theory is rarely, if ever, achieved in the real world of professional practice. That is perhaps as it should be. Constructivist and interpretivist theory assumes that context is critical to professional practice and perhaps the most important implication of that assumption is that any effort to apply a particular model or theory in a particular situation will have to be modified, adjusted, refined, and revised to "fit" the local context. That is certainly true of C-ID models.

In other fields the term "localization" describes a very important concept. For example, in the international business environment localization or "internationalization" is the process of adapting products and procedures as well as marketing efforts to local conditions. Localization can involve anything from making sure the documentation and marketing materials do not violate any cultural sensitivities to redesigning a product so it meets the needs of a different market. There are many famous examples of localization failures that are used to scare participants at localization seminars. One that comes to mind is the effort by Parker Pen to market a ballpoint pen in Mexico using the slogan "It won't leak in your pocket and embarrass you." However, the translator used the wrong verb and in Spanish the slogan said, "It won't leak in your pocket and impregnate you." A similar problem popped up when a new search engine named Dogpile was rarely used in the United Kingdom, as compared to usage patterns in the United States. Dogpile in the U.K. version of English turns out to mean something decidedly unpleasant that dog owners in urban centers have to clean up while walking their dogs.

There are hundreds, if not thousands, of sites on the Internet that help corporations create localized versions of their products and their marketing/documentation materials. There is even a site titled the "Top 10 Fatal Localisation Mistakes" (http://www.seoptimise.com/blog/2008/06/top-10-fatal-localisation-mistakes.html) that covers things like "Translating a Web Site Before Doing Local Market Research." The example used to illustrate what can happen in that case is trying to sell beef in the Indian market.

The examples above are somewhat trivial and easy to fix, but they are humorous and easily understood examples of something that is often much more serious and that can have an enduring impact. If constructivists are correct and context is important, virtually any instructional material will need "localization," just as successful products in one market may need many changes to be successful in another market.

Unfortunately, localization has not always been an accepted and expected activity in education. Developers of many instructional plans and systems considered localization undesirable and went to great lengths to "teacher proof" their materials and force every teacher to use it precisely as the developer planned. Teacher proofing is based on a positivist rather than a constructivist theoretical foundation. The developers who try to teacher proof their materials assume they have discovered the "One Best Way" to use the materials and that adaptations by mere teachers can only reduce the effectiveness of the material. This is a not a position most constructivist educators would accept, and few designers using a C-ID model would invest any effort in teacher proofing their curriculum or materials. I find teacher proofing offensive from a theoretical perspective but Kelly (2004) takes the position that teacher proofing is also a failure in practice:

"Teacher-proofing" does not work
    There have been many attempts over the last three or four decades to bring about curriculum change, most notably those sponsored by the [UK] Schools Council during its lifetime, ... and, most recently, the decision to

change the curricula of all schools to fit the demands of the new National Curriculum.

The most important point to be noted here ... is that we have learned from the experience of these projects and activities about the role of the individual teacher in curriculum change and development. We must especially note the failure of all attempts by the Schools Council to produce "teacher-proof" packages—schemes of work, versions of curriculum, supporting materials and so on of a kind which teachers would accept, use and apply in the precise form that the central planners had in mind. In every case, teachers adapted and used what they were offered in their own ways and for their own purposes. Some project directors were inclined to throw up their hands in despair at this phenomenon, at what they saw, and sometimes described, as "cannibalism." Others went along with it eventually and built into their schemes proper forms of allowance for this kind of personal and local adaptation by teachers.

The implications of this kind of experience for the implementation of forms of centralized control such as the National Curriculum are interesting.... We have here another example of the failure or the refusal of the architects of these policies to take any account or cognizance of the substantial experience and findings of earlier work.

The practice of education cannot be a mechanical, largely mindless activity; it requires constant decisions and judgements by the teacher.... Teaching, interpreted in a purely technicist sense, may be undertaken in a mechanistic manner. If, however, our concern is with education, in the full sense, ... much more than this is required, since education is essentially an interactive process. (p. 9)

This issue of localization has two levels of application in instructional design. At the design level we should not try to create "designer proof" ID models that must be used exactly as prescribed by the creator. Recently there has been an extended discussion in the literature on traditional Instructional Systems Design (ISD) about whether following all the steps in the model actually produces poorer rather than better instruction (Gordon & Zemke, 2000). I will not pursue that debate but I will note that the foundations of C-ID naturally encourage us to adapt ID models to local needs and context. Thus, in the four examples you will read about in this section, you will find many examples of adaptation and revision of the basic C-ID model. While I would not necessarily agree with every change or adaptation I do think that designers should be encouraged to "localize" ID models and also to share with others both the reasons for and the details of those localizations. Reflective analyses of the impact of adaptations and revisions themselves will also be of interest to other designers.

The other level of application has to do with the instructional materials created. A reasonable question to ask a designer is "How much flexibility does your team have when selecting teaching and learning strategies? Are there hard and fast rules that specify what pedagogy you *must* use to teach a certain type of content or a certain type of student?" Constructivists, naturally, will say that there are few, if any, universal answers to such questions. Local context must be considered when selecting both content and pedagogy. However, there have been a number of attempts to automate the process of instructional design (Tennyson & Baron, 1995). The most positivist and empirical of these efforts involves gathering information on what is to be taught to whom, and then provides the designer with a detailed prescription for how instruction should be designed, including the Right pedagogies to be used.

As you read about how teaching and learning strategies were developed in the four design projects included in this section, you will see that the design team did not take the positivist approach. The teams used everything from available research to teacher experience and extended team discussions to select teaching and learning methods, and those methods sometimes changed during the design and development process.

I will end this introduction to the section with a comment about both the design of C-ID models and the design of instruction in general. If design is, to a great extent, a "local" activity then design projects that will be widely shared should keep in mind the need to build in flexibility. If we know that teachers will be adapting and revising both the materials and the way they are used, it behoves design teams to make that process easier and more fruitful by designing material for that purpose. If we were creating children's toys, we should be creating transformer-style toys that can morph into many different things instead of simply being a truck or bicycle or soldier. And, the same goes for C-ID models. Instead of striving to "designer-proof "or "teacher-proof" our creations, we should make the effort to empower designers and teachers by giving them flexible and adaptable tools.

The three chapters in this section illustrate quite different approaches to C-ID. Chapter 19 tells the story of the Brandy Colon's use of the R2D2 C-ID model to create an interactive multimedia resource for teaching students to use a set of critical qualitative research data collection methods. Chapter 20 tells how a group at the University of South Carolina develop a multimedia CD-ROM to teach women about the risks of cervical cancer and the importance of regular PAP smears.

The first two chapters in this section are examples of C-ID work that focus on the process of ID and how ID can be done from a constructivist or participatory perspective. The final chapter is less about process and more about the use of constructivist principles of learning in the design of a course on child welfare. That project was based at the relatively new University of Northern British Columbia and involved the creation of an online course based on constructivist instructional strategies.

REFERENCES

Gordon, J., & Zemki, R. (April, 2000). The attack on ISD. *Training, 37*(4), 42–54.

Kelly, A. V. (2004). *The curriculum: Theory and practice* (5th ed.). Thousand Oaks, CA: SAGE.

Tennyson, R., & Barron, A. (Eds.). (1995). Automating instructional Design: Computer-Based Development and delivery tools. (Series F: NATO ASI Series: Series F: *Computer and Systems Sciences*, Vol. 140. New York: Springer.

CHAPTER 19

CONSTRUCTIVIST INSTRUCTIONAL DESIGN

Creating a Multimedia Package for Teaching Critical Qualitative Research

Brandie Cólon, Kay Ann Taylor, and Jerry Willis

Instructors for quantitative research courses often find that there are many different types of support material for those courses. That is not the case with qualitative courses. Very little support material is available for qualitative research courses. In this chapter we describe the creation of one multimedia package that focuses on one type of qualitative research—critical ethnographic techniques. The package was created to help graduate students learn to use five critical ethnographic techniques: meaning fields, validity reconstruction, role analysis, power analysis, and horizon analysis. We used a Constructivist Instructional Design Model, R2D2, to guide our development work. It is based on an interpretivist epistemology and a constructivist theory of learning. The result was a multimedia instructional package designed and developed for use in courses teaching qualitative

Previously published as *Appreciative instructional design (AiD): A new model.* Paper presented at the 2000 Association for Educational Communication and Technology (AECT) National Conference, Denver, Colorado, October 25–28, 2000 research. The hypermedia, multimedia program called The Critical Researchers Guide to Conducting Qualitative Research (CRIT) has three major components: (1) video cases of middle school settings or sites; (2) definitions and descriptions of qualitative strategies; and (3) application of qualitative techniques. The paper describes revisions and reformulations of the instructional package across the instructional design process.

Webb and Glesne (1992) attest that

> teaching qualitative research methods in colleges of education is challenging and exhilarating precisely because such courses call into question students' taken-for-granted assumptions about so many things: the purpose of research, the uses of method, the nature of knowledge, and what it means to be human. (p. 772)

They add that qualitative research has emerged in education and brought about the birth of new courses, programs, and special interest groups in the area. In the past, few textbooks dealt specifically with qualitative research. Now there are many texts on qualitative research, such as Denzin and Lincoln's (1994) *Handbook of Qualitative Research* and Carspecken's (1996) *Critical Ethnography in Educational Research: A Theoretical and Practical Guide*. However, teaching qualitative research in graduate programs can be difficult for a number of reasons. As a relative newcomer to the group of research methods courses taken by graduate students in education and psychology, the literature is scant on approaches to teaching qualitative methods. It is new in two ways: (1) the actual methods of data collection and analysis and (2) the underlying assumptions and epistemologies are often both complex and contradictory to the content students learned in other courses. According to Web and Glesne (1992) a wide variety of approaches are used and many instructors experience difficulty because the subject matter is complex and unfamiliar.

Even a cursory reading of the qualitative research methodology literature highlights the diversity in the field (Denzin & Lincoln, 1994; Eisner & Peshkin, 1990; LeCompte & Preissle, 1993; Willis, 1995). We will, however, use a general and relatively flexible framework for deciding whether an approach is "qualitative" or not. According to Strauss and Corbin (1990), qualitative research is any research that produces findings not arrived at by means of statistical procedures or other means of quantification. It can refer to research about people's lives, stories, and behavior, but it also can be about organizational functioning, social movements, or interactional relationships. Qualitative research techniques often rely on observation to collect unique data about the problem under study.

Techniques or methods of research are not, however, the only differences between qualitative and quantitative approaches. Qualitative research differs from quantitative research with respect to philosophical foundations, underlying assumptions, and research methods. Although descriptions of qualitative research methods given by different authors vary considerably, most characterizations of qualitative research emphasize participant observation (or more involved actions such as emancipatory and action research methods) and in-depth interviews. Quantitative methods contrast with qualitative research techniques, in which reliance is placed on the research instrument through which measurements are made. Qualitative research usually consists of three components: (1) data, which can come from various sources, (2) analytic or interpretive procedures that are used to arrive at findings or theories, and (3) written and verbal reports. Some researchers gather data by means of interview and observation, documents, books, and videotapes. The data produced are considered to be rich in detail and closer to the informant's perceived world, while quantitative approaches may lead to an impoverishment of data (Carspecken, 1996).

Understanding the major tenets of research paradigms, specifically qualitative, is essential to conducting good research. According to Borg and Gall (1989), qualitative research is more difficult to do well than quantitative research because the data collected are usually subjective and the main measurement tool for collecting data is the investigator. Therefore, before one can conduct qualitative research effectively, extensive training and practice in the methods are necessary.

One way to think about teaching qualitative research is from the perspective of cognitive flexibility theory (Spiro & Jehng, 1990). Qualitative research methods is a subject that is "ill-structured" and "complex." It is ill structured because you cannot teach precise recipes for conducting qualitative research. Many decisions must be made on the fly as the research proceeds. It is complex because the decisions to be made involve considering many aspects, and often from multiple perspectives. Cognitive flexibility theory (Spiro & Jehng, 1990) was created to guide our thinking about how to teach ill-structured, complex content. According to Spiro, Coulson, Feltovich, and Anderson (1998), teaching content such as qualitative research methods via material that was developed using cognitive flexibility theory, and that uses hypertext and hypermedia can be an effective instructional approach. A critical element of this approach is the use of multiple representations that capture the real world complexities of the subject matter, in this case qualitative research.

The increasing interest in qualitative research has, predictably, led to the addition of one or more "qualitative research" courses to the requirements for advanced degrees in education. Instructors for quantitative research courses often find that there are many different types of support material for those courses. That is not the case with qualitative courses. Very little support material is available for qualitative research courses. In this chapter we describe the creation of one multimedia package that focuses on one type of qualitative research—critical ethnographic techniques.

ETHNOGRAPHY AND CRITICAL ETHNOGRAPHY

According to Denzin and Lincoln (1994), ethnography has had many uses and meanings throughout history. The history surrounding ethnography reveals how multiple uses and meanings are brought to each practice. Critical ethnography differs from traditional ethnography in its attempt to link the detailed analysis of ethnography to wider social structures and systems of power relationships in order to examine the origins of oppression. Critical ethnography raises substantive questions about structural relationships. The intention is to go beyond grasping the subject's meanings in order to relate those meanings to wider cultural and ideological forms. Critical ethnography is a widely used technique in critical social research. The involvement and close attention to detail characteristic of ethnography make it useful for rendering visible the invisible, and for revealing anomalies and common-sense notions.

Five Recommended Stages for Critical Qualitative Research

Carspecken (1996) follows Habermas in distinguishing two basic methodological perspectives that can be employed simultaneously within a critical research project: (1) hermeneutic meaning reconstructions and (2) objectivizing studies of social systems. Hermeneutic meaning reconstruction is totally congruent with interpretivist methodologies; one takes the insider's view of a cultural group and reconstructs tacit cultural themes and structures that members commonly employ to interpret the world, judge the world, and construct their social identities. Objectivizing studies of the social system prioritize an "outsider's" view. Here one seeks social structures that help shape and constrain culture.

Carspecken (1996) developed a five-stage scheme for conducting critical qualitative research, where stages 1–3 employ hermeneutic reconstructive techniques and stages 4–5 emphasize the objectivizing stance in one's search for system phenomena. The preliminary steps include creating a list of research questions, a list of specific items for study, and examining researcher value orientations. The five stages are

1. **Stage One: Compiling the primary record through the collection of monological data.** The researcher makes her/himself as unobtrusive as possible within a social site to observe interactions. A primary record is established through note taking, audio taping, and video taping. The information collected is monological in nature because the researcher speaks alone. There is no dialogue with members.

2. **Stage Two: Preliminary reconstructive analysis.** The researcher begins to analyze the primary record as it exists so far. A variety of techniques are employed to determine interaction patterns, their meanings, power relations, roles, interactive sequences, evidence of embodied meaning, and intersubjective structures. This stage is meant to articulate cultural themes that are not observable because they are tacit.

3. **Stage Three: Dialogical data generation.** The researcher ceases to be the only voice in establishing a primary record. The researcher uses special techniques such as interviewing and discussion groups to converse with the subject of study.

4. **Stage Four: Discovering system relations.** The researcher examines the relationship between the social site of interest and other specific social sites bearing some relation to it. System relations are found that are not simply tacit but totally outside the culture of study.

5. **Stage Five: Using system relations to explain findings.** The level of inference goes up as the researcher seeks to explain his/her findings in stages one through four by inference to the broadest system features (pp. 42–43).

These stages were designed to study social action taking place in one or more social sites and to explain this action through examining locales and social systems intertwined with the site of interest. Common subjective experiences and the significance of the activities discovered with respect to the social system at large are assessed (Carspecken, 1996).

This study's emphasis is on the critical ethnographic techniques Carspecken (1996) outlined in stage two of preliminary reconstructive analysis. There are five techniques:

1. Meaning fields,
2. Validity reconstruction,
3. Role analysis,
4. Power analysis, and
5. Horizon analysis

These five techniques for conducting critical ethnographic research were the primary focus of the instructional package developed.

PURPOSE OF THE STUDY

The purpose of this study was to create an instructional product using a hypertext system based on cognitive flexibility theory, about the complex and ill-structured domain of qualitative research and its methods, specifically those detailed in *Critical Ethnography in Educational Research: A Theoretical and Practical Guide* (Carspecken, 1996). A multimedia instructional package was designed and developed for use in courses teaching qualitative research. The hypermedia, multimedia program called *The Critical Researchers Guide to Conducting Qualitative Research* (CRIT) has three major components: (1) video cases of middle school settings or sites; (2) definitions and descriptions of qualitative strategies; and (3) application of qualitative techniques. This final component is comprised of exemplary illustrations of each concept and associated techniques that were used with other qualitative data sets.

METHODOLOGY: CONSTRUCTIVIST
INSTRUCTIONAL DESIGN (C-ID)

Instructional Design and Different Paradigms

Over the years, many instructional design models have been proposed (Bagdonis & Salisbury, 1994). Andrews and Goodson (1995), for example, identified more than 60 models, but descriptions of over 200 models have been published in the educational technology literature. However, most of the ID models are, however, based on behavioral and information processing theories of learning (Dick, 1996). In fact, Walter Dick and Lou Carey developed the most popular ID model in this tradition in 1968. It is now in its fourth version (Dick & Carey, 1996). Dick and Carey called it an *Instructional Systems Design*, or ISD, model because it is based on one form of systems theory, as well as behavioral and information processing theories of learning. "Behaviorism is prominent in the roots of the systems approach to the design of instruction" (Burton, Moore, & Magliaro, 1996, p. 57).

While these models have dominated the field for over 3 decades, they have been criticized recently by a number of scholars. For example, Wilson (1993) felt that rapid change in instructional design is coming from the debate initiated by constructivist theorists. In describing instructional design's state of change, Wilson summarized his objections to the established, ISD, approach this way:

The problem can be simply stated: ID, in its present form, is out of sync with the times.

- Its orientation is behavioristic.
- Its methods are behavioristic.
- Its research base is behavioristic. (p. 1132)

Willis (1995) criticized ISD models because of their behavioral foundation and listed eight characteristics of these ID models that he considered undesirable:

- The process is sequential and linear
- Planning is top down and "systematic"
- Objectives guide development
- Experts, who have special knowledge, are considered critical and central to ID work
- Careful sequencing and the teaching of subskills are important
- The goal is delivery of preselected knowledge
- Summative evaluation is critical
- Objective data are critical (p. 11)

Several other authors have suggested that ID could be based on theories of learning other than behaviorism or information processing. Scholars (Lebow, 1993; Wilson, 1997; Winn, 1992) have offered general suggestions for basing ID on a constructivist foundation. Several instructional designers also have created complete ID models based on constructivist theory. For example, Cennamo, Abell, and Chung (1996) developed the "Layers of Negotiation" model while they were creating a set of video cases for teacher education. They noted that, "Many authors have questioned whether traditional instructional design models are suitable for designing constructivist learning materials" (p. 39). Their model, Layers of Negotiation, is based on the same set of constructivist principles that are the foundations for the instructional material they developed.

The study reported here used another ID model based on constructivist theory-R2D2 (Willis, 1995, 2000).

The R2D2 Model of Instructional Design

The Recursive, Reflective Design and Development (R2D2) model was introduced in a journal article by Willis in 1995. The R2D2 model was the methodology used to guide the process of creating *The Critical Researchers Guide to Conducting Qualitative Research* (CRIT).

Guiding Principles

R2D2 has four overarching principles: (1) recursion, (2) reflection, (3) non-linearity, and (4) participatory design. *Recursion* allows the designers to revisit any decision, product, or process at any time in the design and development of the product, and make refinements and revisions as needed. Recursion, or *iteration*, makes the design process a spiral - the same issues and tasks may be revisited many times across the design and development of a particular instructional product.

Reflection is probably best understood by contrasting it with the opposite principle of design-technical rationality. Design based on a technical-rational approach requires developers to follow a set of pre-defined rules that prescribe what is to be done. Reflective design places less faith in pre-set rules and instead emphasizes the need for the designer to thoughtfully seek and consider feedback and ideas from many sources. For detailed information on the process of reflection in professional practice see Schön (1983, 1987).

The third guiding principle, *non-linearity*, comes from chaos theory (You, 1994). Instead of providing a linear sequence of steps that must be completed in a certain order, R2D2 suggests a set of focal points that need not be approached in any particular predetermined order. Different projects may call for different starting points. For example, the design process need not begin with a detailed plan that requires development of precise objectives at the beginning of the work. Objectives may, instead, emerge over the design process and not be completely set and clear until the end of the project. Thus, the design process commences wherever it is appropriate and progresses as appropriate.

The last principle, *participatory design*, is based on the assumption that the context of use is critically important. Further, the people most familiar with those contexts will be the users. Therefore, they should be involved extensively in all phases of the design and development process (Schuler & Namoika, 1993). In R2D2 the idea of participatory design has been expanded beyond end users to include "experts" in the sense Eisner (1979) meant in his connoisseurship model of educational research. Thus, ID using the R2D2 design model involves a participatory team that guides the process. In education, this team typically includes instructional designers, subject matter experts, teachers, and students. Members of the team are often referred to as *stakeholders*.

These four guiding principles—recursion, reflection, non-linearity, and participatory design—are quite different from the principles that serve as the foundation for many instructional design (ID) models. Most of the existing ID models are based on behavioral and/or information processing theories and proponents of those models have been quite critical of these principles. Merrill (1996), for example, was blunt in his assessment of approaches that try to involve stakeholders in developing a vision of how content should be taught:

> A "visioning activity" is a recipe for disaster in the real world of instructional development. It is a dream of academics who value collaborative approaches to knowledge; but, in practice, it often leads to disaster. There is no doubt that "stakeholders" must have a role in determining "ends" (how the learners will be different as a result of instruction), but when "stakeholders" play a significant role in determining "means" (how those changes in the learners will be fostered), then the result is often ineffective instruction that does not teach. (p. 58)

Merrill (1996) goes on to say that "The consensus of 'stakeholders' often equals poor learning" and he raises doubts about whether students can play a meaningful participatory role in ID since "students are, for the most part, lazy." When instructional design theorists who work from a behavioral, information processing, or cognitive science paradigm have analyzed the R2D2 ID model, they have generally been critical. Dick (1996) compared R2D2 to his model and found it wanting while Merrill rejected outright the guiding principles of R2D2 and similar ID models based on constructivist theory. Merrill has gone so far as to declare that anyone who works from such a different paradigm cannot call themselves instructional designers:

> Those persons who claim that knowledge is founded on collaboration rather than empirical science, or who claim that all truth is relative, are not instructional designers. They have disassociated themselves from the technology of instructional design. We don't want to cast anyone out of the discipline of instructional science or the technology of instructional design; however, those who decry scientific method, and who deride instructional strategies, don't need to be cast off; they have exited on their own. (Merrill, Drake, Lacy, Pratt, & the Utah State University ID2 Research Group, 1996, p. 6)

It is not surprising that scholars such as Merrill are critical of ID models like R2D2 that are based on constructivist theory and interpretivist epistemologies. They are working from behavioral/information processing theories of learning and positivist/postpositivist epistemologies. Constructivist worldviews are based on different foundational beliefs. The four guiding principles of the R2D2 model naturally derive from an interpretivist epistemology and a constructivist theory of learning. They are contrary to postpositivist epistemology and behavioral/information processing theories of learning.

R2D2 is, however, a particularly appropriate ID model for creating instructional materials to teach qualitative research methods. Many forms of qualitative research are based on some of the same guiding principles as the R2D2 model. Most forms of qualitative research, for example, are recursive and non-linear. The purpose, method and form of the research may change and evolve across the study. This generally calls for a reflective approach to research. A qualitative scholar cannot simply follow technical rules and a detailed, pre-established plan that will lead to a "good" study. Finally, several forms of qualitative research are participatory. Participatory action research and emancipatory research in the critical tradition come quickly to mind.

Thus, while R2D2 has been criticized heavily by some ID scholars who practice from a different paradigm, it seems particularly suited to guide the design of instructional materials for teaching qualitative research methods because the its foundations have much in common with a number of approaches to qualitative research. (And, as would be expected, researchers who use postpositivist and positivist paradigms make some of the same criticisms against qualitative research as postpositivist and positivist ID scholars make against R2D2.)

Focal Points

Focal points are convenient ways of organizing thinking about all the things that need to be done in a design project. R2D2 has three focal points: (1) Define, (2) Design and Develop, and (3) Disseminate. This study involved work in the first two focal points. It is important, however, to note that these are not linear stages. We did not complete all the work related to the define focus, for example, before moving on to work in the design and development focus. R2D2 ID is a spiral, which means some tasks were addressed many times across the project.

Define Focus. This focal point addresses the tasks that would, in a traditional instructional design model, be completed before actual design work begins. The *Define focus* in the R2D2 model continues throughout the design process. Early in the process, loose and ill defined ideas about the approaches that might be taken, the characteristics of students who will use the product, and the content to be taught, guide the design process. These ideas gradually evolve, are revised, and emerge from the ongoing work of design and development. The same is true of the basic purpose of the instructional design project. In this study, we began with a general, if somewhat fuzzy, purpose-to create a multimedia program to support teaching critical qualitative research methods.

An important aspect of the *define focus* is the creation of a participatory group. The learners are participating members of the development team rather than objects to be studied. Their input shapes the project's theme and purpose, the topics to be included, the instructional strategies used, and the look and feel of the interface (Schuler & Namioka, 1993). This project's team included experts such as faculty who teach courses in qualitative research, instructional design specialists, and interface design experts. Graduate students, who had completed qualitative research courses and who were instructional technology specialists, were another important part of the team. The end-users (faculty and students) were involved in the entire design process.

Design and Development Focus

The Design and Development focus has three components: (1) preparation tasks, (2) creation, and (3) procedures.

Preparation tasks. These include selection of a development environment, media selection, and selection of instructional strategies. As with most aspects of the R2D2 model, initial, tentative decisions were made to begin the process, but they were always subject to both evolutionary and revolutionary change. Another preparation task is the selection of a development environment. Several criteria were considered when the initial decisions were made about the development environment. Since the product created was a hypermedia package that uses digital video stored on a CD-ROM, an authoring system capable of supporting CD video was required. Another requirement was cross platform compatibility because both Macintosh systems and Windows systems are popular in higher education. CRIT is non-linear and requires a program capable of creating non-linear information landscapes. In addition, we needed an environment that encouraged and supported recursion and reflection. A desirable environment should make it easy for material in all phases of design and development to be revised and edited. After considering all of these requirements, the authoring system SuperCard 2.5 (Allegiant Technologies) was selected. Programs created with SuperCard do not require a viewer or the actual software to run them and these executable files are distributable without paying a royalty. Supplemental software such as graphics programs, video editing packages, and audio editing suites were used as needed.

Two additional activities in this area are *selection of media* and *selection of instructional strategies*. Much of the material on qualitative research methods is in textual form (e.g., textbooks, articles, and book chapters). A medium that facilitates involving the student in the practice of the procedures with appropriate feedback and mentoring is needed (Crandall, 1993; Wolf, 1993). A hypermedia instructional package that presents video clips on the screen that students can use to practice gathering qualitative data meets this need. CRIT includes computer-controlled video and a computer program that provides instruction, guidance, feedback, and opportunities to compare a novice researcher's observations with those of more experienced researchers. A wide range of media are incorporated into CRIT including text, charts, graphics, animation, sound, and video.

Another reason a hypermedia, multimedia environment was selected is its non-linearity. Learners can make many decisions about what material they study and in what order. For this work, an instructional strategy that is well suited to the purpose is Jacobson and Spiro's (1993) cognitive flexibility hypertext. Hypertext environments promote cognitive flexibility - the ability to use knowledge and skills in flexible and innovative ways (Staninger, 1994). The general framework for CRIT was cognitive flexibility theory and within this framework several instructional approaches were available. The video case studies included on the CD-ROM constitute a rich instructional knowledge base the learner can explore and analyze.

The Creation Tasks. The creation tasks address the work involved in actual creation of the product. In the R2D2 model, a prototype is collaboratively developed with the instructional designer, experts in the relevant specialties, and students. For this study, a prototype was developed that included the elements of:

1. Surface Characteristics—screen layout, typography, language, graphics, illustrations, sound;
2. Interface—ook and feel, user interaction, help, support, navigation, metaphors;
3. Scenario—sequence of video cases, options/choices, comparisons;
4. Supporting hypertext and hypermedia instructional content;
5. Instructional strategies.

Drafts of the components of the instructional material were created and evaluated by members of the team (experts and students). The feedback obtained guided a series of revisions and refinements. This was a recursive process that progressed to the final version of the material. CRIT was created by first concentrating on the creation of the components noted above with input from the stakeholders guiding revisions. CRIT is organized as a web of knowledge (i.e., information nodes are linked to others based on their interrelationships). The network's complexity depends on the existing interrelationships.

According to Spiro, Feltovich, Jacobson, and Coulson (1991), "revisiting the same material, at different times, in rearranged context, for different purposes, and from different conceptual perspectives is essential for attaining the goals of advanced knowledge acquisition" (p. 28). CRIT permits students to progress through the information landscape many times from different perspectives. As the product was created, the design team continually revisited the material from different perspectives, as noted earlier, making revisions as needed. The feedback-revise-feedback process occurs across a progressively more "finished" product. It began with components of the product such as the user interface, scenario, and information content. It then progressed to a "single path prototype" (i.e., a version of CRIT that a student progresses through using the one developed path because that is the only path that has been created). From the single path prototype, work progressed to a full alpha version and then to beta versions. The last "beta" version was designated the final product.

Throughout the design and development focus, evaluation plays a critical role. Formative evaluations provide feedback to the development team that improves the product. The participatory design approach is a critical component of the effort to produce a quality instructional package. In addition to the informal evaluations and critiques, there were two formal evaluation strategies: (1) expert and stakeholder appraisal, and (2) student tryouts. Both were used to evaluate many versions of the material across the entire process.

The evaluation strategy for CRIT involved experts and end-users who assisted with two types of evaluation: component feedback and package critique. The R2D2 Model brings students and experts into the development process in the beginning and uses their advice, evaluations, and feedback at every stage to improve the product (Willis, 1995). The information obtained from component feedback guided further design and development. When the single path prototype of CRIT was available, both experts and end-users provided detailed critiques. That process was repeated many times for both alpha and beta versions.

Procedure. CRIT gives students the opportunity to view a number of video cases. They use hypermedia/multimedia software to navigate through the material. Students thus have flexible access to video from classrooms. Viewers see rich visual and verbal cues that can become the basis of their observations. The viewers witness the natural progression of a classroom setting and observe each video case. The program assumes students are already familiar with the particular research techniques they will use. Class discussions and reading assignments in the textbook are common methods of acquiring familiarity. In the program, an expert analysis was done for all video cases regarding the critical qualitative research techniques. Phillip Carspecken, an expert in the field of ethnography, wrote the analyses. The CD-ROM has a SuperCard stack that includes descriptions of techniques, the use of these techniques, practice of these techniques, and reference material. Students can explore the video cases, read the description of the techniques, look at expert analysis of any of the video cases, explore another video case, practice writing an analysis, and look at reference material. With the random access made possible using interactive multimedia, users can juxtapose information in a sequence of their own choosing and examine the content from multiple perspectives (Cennamo, Abell, Chung, Campbell, & Hugg, 1995; Cennamo et al., 1996).

Decisions about the footage that constitutes a video case for the content included in the instructional package were decided by the team. The video cases were pre-existing data of middle-school students in a multicultural, natural classroom setting. The team searched the tape logs for scenes that represented topics discussed in Carspecken's book such as role settings, meaning fields, validity constructions, and power analysis. A compilation of those video clips, graphics, and sound were used to produce the final videotape. Editing was done on a manual-editing system. The final medium of video was CD-ROM because of its portability and cost.

Dissemination

Dissemination was not a part of the study.

RESULTS

The end result of this study was *The Critical Researchers Guide to Conducting Qualitative Research* (CRIT), an interactive multimedia instructional package that supports teaching critical qualitative research techniques and strategies. The focus of CRIT is to introduce qualitative researchers to specific qualitative techniques, provide practice using those techniques, and provide expert guidance. CRIT can support any graduate course on critical qualitative research but it was designed to specifically compliment Dr. Phillip Carspecken's (1996) book, *Critical Ethnography in Educational Research*. Students read and discuss relevant chapters in that book, and then use CRIT to study the methods in a more concrete, hands-on format.

This section of the chapter is divided into three parts: (1) chronology, (2) product overview, and (3) the results of formative evaluation procedures.

Chronology

A working prototype of CRIT was produced after about two months of development, but the actual video used in the program was not developed or selected at that point. Working with experts and end users (graduate students), we selected short segments of video from a library of classroom video at the Center for Information Technology in Education at the University of Houston.

The video clips ultimately selected illustrated a variety of onsite classroom events depicting small group interactions and large group discussions. After the initial selection, a rough edit of the video footage was created. This video was reviewed extensively to identify unnecessary segments. The tape was reviewed by potential users of the product (graduate students taking qualitative research courses) and the rest of the stakeholders. Each individual examined it from his or her perspective.

After the videotape segments were selected, they were incorporated into the prototype. The prototype was a "single path prototype." One path through the material, using one of the video cases, was completed. The development team provided feedback on several versions of this single path prototype, and many revisions were made. When a new version generated no new suggestions for revision, the single path prototype was used as a model to develop all the other paths in the program. The first "full" version prototype was designated the *alpha* version. With feedback and the resulting revisions, the alpha version became the *beta* version, and then Version 1.0. Selecting the video segments required about two weeks and it took about four months to develop the single path prototype and the alpha and beta versions.

Product Overview

When the icon for CRIT is double-clicked, a title screen appears that contains the graphic on the cover of the book, *Critical Ethnography in Educational Research* (Carspecken, 1996). There are two buttons on the title screen (see Figure 19.1), one for the first-time users and one for those who are already familiar with the program.

The "first-time user's" button takes the users to an overview of the program that describes what the program is about and the benefits the learner can expect from using this program (see Figure 19.2).

Although the user has the option to bypass the overview and start the program by clicking the "Experienced User" button (see Figure 19.1), it is important for the user to have an overall understanding of CRIT. Clicking

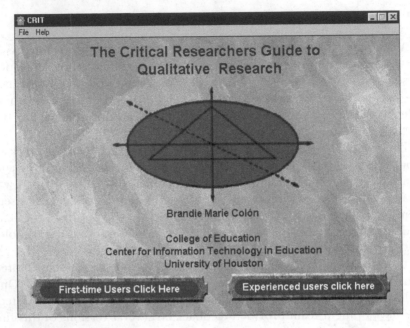

Figure 19.1.   The CRIT Title Screen.

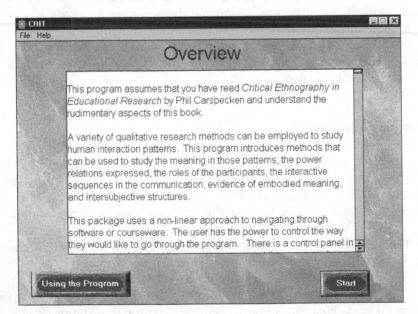

Figure 19.2.   CRIT Program Overview.

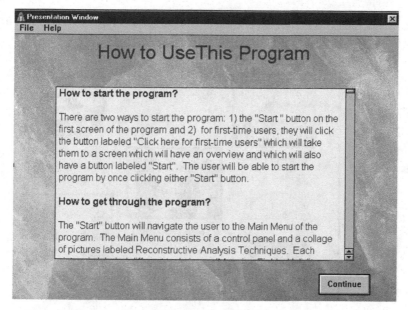

Figure 19.3.   "How to Use This Program."

the first time user button takes a student to a set of directions on "how to use the program" (see Figure 19.3 above).

After reading the directions, the user can navigate through CRIT by clicking the "Start" button. However, the user does not have to read the directions or the overview to start the program as they can always access a help menu with those options. Students also can access the main menu by clicking the "Experienced User" button (see Figure 19.1).

The main menu consists of five graphics. Within each graphic there are text fields labeled Description, Expert, Practice, and Examples (see Figure Figure 19.4).

All of the text fields within the graphics are hyperlinked to the appropriate sections of the instructional material:

1. *Description*: This option includes an introduction and overview of the research method, with some detail on the techniques involved, as well as a conceptual diagram.

2. *Practice*: This alternative lets students use the video clips to practice data collection/analysis using the research method.

3. *Expert*: This choice takes the user to an analysis of the video clip that was done by an expert in the research method. This part of the program is not intended to provide students with an overview

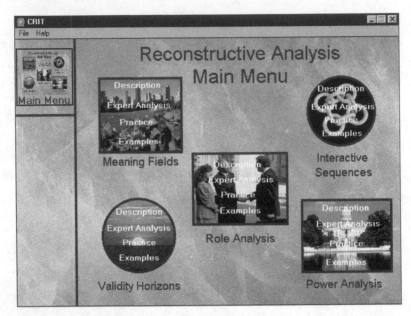

Figure 19.4.   CRIT Text Fields Labelled Description, Expert, Practice, and Examples.

of the "right" way to do the data analysis. Instead, it is treated as an example of how one expert looked at the data.

4. *Examples*: This section includes summaries of research studies that use the research methods covered in the program.

In addition, on the screen shown in Figure 19.4 there is an icon (labeled Main Menu), which is one of the navigational tools featured in CRIT (see Figure 19.4). Once the user starts CRIT, the *main menu* icon appears on all the screens thereafter. Students can click this icon and return easily to the main menu regardless of where they are in the program.

CRIT covers five different critical qualitative research methods: meaning fields, role analysis, interactive sequences, validity horizons, and power analyses. Each of these methods has a description, a video-case example, opportunities to practice the method, and examples of how the method has been used in social science research. The Description section on validity horizons, for example, allows the user to review a detailed description of a validity horizon and view a diagram representing a validity horizon (see Figures 19.5 and 19.6).

Figure 19.7 is an example of the Expert section. It is an analysis of meaning fields for one of the video clips.

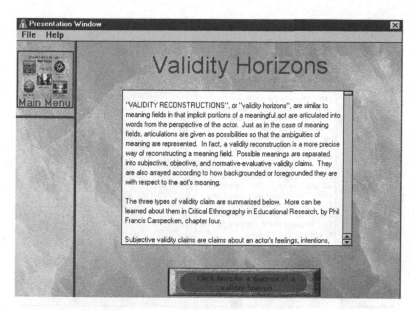

Figure 19.5.   CRIT Validity Horizons Screen.

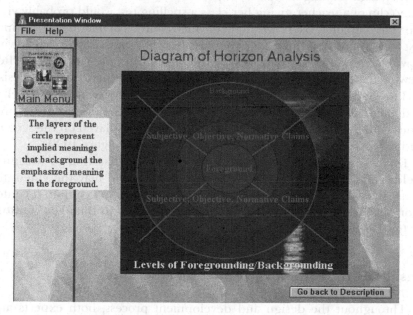

Figure 19.6.   CRIT Diagram of Horizon Analysis.

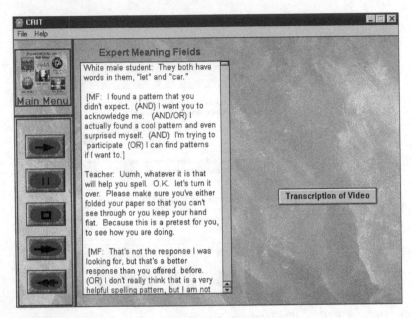

Figure 19.7.   CRIT Example of the Expert Section.

A student looking at how an expert analyzed the validity horizon in a video clip of a teacher giving her class a spelling test would probably view the video clip first. Users have a set of video controls to play, pause, rewind, fast forward, or stop the video. There is also a button to pop up a text window that contains a transcription of the dialog from the video case. After viewing the video, the student can then look at the expert's data analysis.

The Practice section gives students the option of viewing a video clip, and then conducting their own qualitative analysis. The program provides both the case to be studied (on video that is displayed on the screen) and the forms or format for collecting the data (see Figure 19.8).

For example, the user can click on the Practice section of Meaning Fields and select a video clip. The choices are to (a) play the video, (b) pause the video, (c) write some claims, (d) view the transcription, (e) enter more claims, (f) save your work by clicking the "Add to Data File" button, or (g) "Go On to Subjective Claims."

Results of the Formative Evaluation Procedures

Throughout the design and development process, both experts and potential users provided detailed critiques and suggestions for each ver-

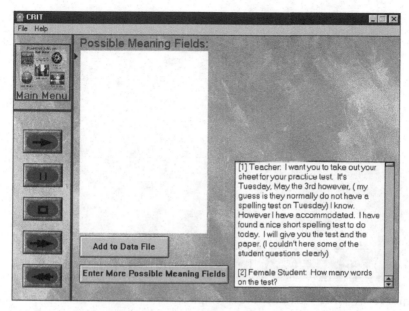

Figure 19.8.   CRIT Practice Session Student Data Collection Field.

sion of the software. An overview summary of input from stakeholders and the changes that resulted from that participation is explored in the next section.

Feedback on and Revisions to Components

Input and feedback from students and experts were important components of the developmental process. The student group (5) that evaluated the program included graduate students who had taken the Ethnography and Cultural Studies course at the University of Houston. Those in the expert group (6) included individuals who had expertise in instructional design, instructional technology, and qualitative research.

In the beginning of the design process, the primary goal was to develop the look and feel of the program, (i.e., the user interface). The goal was to design something that would catch learners' attention, be informative, and at the same time not prescribe a set way of navigating through the program. The first attempt at an interface was a rockscape-style background with contrasting color hotspots pointing to the specific research methods (see Figure 19.9).

This primitive version of CRIT was too "busy" and did not seem to mesh with the subject matter. This was a milestone, however, because it provided a beginning point. Many changes were tried and rejected

Figure 19.9.   CRIT First Interface Attempt.

before the team agreed that we had a pleasing, appropriate background, and a workable user-interface that did not suggest to students
that there was one "correct" or linear path through the various options
of the program.

Discussion with the expert stakeholders suggested that, ideally, the
menu-bar should appear on every screen to provide the user with the
options of viewing video, looking at expert analyses, practicing, looking at
descriptions of techniques and how they were used, or exiting the
program. Making the menu-bar omnipresent provides the user with the
flexibility to "jump" anywhere in the program at any time. The input from
stakeholders suggested that a thematic or metaphoric menu would fulfill
the requirements. Other elements of the program such as the format,
video cases for practice, and explanations of the research method including examples and illustrations of data gathered by others, were arrived at
rather easily. Two of the experts took the lead in suggesting the case format and the options available to students. The navigational options, and
how to present them, were a different matter. Much of the design and
development effort was invested in selection of the video cases and the
design of the navigational options.

The process was discouraging at this point. The attempt at a thematic
menu was a file cabinet labeled with the topics in the program.
Stakeholders felt this theme was too common and that it did not represent

the subject matter of the program. The stakeholders' input caused a major revision to the way the options were laid out. Eventually the menu structure was organized around iconic images that stood for each of the research methods covered. Sources for images included Internet, clip art programs, and photo CDs. After considerable searching, graphics were located that could be used legally in the program that figuratively represented the techniques. After some changes based on feedback from stakeholders, consensus on the graphics used for the main menu was reached. Figure 19.4 shows the graphics used to represent each of the research techniques.

The graphics were chosen, but the task of arranging them on the main menu screen remained. The layout was important because they needed to be arranged so the user could easily navigate through the program with no constraints from the interface. It also was important for the user to have access to all the possible paths from the beginning of the program. For example, the user could click within any of the five figuratively represented techniques and select description, expert analysis, practice, or examples (see Figure 19.4). There was no requirement for students to complete a particular activity before going on to a second activity.

Once the basic structure of the user interface was in place, the next step was to develop first one path and then full versions of the program.

Feedback From and Revisions to the Single Path, Alpha, and Beta Versions

Work on the prototype took about four months. Initial work was a rough mock-up (built in Authorware) of early design ideas. The interface was demonstrated to several users and experts. Feedback was gathered broadly on the general concepts and approach. Significant changes were made and a new version of the prototype included an extended series of design walkthroughs. Feedback on these walkthroughs led to a number of revisions.

Five individuals with expertise in instructional technology and design used the single path prototype and offered detailed suggestions. Also, four graduate students who were potential users also used the prototype and critiqued it. These individuals were invited to observe, explore, construct, and evaluate their interpretations and opinions of the program. The process of using CRIT was the focus of evaluation in these tryouts. The results of the prototype, alpha, and beta tryouts of CRIT by experts and potential users are combined here. Data came from open interviews, informal discussions, and a short questionnaire. Each participant spent about an hour on each tryout. Many changes were made as a result of feedback. These changes are described below in the three categories: (1) conceptual, (2) cosmetic, and (3) media and implementation.

Conceptual. Conceptual suggestions were about the major ideas or themes within the program. Below are some of the revisions made based upon feedback from various stakeholders.

The main menu, or "homepage," is the entry point to the content CRIT presents. *Main menu* implies some type of top-level index. A common assumption about interactive instructional programs based on traditional paradigms is that there is a single starting point to begin the program. A second common assumption is that there is a sequential order for content presentation.

The students who tried the menu-bar liked it, but one of the experts suggested metaphorically presenting the information. Using feedback from one of the expert stakeholders, a file cabinet was used as a main menu. Other experts objected to the file cabinet, commenting that the metaphor implied a linear structure. Taking that into consideration, graphics that would figuratively represent the reconstructive analysis techniques featured in the program were selected. This involved a number of iterations as images were considered, critiqued, and then rejected.

Cosmetic Revisions. Numerous changes were made that, while cosmetic in nature, contributed to the "finished" or "professional" look of the product. For example, all of the individuals in the tryout-sessions had difficulty reading the text on the screen because the background was transparent. Burgundy text on a sand-marble background was originally used. As a result of the comments, all text backgrounds were changed to white opaque backgrounds (see Figure 19.2). Other cosmetic revisions included changing the "busy" and inconsistent layouts of several screens, and improvements in the Overview section that involved dividing it into two parts. Dividing the section into two parts reduced the amount of text a person had to read in a section and made the instructions for use clearer and easier to understand.

Other changes included adding a video counter to allow students to determine exactly where they were in a video clip as they wrote their analyses. This caused another problem, however. The cursor "jittered" on the screen and was distracting enough to render the video counter unusable. Instead, a full-text transcription of the video was provided in place of the counter. The user has the option to click on the button for a verbatim transcript of each video case.

Finally, experts and users found many spelling and grammatical errors that were corrected as they were identified. The subject matter expert reviewed the content and a number of changes were made based on that review. Also, many of the student users felt the text content was too verbose and cumbersome. The content matter expert revised some of the content focusing on cutting down the amount of text.

Media and Implementation Changes. The original video was on S-video tapes. It was digitally edited and saved using MPEG compression. The size of the video files range from 250 to 350 megabytes; the Authorware program or file is 4 megabytes. With three video clips and the Authorware program, storage capacity became a problem. In order for the video clips and the program to fit on a standard CD-ROM, the video clips were saved in "thousands" of colors, instead of millions of colors. That reduced the size of the video files and made it possible to put CRIT on a CD-ROM.

SUMMARY OF RESULTS

The overall reaction from stakeholders and students was that this type of program is valuable for learning the content. All believed the final version of the instructional package was easy to use and was a worthwhile learning experience. The match between content intended for teaching advanced qualitative research techniques, and teaching strategy (video cases) was considered appropriate.

Specific suggestions and areas that would benefit from revision were offered by many of the people who participated in the design and development process. The CRIT video component was noted as an area of potential strength in its representation of techniques. Some participants questioned the length of the video clips and the clips were consequently reviewed. More video data was added so students could see more context and develop a better feel for the situation being portrayed. Video cases in the final version ranged from 2.5 to 3 minutes.

CRIT's navigational component was another strength cited by the stakeholders. In observing some of the tryouts, the instruction was interactive and engaging; learners found themselves engrossed. More knowledgeable learners were comfortable with the navigation as well. Also, learners with a richer and more extensive base of subject-matter knowledge seemed to quickly grasp the "big picture" of the content and the program. The novelty and appeal of working with authentic cases was appealing to the students and was also considered a strength.

One student expressed concerns about the content of the video and the need for the cases to be more ethnically diverse. A range of ethnic backgrounds is represented in the cases-Anglo, African American, Hispanic-and most of the participants felt the video cases were diverse in terms of student representation.

The concept of practicing the qualitative techniques in real-time was highly praised, as well as the user-friendliness of CRIT. Both experts and student evaluators thought the final version was effective in helping students better understand the qualitative research techniques.

If the reactions of stakeholders to CRIT were distilled to a few comments, most would focus on the positive benefits of authentic practice and video cases. The video helped students visualize concepts, and the ease of navigation and logical flow to and from different parts of the program allowed students to concentrate on learning. Students were particularly pleased with the opportunity to see actual video cases to which they could apply the research skills they were learning.

DISCUSSION

This study illustrates two "trends" in education today. The first is the creation of electronic support materials for all types of content—from kindergarten to graduate school. When instructional designers create electronic content, they often use instructional strategies and approaches that come primarily from the behavioral side of psychology. Some authors even feel that behavioral approaches dominate the educational uses of technology today. CRIT illustrates the point that other worldviews—critical theory and constructivism/interpretivism—can be used to make decisions both about what content should be included and the instructional strategies that will be used. Information technology is a flexible and fluid medium, and in the case of CRIT, it became a powerful means of helping students learn some admittedly complex and difficult research methods.

The second "trend" this study represents is the move away from linear models of instructional design (ID) based on behavioral psychology and an increasing interest in ID models based on constructivist as well as chaos theory. The R2D2 model is non-linear, recursive, reflective, and participatory. Traditional ID models are linear, tend to emphasize the role of the designer as THE expert, and frequently involve very little participation on the part of stakeholders. ID is a well-established field within educational technology, and perhaps because it has its origins in the behavioral psychology in the 1960s and 70s, it has been slower to seriously consider other foundations. Other areas in education and the social sciences have already begun to incorporate epistemologies, methodologies, and teaching/learning strategies from other paradigms. ID, and educational technology in general, has been much slower to adopt other perspectives. This study illustrates how an "alternative" approach to teaching "alternative" content would be supported by material developed using an "alternative" ID model.

The ID model used violated many of the "rules" of traditional design. The designer played a collaborative and facilitative role rather than an expert role. Goals and objectives, as well as content and teaching strategies, were not decided in advance by the expert. The act of decision-making was based on the understanding, beliefs, and values of the stakeholders. The process that guided the design and development of CRIT included negotiating a set of shared understanding, beliefs, and values.

In addition, there were no prescribed linear steps; rather questions were addressed in a spiral fashion. The same issues were addressed many times, but often at different levels or with different perspectives because the team developed better understanding of the project as design and development progressed.

Using R2D2, decisions were made but they were never final. The R2D2 model, however, may not be for everyone. This model calls for beginning design and development with "fuzzy," ill-defined goals and objectives that change as work progresses. This can be frustrating for designers who traditionally use a linear model. Frustration arises from constant revision and modifications of the program. Design and development may take less time and effort with a traditional design model that starts with a specific set of prescribed objectives. However, the CRIT program likely would not have been as creative or effective if we had used that type of design model. Constant feedback and input from stakeholders made this project a collage of ideas and creativity. The program created using the R2D2 model is quite different from the one that would have been created using a traditional linear model.

In summary, we feel CRIT is a significant contribution to the resource base for teaching qualitative research methodology. It is more an example of what is needed, however, than a solution. There are hundreds of qualitative data collection and analysis methods. Only a few of them were addressed in CRIT. We need many more of this type of instructional resource. Think, for example, of a CD-ROM or Web-based resource that helps students use any of six or seven different formats to conduct interviews and analyze the results, or to conduct participant observer studies in school classrooms, or collect an analyze historical data on racial biases in local government decisions. Qualitative research methodology is not as cut and dried as many forms of quantitative research, but that does not mean that "anything goes." Becoming a scholar who can conduct strong qualitative research is not easy, and courses that support that goal can benefit from a large library of technology-enhanced instructional resources. We believe R2D2 is a flexible model for guiding the design of such instructional materials, and recommend it to you for consideration.

REFERENCES

Andrews, D. H., & Goodson, L. A. (1995). A comparative analysis of models of instructional design. In G. Anglin (Ed.), *Instructional technology. Past, present, and future* (pp. 161–182). Englewood, CO: Libraries Unlimited.

Bagdonis, A., & Salisbury, D. (April 1994). Development and validation of models in instructional design. *Educational Technology, 34*(4), 26–32.

Borg, W., & Gall, M. (1989). *Educational research: An introduction.* White Plains, NY: Longman.

Burton, J. K., Moore, D. M., & Magliaro, S. G. (1996). Behaviorism and instructional technology. In D. H. Jonassen (Ed.), *Handbook of research for educational communications and technology* (pp. 46–73). New York: Simon & Schuster Macmillan.

Carspecken, P. F. (1996). *Critical ethnography in educational research: A theoretical and practical guide.* New York: Routledge.

Cennamo, K., Abell, S., Chung, M., Campbell, L., & Hugg, W. (1995). *A "Layers of Negotiation" model for designing constructivist learning materials.* Proceedings of the Annual National Convention of the Association for Educational Communications and Technology (AECT), Anaheim, CA, 95, pp. 32–42.

Cennamo, K. S., Abell, S. K., & Chung, M. L. (1996, July–August). A "Layers of Negotiation" model for designing constructivist learning materials. *Educational Technology,* 39–48.

Crandall, J. (1993). Professionalism and professionalization of adult ESL literacy. *TESOL Quarterly, 27*(3), 497–515.

Denzin, N., & Lincoln, Y., (Eds.). (1994). *Handbook of qualitative research.* Thousand Oaks, CA: SAGE.

Dick. W. (1996). The Dick and Carey model: Will it survive the decade? *Educational-Technology-Research-and-Development, 44*(3), 55–63.

Dick, W., & Carey, L. (1996). *The systematic design of instruction* (4th ed.). New York: HarperCollins.

Eisner, E. (1979). *The educational imagination: On the design and evaluation of school programs.* New York: McMillan.

Eisner, E.W., & Peshkin, A. (1990). *Qualitative inquiry in education: The continuing debate.* New York: Teacher's College Press.

Jacobson, M., & Spiro, R. (1993). *Hypertext learning environments, cognitive flexibility, and the transfer of complex knowledge: An empirical investigation.* Center for the Study of Reading Technical Report. College of Education, University of Illinois at Urbana-Champaign.

Lebow, D. (1993). Constructivist values for instructional systems design. Five principles toward a new mindset. *Educational Technology, Research, and Development, 41*(3), 4–16.

LeCompte, M. D., & Preissle, J. (1993). *Ethnography and qualitative design in educational research* (2nd ed.). San Diego, California: Academic Press.

Merrill, M. (1996, July-August). What new paradigm of ISD? *Educational Technology, 36*(6), 57–58.

Merrill, M. D., Drake, L, Lacy, M, Pratt, J., & the Utah State University ID2 Research Group. (1996). Reclaiming instructional design. *Educational Technology*, *36*(5), 5–7. Retrieved from http://www.coe.usu.edu/it/id2/reclaim.html

Schön, D. (1983). *The reflective practitioner: How professionals think in action*. New York: Basic Books.

Schön, D. (1987). *Educating the reflective practitioner*. San Francisco: Jossey-Bass.

Schuler, D., & Namoika, A. (1993). *Participatory design: Principles and practices*. Hillsdale, NJ: Erlbaum.

Spiro, R., Coulson, R., Feltovich, P., & Anderson, D. (1988). Cognitive flexibility theory: Advanced knowledge acquisition in ill-structured domains. Center For the Study of Reading. *Technical Report* No. 441, College of Education, University of Illinois at Urbana-Champaign.

Spiro, R., Feltovich, P., Jacobson, M., & Coulson, R. (1991). Cognitive flexibility, constructivism, and hypertext: Random access instruction for advanced knowledge acquisition in ill-structured domains. *Educational Technology*, *31*(4), 24–33.

Spiro, R., & Jehng, J. (1990). Cognitive flexibility, constructivism and hypertext: Theory and technology for the nonlinear and multidimensional traversal of complex subject matter. In D. Nix & R. J. Spiro (Eds.), *Cognition, education, and multimedia: Exploring ideas in high technology* (pp. 163–205). Hillsdale, NJ: Erlbaum.

Staninger, S. (1994, July–August). Hypertext technology: Educational consequences. *Educational Technology*, 51–53.

Strauss, A., & Corbin, J. (1990). *Basics of qualitative research: Grounded theory procedures and techniques*. Newbury Park, CA: SAGE.

Webb, R. B., & Glesne, C. (1992). Teaching qualitative research. In M. D. LeCompte, W. L. Millroy, & J. Preissle (Eds.), *The handbook of qualitative research in education* (pp. 771–814). San Diego, CA: Academic Press.

Wittgenstein, L. (1958). *Philosophical investigations* (3rd ed.). (G. Anscombe, Trans.). New York: Macmillan.

Willis, J. (1995). A recursive, reflective instructional design model based on constructivist-interpretivist theory. *Educational Technology*, *35*(6), 5–23.

Willis, J. (2000). The maturing of constructivist instructional design: Some basic principles that can guide practice. *Educational Technology*, *40*(1), 5–16.

Wilson, B. G. (1993). *Constructivism and instructional design: Some personal reflections*. Proceedings of Selected Research and Development Presentations at the Convention of the Association for Educational Communications and Technology Sponsored by the Research and Theory Division. 15th, New Orleans, Louisiana, January 13–14, 1993.

Wilson, B. (1997). Reflections on constructivism and instructional design. In C. Dills & A. Romiszowski (Eds.), *Instructional development paradigms*. Englewood Cliffs, NJ: Educational Technology.

Winn, W. (1992). The assumptions of constructivism and instructional design. In T. Duffy & D. Jonassen (Eds.), *Constructivism and the technology of instruction: A conversation* (pp. 177–182). Hillsdale, NJ: Erlbaum.

Wolf, M. (1993). Mentoring middle-aged women in the classroom. *Adult Learning*, *4*(5), 8–9.

You, Y. (1994). What can we learn from chaos theory? An alternative approach to instructional design. *Educational Technology Research and Development, 41*(3), 17–32.

CHAPTER 20

A CERVICAL CANCER CD-ROM INTERVENTION FOR COLLEGE-AGE WOMEN

Lessons Learned From Development and Formative Evaluation

**Alexandra Evans, Elizabeth Drane,
Karl Harris, and Tara Campbell-Ray**

Despite the decline of incidence rates of cervical cancer over the past 50 years, data suggest an increase of precervical cancer among college women. This increase may be the result of the high prevalence of the Human Papillomavirus (HPV) among this population. To increase cervical cancer prevention behaviors, we developed a computer-based intervention targeting women ages 18 to 24. Through the use of story scripts, role models, and demonstrations (both animated and videotaped procedures), women learn the importance of regular Pap smear screenings, personal risks for developing cervical cancer, the meaning of and dealing with abnormal Pap smears, communication with health care providers, and sexual practices that impact the transmission of HPV. Results from the formative evaluation support the

Previously published in *Health Promotion Practice* (2002), *3*(4), 447–456.

viability of CD-ROM interventions in health education and emphasize the importance of addressing women's cognitive and affective perceptions surrounding cancer, sexual health, and gynecological exams. Lessons learned from the development process are presented.

In this chapter we share our experiences with the development and formative evaluation of an intervention strategy that uses a relatively new technology (CD-ROM) to address a relatively new understanding of a health risk—Human Papillomavirus (HPV), a sexually transmitted disease—as the cause of cervical cancer.

BACKGROUND

Carcinoma of the cervix is one of the more common malignancies in women, accounting for 15,700 new invasive cases, 4,900 deaths, and 65,000 new in situ cases in the United States each year (American Cancer Society, 1998; Greenlee, Murray, Bolden, & Wingo, 2000). Due to the widespread use of the Papanicolaou (Pap) smear in the United States, the incidence of cervical cancer among American women has declined significantly over the past 50 years and is now considered a preventable and treatable disease.

However, as overall rates of cervical cancer have decreased, the incidence of precancer among younger women has increased at a rate of approximately 3% a year since 1986 (Larsen, 1994). It is not unusual to discover precancerous cervical intraepithelial neoplasia (CIN) in women in their late teens and early 20s (Moore, 1997). This increase in incidence among younger women may reflect the increasing prevalence of HPV, a sexually transmitted virus believed to be the causal factor of cervical cancer (National Institutes of Health [NIH], 1996). Epidemiologists suggest that up to 75% of the people who are sexually active may have been exposed to HPV, with peak prevalence rates of HPV found in women from 22 to 25 years old (Koutsky, 1997; NIH, 1996).

Because cervical cancer is most treatable when detected early, the American College of Obstetricians and Gynecologists, the American Cancer Society, and the National Cancer Institute recommend that all sexually active women and women over the age of 18 receive annual Pap tests (American Cancer Society, 1993; American College of Obstetricians and Gynecologists, 1996; NIH, 1996). However, despite the fact that routine Pap smears can prevent nearly all deaths from cervical cancer, it is estimated that fewer than 50% of women in the United States have had a Pap smear within the past year (Martin, Calle, Wingo, & Heath, 1996). In a more recent study, 66% of a sample of college women reported ever having had a Pap smear (Harris, 2000).

Lack of knowledge or perceptions of low levels of risk for cervical cancer may be two reasons that women do not obtain annual Pap smears. Although the majority of women know who should get a Pap smear and how often this test should be performed (Massad, Meyer, & Hobbs, 1997), there is a general lack of knowledge surrounding the risk factors, namely HPV, for this type of cancer (Hasenyager, 1999; Linnehan & Groce, 1999). In recent surveys, 56% to 70% of college women were unable to name the cause of cervical cancer (Harris, 2000; Yacobi, Tennant, Ferrante, Naazeen, & Roetzheim,1999), and 63% of a random sample of college students had never heard of HPV (Yacobi et al., 1999). In addition, research conducted with college age women indicate that, in general, young women underestimate their risk of becoming infected with any sexually transmitted disease (STD), and HPV in particular (Burak & Meyer, 1998; Kaiser Family Foundation, 1998). Therefore, there is a need for more education about cervical cancer and HPV among college-age women.

The more commonly used method to disseminate information about cervical cancer and Pap smears has been didactic in nature (e.g. educational pamphlets and other text-based media and information from health care providers). Didactic methods are limited in their ability to produce behavior changes, especially in regard to behaviors that are considered sensitive, such as sexuality behaviors (Kirby, Short, & Collins, 1994). In addition, discussions with health care providers, which provide another common method for delivery of cervical cancer information, are also limited in their effectiveness. Oftentimes, physicians are under time constraints, or women are uncomfortable asking about personal or potentially sensitive information (Kavanagh & Broom, 1997). Therefore, an exploration of the use of alternative methods to disseminate this type of sensitive information is warranted.

To test the effectiveness of an alternative delivery system, we developed a theory-based multimedia CD-ROM intervention for women ages 18 to 24. The goal of the CD-ROM was to increase young women's cervical cancer prevention behaviors by providing women with information about issues surrounding cervical cancer and Pap smears, such as the natural development of cervical cancer, risk factors, and specific details about the Pap smear procedure. For this intervention, cervical cancer prevention came to be defined as strategies to reduce risk from infection with HPV (i.e., condom use), obtaining regular Pap smears, and strategies for dealing with abnormal Pap smear results. Moreover, the intervention was designed to empower women to make informed decisions surrounding cervical cancer prevention and to provide them with effective emotional and behavioral responses to the Pap smear procedure and its results. To our knowledge, this is the first CD-ROM that specifically targets cervical cancer prevention behaviors among women of this age group.

Women of college age were selected as the priority audience for our intervention for several reasons. First, due to their sexual practices and the biological immaturity of their cervical anatomy, they are at an increased risk for HPV infection, which places them at greater risk for cervical cancer (Bosch & de Sanjose, 2002). Second, young women are more likely to be exposed to cofactors (such as cigarette smoking or infection with other STDs), which increase their risk of precancerous lesions (Bauer et al., 1993; Kataja et al., 1993). Lastly, college-age women are likely to have experience with computers (Harrington, 1990) and can therefore be expected to appreciate this type of intervention for the delivery of health information.

The use of CD-ROM interventions in the health care field is becoming more popular as the technology improves and development costs decrease (Skinner, Siegfried, Kegler, & Strecher, 1993). A review of computer-based programs for health purposes illustrates a broad use of these programs in a variety of fields such as HIV and AIDS (Duncan, Duncan, Beauchamp, Wells, & Ary, 2000; Evans, Edmundson-Drane, & Harris, 2000; Noell, Ary, & Duncan, 1997; Seidner, Burling, & Marshall, 1996), diabetes (Castalini, Saltmarch, Luck, & Sucher, 1998), and drugs and alcohol (Kinzie, Schorling, & Siegel, 1993). Many of these programs have been shown to be effective, especially for increasing knowledge and changing attitudes. However, computer-based programs specifically designed for facilitating behavior change are still scarce.

The purpose of our chapter is to describe the development and the formative evaluation of the interactive, theory-driven CD-ROM intervention. Within our description, we convey the important role of women as both our informers and our priority population. Our "Lessons Learned" addresses both issues surrounding CD-ROM interventions and issues related to the development of interventions designed to change cervical cancer screening behaviors.

METHOD

The intervention, titled "Taking Care of Business: An Inside Look at the Pap Smear," took approximately a year to develop. The development occurred as three interrelated and cyclical processes: preliminary input, prototype development, and formative evaluation. Figure 20.1 illustrates the flow of the intervention development.

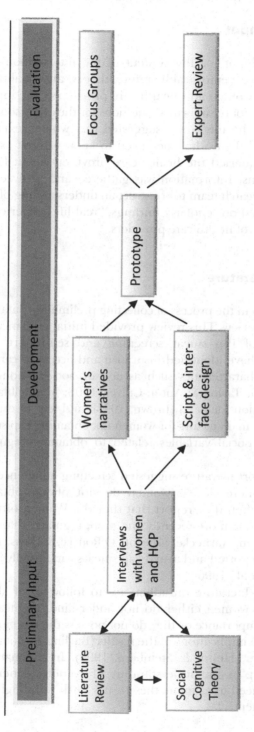

Figure 20.1. Development process.

Preliminary Input

Preliminary input for the content of the intervention stemmed from three main sources: empirical literature, theory, and qualitative data from interviews with women and health care providers. The input process was iterative, as each source informed questions for the other sources. For example, data from the interviews suggested that women expect health care providers to tell them what they need to know without asking questions; thus, we then queried the health care providers about how to facilitate such interactions. Information was gathered and analyzed concurrently, allowing the research team to construct an understanding of cervical cancer prevention based on scholarly findings, "real life" experiences of women, and the insights of health care providers.

Empirical Literature

Our first step in the process of collecting preliminary data was an exhaustive literature review. This review provided initial guidance to the specific determinants of Pap smear screening and self-report and adherence issues. Studies have identified insurance and access to care issues; certain demographic characteristics such as age or socioeconomic status (SES) (Howe, Delfino, Taylor, & Anton-Culver, 1998; Mandelblatt et al., 1999); lack of recognition for need (Brown, 1996); and fear of cancer (Lerman & Rimer, 1995) as main factors for women not obtaining Pap smears. However, data on psychosocial variables related to obtaining regular Pap smears remain limited.

The self-report literature suggested screening adherence reports should be examined more closely; women do not always obtain timely Pap smears, even when they report that they do. Women often "remember" having Pap smears more recently and more regularly than clinical records indicate (Sudman, Warnecke, Johnson, O'Rourke, & Davis, 1994); the difference between reported and actual Pap smears can actually average nearly 2 years (Paskett et al., 1996).

The clinical literature on adherence to follow-up of abnormal results suggested that women either do not understand the meaning of the Pap smear and its importance or they do not possess the emotional coping skills to deal with the implications of the results (Barling & Moore, 1996; Miller, Roussi, Altman, Helm, & Steinberg, 1994). In summary, the literature suggested a gap in our understanding of women's experiences with Pap smears and a need to examine these issues further through personal interviews with women.

Theoretical Framework

Social Cognitive Theory (SCT) provided the primary theoretical framework for the intervention. SCT addresses the psychosocial, behavioral, and environmental correlates of health behaviors and the methods of promoting behavior change (Bandura, 1986). The social construction paradigm (Berger & Luckmann, 1967) guided the interviews and provided a framework to understand from whom and how women conceptualize Pap smears and cervical cancer. Using Intervention Mapping techniques (Bartholomew, Parcel, Kok, & Gottlieb, 2000), the most salient and amenable determinants of cervical cancer screening behavior were identified from the literature and selected to be addressed in the intervention. These constructs included self-efficacy, outcome expectations and expectancies in terms of getting a Pap smear, observational learning, reinforcements, and emotional coping responses. Knowledge was an additional construct included in the intervention (see Table 20.1). Consequently, the selected constructs guided the development of questions for the personal interviews. For example, SCT's construct of outcome expectations suggests that, when it comes to obtaining an annual Pap smear, women who have more positive outcome expectations—compared with women who have negative outcome expectations—about the issue will be more likely to get a Pap smear. Thus, we asked the women in the interviews about their outcome expectations and what strategies could be used to increase the number of their positive outcome expectations.

Personal Interviews With Women

After obtaining permission from our university's Institutional Review Board, we conducted a series of interviews with women. The purpose of the interviews was twofold. First, we wanted to gain a better understanding of women's experiences, and second, we wanted to gather women's stories to illustrate the information conveyed in the intervention (i.e., selected women were videotaped, and their videotaped testimonials were included in the CD-ROM intervention).

Snowball sampling seeking maximum variation was used to identify women to interview. At first, key informants were selected from college-age women who wanted to share their experiences. After each woman was interviewed, we asked her if she knew of anyone who had an experience different from her own who might be willing to share those experiences. Using this method, we were able to interview 32 women with experiences ranging from never having had a Pap smear to having had a hysterectomy due to cervical cancer. To get a more comprehensive contextual framework surrounding cervical cancer preventive behaviors, we decided to interview women of all ages instead of restricting our interviewees to women of our target age range. The 32 women ranged in age (18 to 56), SES (very low to very high), ethnic background (Black, Hispanic, Asian, and White), educational level (high school graduate to PhD level), sexual experiences (virgin to multiple partners), and Pap smear experiences. All women received a small gift for their participation. We found that no woman turned down our request for an initial interview.

All interviews were tape-recorded, transcribed, and coded. As part of the interview process, we asked each participant if she would be willing to be videotaped in a second interview so that parts of her story could be included in the CD-ROM.

Responses from the interviews supported our original choice of theory in that many SCT constructs were indicated as important factors for obtaining Pap smears. From the interviews, we learned again that large gaps exist in women's knowledge, even among women who routinely have Pap smears or have had treatment for gynecological problems. It also became clear that the experiences surrounding cervical cancer screening are emotionally charged for many women. Women reported that they rarely share their experiences or their lack of knowledge. Consequently, women do not know how they learned what they knew, nor do they have other women's experiences to guide them through the gynecological health process (Harris, Evans, Edmundson-Drane, & Diase, 1999). At the same time, they expressed a strong desire to talk about these issues with other women.

For the purpose of our intervention, the interviews demonstrated the need to include both cognitive and affective components. In addition, the interviews supported the notion that the intervention needed to include actual videotaped testimonials of women's stories.

Interviews With Health Care Professionals

Interviews with health care professionals (HCP) experienced in the field of adolescent gynecology provided additional information about practical issues regarding Pap smear exams. On videotape, the health care professionals discussed their experiences with women's Pap smears, their perceptions of women's responses to information, and methods to facilitate women's participation in the Pap smear process in order to have a more comfortable and informative experience (i.e., how to ask questions). Information from these interviews was used to develop sections related to interacting with the health care provider, the mechanics of a Pap smear, and interpretations of the Pap smear results.

Table 20.1.   Description of the CD-ROM Intervention

Section Title	Description of Content	SCT Constructs
Scheduling your Pap smear	Emphasizes the importance of a Pap smear, potential consequences of not getting Pap Smears, tips for scheduling Pap smears, includes (dis)advantages of obtaining yearly Pap smears (as compared to once every 3 years)	Self-efficacy, physical outcome expectations and expectancies, role modeling, behavioral capability
Doctor-patient communication	Emphasizes the importance of communication during a doctor's visit, offers tips for asking questions	Self-efficacy, knowledge, physical and social outcome expectations and expectancies, role modeling, behavioral capability
Life cycle of a cancer cell	Presents information on how cancer spreads, describes natural progression to cervical cancer	Knowledge, physical outcome expectations and expectancies
Anatomy of a Pap smear	Allows user to view real Pap smear procedure, illustrates the procedure using 3-D modeling	Knowledge, physical and self-evaluative outcome expectations and expectancies, behavioral capability
Understanding your test results	Explains meaning of abnormal test results, suggests strategies for dealing with abnormal results	Knowledge; physical, social, and self-evaluative outcome expectations and expectancies
Risk factors for cervical cancer	Explains risk factors for cervical cancer with focus on HPV, describes how HPV is transmitted, offers risk reduction techniques	Knowledge; physical, social, and self-evaluative outcome expectations and expectancies; self-efficacy

Note: SCT = Social Cognitive Theory; HPV = Human Papillomavirus.

Prototype Development

Using the preliminary information, a storyboard was created by the development team, which consisted of behavioral scientists, programming specialists, and a graphic artist. The behavioral scientists were responsible for assuring accuracy of information, theoretical concepts, instructional design, and evaluation. The programming specialists crafted the story lines and artistic components.

The development of the intervention was guided by several principles. We created a story in which segments of information were interwoven, with the intent that the story would help the women retain information (Schank, 1995). In an effort to make the topic more approachable and easier to discuss, a lighthearted tone in the intervention was used. In addition, because our preliminary interviews suggested women's emotions toward Pap smears and Pap smear results could influence women's screening behaviors, we decided that the affective component of gynecological screenings needed to be addressed. To decrease possible bias against those women less comfortable with computers, the CD-ROM was programmed to be very user-friendly, requiring very little computer knowledge. A short tutorial was included at the beginning of the program.

The script and interface were designed using software such as Director and Adobe Photoshop. Music and other features such as clip art came from sources in the public domain with a couple of exceptions (e.g., animation sequences), which were contracted out. Vignettes for the overall story line were videotaped using local actors at local settings. In addition, women who agreed during the interviews to be videotaped were taped for the personal testimonials that were inserted in the program. Although most video segments used for the CD-ROM portrayed women of college age, some older women with particularly salient testimonials were included as well.

Description of the CD-ROM Intervention

The final version of the CD-ROM consists of six sections, each of which provides scenarios that can be viewed individually. The sections include "Anatomy of a Pap Smear, Understanding Your Test Results, Scheduling Your Pap Smear, Health Care Provider-Patient Communication, Risk Factors for Cervical Cancer," and "Life Cycle of a Cancer Cell" (see Table 20.1). For validity purposes, the prototype of the intervention was reviewed by a team of experts in the field of adolescent sexuality.

Each section begins with a short vignette that is related to the overall story line. The vignette introduces the main topics for the specific section and is followed by a variety of theoretical strategies addressing both cognitive and affective domains. For example, the section titled "Anatomy of a Pap Smear" begins with a short story vignette that shows a woman being questioned by a friend about the Pap smear (e.g., what is the purpose of a Pap smear, how is it performed, etc.). This segment is followed by a vignette showing a physician performing a Pap smear. While the physician is performing the test, he talks the woman through the procedures and answers her questions. This vignette is followed by graphical information describing the types of instruments commonly used to perform the test and the types of questions to ask health care providers to obtain the most accurate results.

FORMATIVE EVALUATION AND LESSONS LEARNED

The formative evaluation was designed to determine both the acceptability of the CD-ROM program by college-age women and to direct future versions of this intervention. We conducted four focus groups with 32 women. Two focus groups consisted of women ages 18 to 24, and two focus groups consisted of older women. Although we were interested in the younger women's acceptance of the intervention, we also wanted to obtain input from the older women regarding the marketability of a gynecological health intervention in CD-ROM format for their daughters. The women selected for the focus groups ranged in ethnic background, educational level, and SES.

Women were recruited using fliers displayed at a university campus, in employment offices, libraries, and laundromats. All participants were paid $50 in cash at the end of the focus group session. Two trained focus group leaders presented structured questions after group interaction with the CD-ROM. Focus group discussions were tape-recorded, transcribed, coded, and analyzed.

The following findings address the viability of using CD-ROM technology for this type of health education intervention and concern the women's responses to content and style of delivery. It is interesting to note that, although the focus groups consisted of women of different ages with varying ethnic, computer literacy, and educational backgrounds, their responses to the CD-ROM were, in general, unanimous.

VIABILITY OF CD-ROM HEALTH INTERVENTIONS

The CD-ROM intervention proved to be an acceptable method of delivery, even though few of the women had previous experience with CD-ROMs. The main reported benefits of the CD-ROM were the ability to navigate large amounts of information and the nontext format. The most effective components were the pelvic exam footage and 3-D modeling, the role modeling of interactions with the health care provider, and women's testimonials.

The participants did not appreciate the interactive pieces that were designed to allow women to practice certain skills.

An unexpected result from the formative evaluation indicated that our intervention prompted much discussion, even among women who are not accustomed to discussing gynecological health. The intervention challenged participants' interpretations of their own experiences (by introducing a sexually transmitted disease as a risk factor for cervical cancer). In addition, the CDROM presented images of real women discussing their own experiences, effectively giving them formal language and words where there may have been only images and impressions before. This combination of factors may have inspired our participants to share and explore their own experiences and beliefs.

Content and Features

The focus group responses to the information presented in the CD-ROM were consistent with the preliminary interviews. Women had little knowledge of HPV, cervical cancer, or the purpose of Pap smears. We learned, however, that women were not prepared to accept the information included in the CD-ROM. We did not properly address their prior knowledge, especially about HPV as a risk factor for cervical cancer. Even though the CD-ROM discussed the risk factors for cervical cancer, it did not clearly emphasize that this cancer is different from other cancers. After participating in the CD-ROM, the women maintained that the intervention also should have addressed the environmental, genetic, and dietary risk factors for cervical cancer, even though the literature suggests that these factors are not significantly associated with cervical cancer.

According to the focus groups, the experiences shared by women on the CD-ROM through personal testimonials were the most powerful and positive feature of the intervention. Indeed, the women wanted all the information to be presented by women. Although we had included vignettes with both a "sensitive" male and a female health care provider, upon first viewing, the women only wanted to see and receive information from other females.

The intervention design allowed participants to select any topic in any order. However, the screen interface appeared linear, in that the available topics were presented in a list-wise fashion. The focus group members were reminded they could select topics in a random order, but they chose topics in the order appearing on the screen. In this case, the information about risk factors for cervical cancer (i.e., HPV) was viewed at the end, which seemed to be problematic.

Strong Affective Response

Women responded with disbelief that an STD could cause cancer. Many expressed anger when they began to reflect upon their own experiences with health care providers. They were angry and shocked because they believed the health care community informed them inadequately about Pap smears, cervical cancer, and their risk behaviors. At the same time, they expressed embarrassment that they had not been able to ask the types of questions they needed to ask in order to understand their gynecological health.

Style

The use of a lighthearted tone in some of the sections was interpreted as trivialization of a serious health concern. The reasons for our approach were related to the social construction of cervical cancer and STDs. Although cervical cancer precursors (i.e., abnormal Pap smears and dysplasia) are treatable and preventable, the preliminary interviews indicated that women construct cervical cancer similar to other cancers. STDs remain marginalized as they are often constructed as diseases other people acquire. Furthermore, the women we interviewed were uncomfortable talking about Pap smears and STDs, so we wanted to provide women with a more "comfortable" language. However, the type of humor included in the intervention was not effective in accomplishing those goals and, in fact, had a negative effect on the women.

Overarching Theme/Metaphor

Currently, there is a debate within the instructional design community concerning the use of metaphor. The debate centers on whether or not a metaphor detracts from or enhances learning. A key argument is that instructional design is more effective when focused on content. Our experience with the use of a metaphor taught us that some metaphors do detract from learning and lead the learner to focus on evaluating the metaphor instead of the information. We created a story in which segments of information could be interwoven, hoping that the story would help the participants retain the information. The script was professionally performed but did not appeal to all viewers. The participants' inability to accept the story line seemed to contribute to their inability to accept the information presented in the intervention.

Organizational Structure of the Development Team

We found that decision making during the development phase was difficult with our multidisciplinary team. After talking with several colleagues, we believe this experience was not unique. Graphic artists and computer technology design specialists usually have strong opinions about what they think is essential to the project. Their perspective is, by definition, distinct from behavioral science, health, and medicine. Even among teams with excellent communication, the behavioral/medical team members are, for the most part, at the mercy of the artistic developers, simply because they do not have the skills to override changes made by the artistic members of the design team. We recommend an organizational structure that makes an effort to build rapport and consensus between artists and scientists. At the same time, the organizational structure should allow for final decisions, particularly those related to applications of theory and content, to be made by the health/behavioral science members of the team.

CONCLUSION

To evaluate the effectiveness of a new method to increase cervical cancer prevention behaviors among women of college age, we developed a CD-ROM intervention. Our experiences with the development and formative evaluation of the program provided us with several important lessons learned. First, an alternative method for providing information in regard to cervical cancer prevention is both needed and appreciated by women. In accordance with results from other computer-based program evaluations, using CD-ROM technology to provide sensitive information was found very acceptable (Evans et al., 2000; Fernandez, 1995; Paperny & Starn, 1989; Reis & Tymchyshyn, 1992; Thomas, Cahill, & Santilli, 1997). However, certain aspects of the intervention were more effective than others. For example, the vignettes of the women's personal stories were very effective, whereas the interactive activities were not. More research is needed to learn what strategies can be most effectively used in computer based interventions for this priority population.

Our formative evaluation also emphasized the adage, "pretest, pretest, pretest." Our knowledge of the interview responses did not necessarily translate into knowledge of how women received the intervention. Because development of interactive computer interventions is both costly and time consuming, developers need a clear understanding of their target population and need to appeal to their preintervention knowledge, attitudes, and behaviors before presenting a different way to think about a health concern. For the development of this particular CD-ROM, we did not thoroughly pretest the different sections, colors, metaphor, and so forth. at each development phase. Women in the focus groups suggested several "small" changes, which actually translated into large changes in terms of time and money. Therefore, pretesting of all aspects and at different phases of the intervention is necessary and costeffective.

The focus group responses affirmed that emotion is an integral component of health behavior that cannot be neglected by health educators. Health educators need to do a better job of helping women integrate information about Pap smears and cervical cancer into their current cognitive structures about cancer, about who contracts sexually transmitted diseases, and about a woman's role in her own gynecological health care. Essentially, the participants were being asked to rethink their own experiences and the experiences of loved ones who had been treated for precancerous conditions. Through the inclusion of testimonials, CD-ROMs are effective tools to address affective information.

Although this intervention has not been formally evaluated, the results from the formative evaluation were very promising. We are currently revising the CDROM, incorporating many of the suggestions made by women who viewed the prototype of the intervention. Some of the revisions include presenting the information in a nonlinear format, including more women's stories, revising the metaphor and style of presentation, and presenting the information about the link between HPV and cervical cancer using different methods.

Distribution and Evaluation of the CD-ROM

After completion of the revisions, we will use a multipronged approach to distribute and formally evaluate the CD-ROM. One approach will be to provide free copies of the CD-ROM to Planned Parenthood sites in larger metropolitan areas, asking for their feedback on it, as well as their opinions of the process of having clients view the CD-ROM during their visits. Our evaluation will be designed so that we can compare outcome expectations, self-efficacy, knowledge, and behavioral skills of women who have reviewed the CDROM to those of women who have not. Another approach will be to apply for funding for audience specific pilot evaluations from HMOs. Our strategy is to approach the medical director or the director of disease management with a pilot proposal where we supply the CDs and the HMO incorporates them into an area's efforts and tracks and assesses results.

The last prong of the distribution strategy is a direct appeal to college-age women via the World Wide Web. We are currently in the process of reformatting the CD-ROM to a Web site that will be linked to popular teen Web sites such as grrrl.com and goosehead.com, as well as to Web sites known to be geared toward women, such as ivillage.com. The visitors will view parts of the CD-ROM, be asked to complete a survey, and be provided with a place to order the complete CD-ROM if they are interested.

This study represents one of the many examples of how interactive technology can be used for health promotion. As the software and hardware become more user-friendly and affordable, health educators need to continue to explore methods to use these technologies most effectively.

ACKNOWLEDGMENT

This chapter was supported in part by NCI/NIH grant #2R25CA57712-06 (Behavioral Science Education Cancer Prevention and Control). Please direct correspondence concerning the article to Alexandra Evans, Department of Health Promotion, Education, and Behavior, Norman J. Arnold School of Public Health, University of South Carolina, HESC 215, Columbia, SC 29208. Telephone: (803) 777-4862; fax: (803) 777-6290; e-mail: sevans@sc.edu.

REFERENCES

American Cancer Society. (1993). Summary of American Cancer Society recommendations for the early detection of cancer in asymptotic people. CA, 43, p. 45.

American Cancer Society. (1998). Cancer Facts & Figures, 1998. Atlanta, GA: Author.

American College of Obstetricians and Gynecologists. (1996). Guidelines for Women's Health Care. Washington, DC: Author.

Bandura, A. (1995, March). Moving into forward gear in health promotion and disease prevention. Paper presented at the annual meeting of the Society of Behavioral Medicine, San Diego, CA.

Barling, N. R., & Moore, S. M. (1996). Prediction of cervical cancer screening using the theory of reasoned action. Psychological Reports, 79, 77–78.

Bartholomew, L. K., Parcel, G. S., Kok, G., Gottlieb, N. H. (2000). Intervention mapping: A process for designing theory and evidence-based health promotion programs. Mountain View, CA: Mayfield.

Bauer, H. M., Hildesheim, A., Schiffman, M. H., Glass, A. G., Rush, B. B., Scott, D. R., et al. (1993). Determinants of genital Human Papillomavirus infection in low-risk women in Portland, Oregon. Sexually Transmitted Diseases, 20(5), 274–278.

Berger, P. L., & Luckmann, T. (1967). The social construction of reality: A treatise in the sociology of knowledge. Garden City, NY: Doubleday.

Bosch, F. X., & de Sanjose, S. (2002). Human Papillomavirus in cervical cancer. Current Oncology Report, 4(2), 175–183.

Brown, C. L. (1996). Screening patterns for cervical cancer: How best to reach the unscreened population. *Journal of the National Cancer Institute Monographs, 21,* 7–11.

Burak, L. J., & Meyer, M. (1998). Factors influencing college women's gynecological screening behaviors and intentions. *Journal of Health Education, 29*(6), 365–370.

Castalini, M., Saltmarch, M., Luck, S., & Sucher, K. (1998). The development and pilot testing of a multimedia CD-ROM for diabetes education. *The Diabetes Educator, 24,* 285–296.

Duncan, T. E., Duncan, S. C., Beauchamp, N., Wells, J., & Ary, D. V. (2000). Development and evaluation of an interactive CD-ROM refusal skills program to prevent youth substance use: "Refuse to Use." *Journal of Behavioral Medicine, 23,* 59–72.

Evans, A., Edmundson-Drane, E., & Harris, K. K. (2000). Computer-assisted instruction: An effective instructional method for HIV prevention education? *Journal of Adolescent Health, 26,* 244–251.

Fernandez, M. (1995). *Evaluation of an interactive videodisc on breast and cervical cancer prevention and early detection for Hispanic women.* Unpublished doctoral dissertation, University of Texas, Houston.

Greenlee, R. T., Murray, T., Bolden, S., & Wingo, P. A. (2000). Cancer statistics. *CA Cancer Journal, 50,* 7–33.

Harrington, J. (1990). Changes in the computer background in incoming freshmen. *Collegiate Microcomputer, VIII,* 147–154.

Harris, K. K. (2000). *Psychosocial and contextual correlates of women's gynecological health screening.* Unpublished dissertation, University of Texas, Austin.

Harris, K. K., Evans, A. E., Edmundson-Drane, E., & Diase, M. (1999, November). *Women's voices in cervical cancer prevention.* Paper presented at the American Public Health Association, Chicago.

Hasenyager, C. (1999). Knowledge of cervical cancer screening among women attending a university health center. *Journal of American College Health, 47,* 221–224.

Howe, S. L., Delfino, R. J., Taylor, T. H., & Anton-Culver, H. (1998). The risk of invasive cervical cancer among Hispanics: Evidence for targeted preventive interventions. *Preventive Medicine, 27,* 674–680.

Kaiser Family Foundation. (1998). *Kaiser Family Foundation/MTV/TEEN PEOPLE national survey of 15-17 year olds: What teens know and don't (but should) about sexually transmitted diseases.* Menlo Park, CA: Author.

Kataja, V., Syrjanen, S., Yliskoski, M., Hippelinen, M., Vayrynen, M., Saarikoski, S., et al. (1993). Risk factors associated with cervical Human Papillomavirus infections: A case-control study. *American Journal of Epidemiology, 138*(9), 735–745.

Kavanagh, A. M., & Broom, D. H. (1997). Women's understanding of abnormal cervical smear test results: A qualitative interview study. *British Medical Journal, 314,* 1288–1391.

Kinzie, M. B, Schorling, J. B., & Siegel, M. (1993). Prenatal alcohol education for low-income women with interactive multimedia. *Patient Education and Counseling, 21,* 51–60.

Kirby, D., Short, L., & Collins, J. (1994). School-based programs to reduce sexual risk behaviors: A review of effectiveness. *Public Health Reports, 109,* 339–360.

Koutsky, L. (1997). Epidemiology of genital Human Papillomavirus infection. *The American Journal of Medicine, 102*(5A), 3–8.

Larsen, N. S. (1994). Invasive cervical cancer rising in young White females. *Journal of the National Cancer Institute, 86*(1), 6–7.

Lerman, C., & Rimer, B. K. (1995). Psychological impact of cancer screening. In R. T. Croyle (Ed.), *Psychosocial effects of screening for disease prevention and detection* (pp. 65–81). New York: Oxford University.

Linnehan, M. J. E., & Groce, N. E. (1999). Psychosocial and educational services for students with genital Human Papillomavirus infection. *Family Planning Perspectives, 31*(3), 137–141.

Mandelblatt, J. S., Gold, K., O'Malley, A. S., Taylor, K., Cagney, K., Hopkins, J. S., et al. (1999). Breast and cervix cancer screening among multiethnic women: Role of age, health, and source of care. *Preventive Medicine, 28,* 418–425.

Martin, L. M., Calle, E. E., Wingo, P. A., & Heath, C. W. J. (1996). Comparison of mammography and Pap test use from the 1987 and 1992 national health interview surveys: Are we closing the gaps? *American Journal of Preventive Medicine, 12*(2), 82–90.

Massad, L. S., Meyer, P., & Hobbs, J. (1997). Knowledge of cervical cancer screening among women attending urban colposcopy clinics. *Cancer Detection & Prevention, 21*(1), 103–109.

Miller, S. M., Roussi, P., Altman, D., Helm, W., & Steinberg, A. (1994). Effects of coping style on psychological reactions of low-income, minority women to colposcopy. *Journal of Reproductive Medicine, 39*(9), 711–718.

Moore, G. (Ed.). (1997). *Women and cancer: A gynecological oncology nursing perspective.* Sudsbury, MA: Jones & Bartlett.

National Institutes of Health. (1996). Cervical cancer. NIH Consensus Statement 1996 April 1-3. *Journal of the National Cancer Institute, Monographs 21,* 1–38.

Noell, J., Ary, D., & Duncan, T. (1997). Development and evaluation of a sexual decision-making and social skills program: "The Choice Is Yours Preventing HIV/STDs." *Health Education and Behavior, 24,* 87–101.

Paperny, D., & Starn, J. (1989). Adolescent pregnancy prevention by health education computer games: Computer assisted instruction of knowledge and attitudes. *Pediatrics, 83,* 742–752.

Paskett, E. D., Tatum, C. M., Mack, D. W., Hoen, H., Case, L. D., & Velez, R. (1996). Validation of self-reported breast and cervical cancer screening tests among low-income minority women. *Cancer Epidemiology, Biomarkers and Prevention, 5,* 721–726.

Reis, J., & Tymchyshyn, P. (1992). A longitudinal evaluation of computer assisted instruction on contraception for college students. *Adolescence, 27,* 803–811.

Schank, R. C. (1995). *Tell me a story: Narrative and intelligence.* Evanston, IL: Northwestern University Press.

Seidner, A. L., Burling, T. A., & Marshall, G. D. (1996). Using interactive multimedia to educate high risk patients about AIDS and sexually transmitted diseases. *Computers in Human Services, 13,* 1–15.

Skinner, C. S., Siegfried, J. C., Kegler, M. C., & Strecher, V. J. (1993). The potential of computers in patient education. *Patient Education and Counseling, 22,* 27–34.

Sudman, S., Warnecke, R., Johnson, T., O'Rourke, D., & Davis, A. M. (1994). Cognitive aspects of reporting cancer prevention examinations and tests. *Vital and Health Statistics, 6,* 1–20.

Thomas, R., Cahill, J., & Santilli, L. (1997). Using an interactive computer game to increase skill and self-efficacy regarding safer sex negotiation: Field testing results. *Health Education and Behavior, 24,* 71-86.

Yacobi, E., Tennant, C., Ferrante, J., Naazeen, P., & Roetzheim, R. (1999). University students' knowledge and awareness of HPV. *Preventive Medicine, 28,* 539–541.

CHAPTER 21

FROM PEDAGOGY TO TECHNAGOGY IN SOCIAL WORK EDUCATION

A Constructivist Approach to Instructional Design in an Online, Competency-Based Child Welfare Practice Course

Gerard Bellefeuille, Robert R. Martin, and Martin Paul Buck

This chapter documents the design and pilot delivery of a computer-mediated, competency-based child welfare practice course founded on constructivist instructional principles. It was created in 2003 as part of the University of Northern British Columbia (UNBC) Social Work program's child welfare specialization stream. Offered to learners via the Internet using Web-based tools and resources, the course expanded access to the child welfare specialization option for social work students studying at UNBC's three rural regional campuses. The article examines emerging teaching and learning options across four components of online course development and delivery. The background of the specialization stream in the social work program is

Previously published in *Child & Youth Care Forum* (2005), 34(5), 371–389.

Constructivist Instructional Design (C-ID): Foundations, Models, and Examples, pp. 493–512

reviewed, constructivist instructional design theory is summarized, and a rationale for adopting this approach is discussed. This is followed by a brief report on the findings of the formative evaluation of the pilot delivery. Finally, the evolving underpinnings of online instruction are considered, including shifts in the roles of learners and instructors and the role of pedagogy in an evolving educational paradigm.

The creation of the province of British Columbia's Child Welfare Specialization Stream was the result of a special funding initiative between several BC post-secondary institutions and the provincial Ministry for Children and Family Development (MCFD) responsible for child protection services. The course reviewed in this article was created in response to the recommendations in the 1995 report of the Gove Inquiry Into Child Protection. This review was mandated to examine causal factors in the death of Mathew John Vaudreuil, a child in care, and to recommend remedial actions to prevent similar tragedies from occurring in the future. The Gove Inquiry reported that MCFD training for child protection workers was inadequate and recommended that measures be taken to strengthen the province's Bachelor of Social Work and other relevant human services degree programs in the specialized area of child protection practice (Gove, 1995).

As part of the partnership initiative, a consortium of BC schools of social work and child and youth care agreed to develop a competency-based curriculum. The challenge was to satisfy the ministry's needs for competency-ready workers while maintaining the School's anti-oppressive critical learning stance—a key accreditation standard—for the students. This stance is rooted in the conviction that many traditional practice strategies actually contribute to oppression in society and that many forms of social work education actually mask the oppressions that need to be challenged.

The integration of an online competency-based child welfare practice course into the UNBC social work program generated considerable discussion. The epistemological underpinnings of the instructional design, the technology-mediated learning environment, and content in a child welfare specialization stream within a structural approach to social work education all generated vigorous academic debate within the department.

EPISTEMOLOGY, INSTRUCTIONAL DESIGN, AND COMPUTER-MEDIATED LEARNING

Constructivist epistemology is based on the assumption that reality is not an external absolute, but a composite constructed from the knower's mental activity and previous base of experience. Proponents of constructivist educational theory argue that humans "assemble" their own individualized versions of reality in an attempt to know and make sense of their world (Duffy & Jonassen, 1992; Ricks, 2002; Woolfolk, 1993). A 2004 edition of the E-Learning Guild's online journal describes the impact of this perspective on approaches to instructional design for online courses:

> The constructivist view is that learning is an active process of constructing knowledge, where the learners are doing the construction. Learning is not acquisition of knowledge. Learning is a change in meaning, ideas, or concepts, constructed from prior knowledge and experience. The instructor's job is not to instruct as such, but to support the construction process, mainly by creating an environment in which the construction can take place. Technology, especially computer technology and the Web, offers many resources that have proven successful over the last two decades as supplements to constructivist practices in the classroom, in adult education, and in distance learning.

Constructivist reality is based in one's own interpretation of what exists, obtained through reasoning about one's personal experiences, beliefs, and perspectives. Knowledge and reality do not have an objective or absolute value or, at the very least, we have no way of knowing this reality. Our way of knowing continues to be transformed by our learning.

Mezirow (1991) argues that transformative learning

> begins when we encounter experiences, often in an emotionally charged situation, that fail to fit our expectations and consequently lack meaning for us, or we encounter an anomaly that cannot be given coherence either by learning within existing schemes or by learning new schemes. (p. 94)

A case study presentation followed by re?ective online discussion and journaling can be an effective way of constructing new meaning. He goes on to state,

> Reflection is involved in problem solving, problem posing and transformation of meaning schemes and perspectives. We may reflect on the content of a problem, the process of our problem solving or the premise upon which the problem is predicated. Content and process reflection can play a role in thoughtful action by allowing us to assess consciously what we know about taking the next step in a series of actions. Premise reflection involves a movement through cognitive structures guided by the identifying and judging of presuppositions. Through content and process reflection we can change our meaning schemes: through premise reflection we can transform our meaning perspectives. Transformative learning pertains to both the transformation of meaning schemes through content and process reflection, and the transformation of meaning perspectives through premise reflection. (p. 117)

Von Glasersfeld (1995) argues that, "From the constructivist perspective, learning is not a stimulus-response phenomenon. Knowledge is the result of active learning, which is a generative process. It requires self-regulation and the building of conceptual structures through reflection and abstraction" (p. 14). This means that learners constantly relate new information to what they already know (Jonassen & Mayes, 1993), resulting in the "social construction" of new personalized meaning. Because of this integration of prior learning, constructivist instructional design approaches promote a more open-ended, cumulative learning experience. Differences in learning styles, personal values, and cultural variables can be flexibly accommodated while still staying true to the spirit of course learning outcomes in a competency-based curriculum.

Computer-mediated instructional design and its Web-based interactive options offer environments that, we would argue, are more inherently constructivist in nature. Many communication tools are available at a click of the mouse (Dede, 1996; Dodge, 1996; Eastmond & Ziegahn, 1996). Online learners can take a virtual tour of another country, gain entry to a virtual seminar with visiting lecturers, or search libraries around the world from the comfort of their homes or offices.

As Harasim (1996) states in *Computer Networking and Scholarly Communication in the Twenty-First-Century University,* a transformed paradigm accentuates "the focus from knowledge transmission to knowledge building" (p. 205). From this philosophical vantage point, the role of the course instructor shifts from one of knowledge transmitter to one of a facilitator who provides ample opportunities for skill and knowledge acquisition through online interaction and meaning-making. The role of the learner also shifts from a receptor of knowledge from an "expert" to one of taking on greater responsibility for their own learning (Lefrere, 1996). The somewhat clichéd expression for this phenomenon, now overused but nonetheless accurate, is the concept of shifting from "the sage on the stage, to the guide on the side."

APPLYING CONSTRUCTIVIST PRINCIPLES TO WEB-BASED INSTRUCTIONAL DESIGN

The World Wide Web was born in 1994 with the introduction of Mosaic, the first full-featured Web browser software program which went on to become Netscape Navigator. In the process of adapting to the opportunities made available by the Web, colleges and universities began to shift some of their focus from classroom instruction to the development and delivery of online and "blended" courses. Initially, early adopters in the academic community tended to replicate lecture-style delivery models online, and the results were sometimes not well received. Carol Twigg, Executive Director for the Center for Academic Transformation at Rensselaer Polytechnic Institute, reported in 2001 that, "The vast majority of [on-line] courses are organized in much the same manner as are their campus counterparts." This is gradually changing, however, with the evolution of more full-featured learning management system (LMS) software that allow creation of learning environments that reflect and enable, to varying degrees, the broader application of constructivist principles.

The relationship between computer-mediated learning and constructivist instructional design is based on the notion that an expanding "toolbox" of electronic product and process options provides students with enhanced access to information with which to develop knowledge and skills, do research, and test their personal ideas in a broader interactive milieu. Online learning includes high-quality interpersonal interaction (Harasim, Hiltz, Teles, & Turoff, 1995), allowing students to present and test their growing competence across broader audiences when freed of place and time constraints. It can also expose them to opinions of a more diverse group of people in the real world beyond the classroom, school, and local community, all conditions optimal for constructivist learning (The Cognition and Technology Group at Vanderbilt, 1993; Jonassen, Peck, & Wilson, 1999; Reigeluth, 1991; Winograd & Flores, 1986; Winfield, Mealy, & Scheibel, 1998).

The instructional design framework selected to guide the development of the competency-based child welfare practice course, presented in Table 21.1, was adapted from the work of several educational scholars including Collins, Brown, and Newman (1989), Bednar, Cunningham, Duffy, and Perry (1998), Duffy and Jonassen, (1992), Lebow, (1993), Woolfolk, (1993), and Willis, (1995). The specific instructional design principles establish a rich context within which meaning can be negotiated and ways of understanding can emerge and evolve (Hannafin, Land, & Oliver, 1999; Lebow, 1993).

Learning management systems allow for activities involving learners, instructors, and online guests to be designed as synchronous (in real time/ at the same time) or asynchronous (in delayed time or not at the same time) events. An example of a synchronous event would be a group of learners exchanging dialogue with an instructor in an online chat room where they interact with one another using text chat or audio-conferencing features, accessing Web pages, or downloaded digital resources to assist the discussion, on a Wednesday evening at 7 P.M. in their local time zone. This might be thought of as an alternative to a small working group meeting in a breakout room in a face-to-face learning environment. In the online environment, however, it is possible for a learner working an evening shift to take an hour break to participate in the online chat and then go back to work having applied some of their work-related context of the day to the online interaction. An example of an asynchronous event would be the same online discussion taking place over a 3-day time frame via "threaded" e-mail postings where the instructor and other resource people logging in and out to read others' entries and add their own as time schedules permit and relevant opinions are formed.

One major advantage of Web-based learning is that it allows for self-paced learning and re?ection, two central premises of constructivist epistemology. Students can interact at their convenience and have the flexibility to take courses without physically walking into a classroom. Table 21.1 illustrates how key variables were integrated across five foundational areas of course and program design. By drawing on the strengths of a Web-based approach to learning online, constructivist principles can be applied to curriculum development projects using a variety of creative strategic approaches.

THE USE OF WeBCT AS AN E-LEARNING PLATFORM

The learning management system (LMS) known as Web Course Tools or WebCT was selected as the e-learning platform. UNBC had a license for the software, training was available for instructors, and there was a growing community of peers at other institutions willing to share what they had leaned. The software was developed in the mid-1990s by a team led by Dr. Murray Goldberg at the University of British Columbia. It contains a variety of course development and delivery features including communication tools such as the bulletin board (now called discussions); private and group e-mail; online "chat" which allows learners to send text back and forth in "real time;" streaming audio and video file transfers; online activity tools such as quizzes, short answer assignments, self-tests, surveys, and tools for organizing course content and developing student presentations using a Web browser.

Students can interact with other students, teachers, and professionals in communities far from their classrooms. The tools can also provide students access to many different types of online resources that can help them understand both their own culture and the cultures of other participants. Options for facilitating open dialogue, focused discussion on key points, and guiding debate between individuals and small working groups or within larger class cohorts enable a diverse range of learning and teaching strategies. When competently designed and led by a course instructor who is acquainted with Web-based delivery tools and techniques, these activities can be used to facilitate individual learners' social construction of meaning in new and different ways. Internet search

Table 21.1. A Constructivist Instructional Design Framework

Assumptions	Values	Instructional Design Principles	Instructional Strategies	Exemplars of a Constructivist Learning Environment
Individuals interpret and construct meaning based on their experiences and evolved beliefs	Collaboration	Emphasis on the *affective* domain of learner	Interactive	Embedding skills and knowledge in holistic and realistic contexts
	Personal autonomy	Instruction *personally relevant* to the learner	Experientia	Scaffolding and coaching of knowledge
	Generativity	Help learners develop skills, attitudes, and beliefs that support *self-regulation* of the learning process	Independent	Authentic learning tasks
	Reflectivity	Context offers balance and control of learning environment with promotion of personal autonomy	Direct and indirect strategies	Multiple perspective building, and multiple representations
	Active engagement	Embed reasons for learning into the learning activity		Collaborative learning activities
	Personal relevance	Strategically explore errors		
	Pluralism			

engines make it possible for instructors to tailor resources for assignments and learning activities to quickly adapt to diverse cultures, belief systems, and learning styles. Group learning exercises integrate the experiential bases of participants with personal interest areas, providing high levels of motivation when combined with fast access to a world of information. A side bene?t has been the requirement that learners expeditiously develop a set of critical evaluative skills to assess and determine the worth of the many resources they can tap into online. While this skill set can be challenging to acquire, it is fast becoming a basic literacy item in the learner's toolbox for the digital age.

THE INSTRUCTIONAL DESIGN OF A COMPETENCY-BASED CHILD WELFARE PRACTICE ONLINE COURSE

Initially a traditional instructional design template was used to break down the course into twelve separate but related online learning modules. A literature review was conducted to identify the main knowledge domains of child welfare practice. British Columbia's Ministry of Children and Family Development (MCFD) had created a comprehensive list of core competencies for effective child protection practice. Once the main knowledge domains were established, these competencies were reviewed and functionally integrated among the twelve learning modules illustrated in Figure 21.1.

WeBCT COURSE MANAGEMENT AND COMMUNICATION TOOLS

The competency-based ("demonstrate that you can follow the rules") focus of the child welfare practice course, combined with the structural perspective ("critically thinking outside of the box") of the Social Work program, presented a considerable challenge to the instructional design team. Following the principle that learners should progress at their own pace, while at the same time maintaining some degree of group cohesion, learners were given one week to work through each separate learning module. Individual learning styles, life situations, and approaches to completing required components resulted in a diverse range of "routes to learning."

Each module focused on mastery of specific competencies based on a particular knowledge domain. The emphasis was as much on process as on arriving at the "right" answers, although correct answers were provided along with feedback on incorrect choices, in content areas where a right/wrong conclusion was required. Preferred answers were provided as

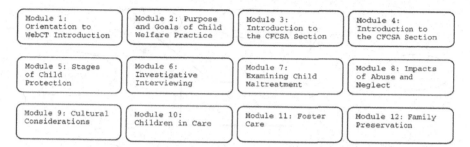

Module 1: Orientation to WebCT Introduction	Module 2: Purpose and Goals of Child Welfare Practice	Module 3: Introduction to the CFCSA Section	Module 4: Introduction to the CFCSA Section
Module 5: Stages of Child Protection	Module 6: Investigative Interviewing	Module 7: Examining Child Maltreatment	Module 8: Impacts of Abuse and Neglect
Module 9: Cultural Considerations	Module 10: Children in Care	Module 11: Foster Care	Module 12: Family Preservation

Figure 21.1.   Online learning modules and main child welfare practice knowledge domains.

feedback to the learners following their completion of other assignments where there was more than one correct solution or where personal initiative to "go another step" could enhance their knowledge and skill acquisition in the competency area. The majority of participants were practitioners with previous experience in human services work environments. The constructivist approach enabled them to draw on the richness of that experience and develop a shared pool of knowledge and skill as they completed exercises and assignments with others in their learning community. Learners were required to work through a variety of online activities supported by Web-based content from predefined online sites. The activities were accessed and supported by a range of WebCT tools illustrated in Figure 21.2.

INSTRUCTIONAL FORMAT

Each learning module followed the same instructional sequence, shown in Figure 21.3, to make it easier for learners to navigate through the course content and develop con?dence with the online learning process.

COURSE HOME PAGE

The "click to enter" icons illustrated on the course home page in Figure 21.4 provided easy and immediate access to the organizing components of the online classroom. These included course modules, a course outline, communication and management tools, learner grade profile, selected resource materials, and a self-help guide. Learners were shown how to use multiple screens on their desktop by "toggling" back and forth from one set of features to another to increase the ease of moving from one tool to the next.

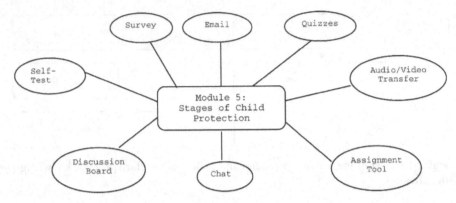

Figure 21.2.   WebCT course management and communication tools.

Table of Contents: Module 5
1.   Introduction
2.   Learning Outcomes
3.   Related Child Welfare Competencies
4.   Required Readings
• 5.   What are Practice Standards
•   5.1 Practice Standards: Definition
•   5.2 Stages of Child Protection
•   *Activity # 5-1 (Discussion Board)*
• 6.   Risk Assessment
•   *Activity # 5-2 (Assignment Tool)*
•   *End of Module Quiz   (Quiz Tool)*

Figure 21.3.   Standard structure of learning modules.

INSTRUCTIONAL STRATEGIES

The learning activities in Web-based environments play a fundamental role in determining learning outcomes. They determine how the learners will engage with the course materials and the kinds of knowledge construction that will take place (Wild & Quinn, 1997). In this course design, a mix of constructivist and traditional (objectivist) instructional strategies were integrated. This blending of strategies and approaches allowed the course instructor to maintain a flexible but critical learning stance while teaching to the specific child welfare practice competencies.

The concept of "scaffolding" is introduced visually in Figure 21.5. Scaffolding enables the instructor to draw to a greater extent on individual experiential storehouses and the willingness of learners to initiate more independently as their growing knowledge and skills bases expand. As

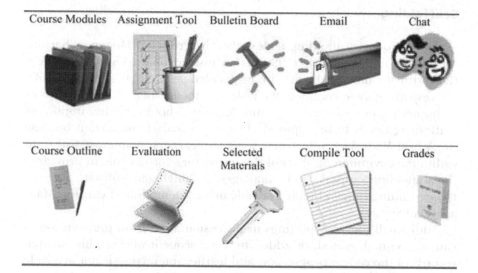

Figure 21.4.   Home Page features.

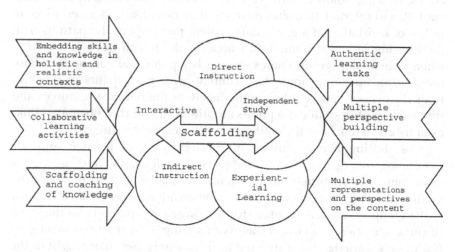

Figure 21.5.   Mixed online instructional design framework.

their confidence and competence in using WebCT tools and methods grew, instructor options expanded in the online learning environment. Learners were able to use the knowledge and skills acquired in previous modules to "push the envelope" and try more advanced features in subsequent sections of the course.

Scaffolding

The notion of scaffolding is described by Presseley, Hogan, Wharton-McDonald, and Mistretta, (1996) as an instructional technique wherein the instructor models the activity in detailed steps, then gradually shifts the responsibility to complete the task onto the learner. In the contexts of computer-supported teaching and learning, however, the notion of scaffolding needs to be adjusted. This is particularly important because the World Wide Web offers new ways for learners to interact with and within the environment. Parts of the instructor's former role in providing the scaffolding support can be managed by "intelligent software agents," the expanding range of tools available in a contemporary Learning Management System.

Additionally, learner postings in discussion areas from previous weeks can be revisited, edited, or added to in subsequent weeks. This "written record" of the course progression and learner interaction is not available in the lecture-based face-to-face environment. The incremental building block approach that it affords is a key strength of the well-designed online course offering. Another is the capacity to change directions in midstream via URLs to relevant Web sites or immediate downloads of a new piece of policy or legislation, if a particularly salient point opens the path to critical learning in another important area. In the fourth week of a course, when a discussion group comes upon a key point that can be foreshadowed as an upcoming important content area, participants in synchronous discussions can be taken immediately to the relevant resources and discussion outlines while the points are still fresh and the conceptual links established to reinforce the connection between the two areas.

This shifting context allows innovative forms of learner supports depending on learners' knowledge and skill bases and instructor availability. In the online environment, the challenge is to replicate the quality learning aspects of verbal communication between an instructor and a group of learners in a face-to-face classroom and recreate the value of those activities in a virtual classroom setting. Constructivist principles posit that knowledge is constructed by a student rather than taught to the student. The process of knowledge construction is viewed as deeper than the traditional approach (e.g., the student is more actively engaged in solving meaningful problems). This transition lends itself to a more learner-centered approach to attaining learning outcomes. Content review, conceptual debate in online dialogue venues, and summarization of key learning points by the facilitator lend themselves well to the activity of scaffolding. The orientation module illustrated in Figure 21.6 was constructed for this purpose wherein students were introduced to the

Go to the discussion board using the multiple windows technique
and click on the "Welcome to Socw 422" link to access all of the
messages posted to this link. You will find an initial posting
from the course instructor entitled "Welcome" (refer to Figure
1.2).

Topic	Unread	Total	Status
All	197	1345	
Main	13	91	public, unlocked
Welcome to Socw 422	0	6	public, unlocked

Figure 21.6.   Page 5 of course orientation manual.

various components of the course and guided through a series of activities
to gain confidence in the online instructional process.

Direct Instruction Strategy

Direct instruction was used to provide guided readings in advance of
interactive group learning activities. It also was used to promote the
acquisition of knowledge and practice skills through repetitive activities
(e.g., completion of small tasks to build familiarity with legislation, multiple choice definition quizzes, and word puzzles). See Figure 21.7 for an
example.

Indirect Instruction Strategy

In contrast to direct instruction, indirect instruction is mainly student-centered and seeks a high level of student involvement in examining, exploring, drawing inferences from data, or forming hypotheses. It
takes advantage of students' personal interests, life experiences, and
natural curiosity by encouraging them to generate alternative solutions.
Within this approach the role of the instructor is one of facilitator, supporter, coach, and resource person (Martin, 1983). Brown and Voltz
(2005) elaborate on instructional design implications for using such
strategies:

> rich learning activities allow students to learn with computers rather than
> from computers. The change in learning context affects the student–teacher
> relationship, which becomes a multifaceted interaction among student,

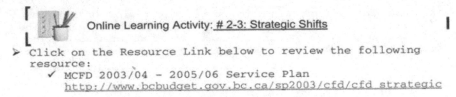

Figure 21.7.   Example of a directed reading strategy.

online materials, the broader community of Internet users, and, in many cases, teachers as facilitators and mentors.

The WebCT discussion board area was used to encourage students to think and talk about what they have observed, heard, or read. In one case learners were asked to post responses to the following question: "What are your beliefs about the power of a court with regards to parents' jurisdiction in detaining a pregnant woman against her will in order to protect her unborn child from conduct that may harm the child?" This is an inferential open-ended question to which there is no correct answer. It required students to make inferences and encouraged critical thinking about their beliefs and values, the mores and laws of our time, and the rules of the court.

Experiential Learning

Experiential learning is inductive, learner-centered, activity-oriented, and promotes concept attainment through experiential practice. Instructional scenarios (e.g., case studies) that drew on participants' former work experiences and associated problem-solving skills were frequently used to promote the use of critical and evaluative thinking. The ambiguous or open-ended nature of the case studies encourages students to contemplate multiple perspectives in light of the different stakeholders in the situation. The case studies were based on familiar, real-life scenarios in the context of northern British Columbia. Learners were encouraged to share current work experiences and problem-solving strategies in their explorations of the case studies. For example in Course Module 4, learners were asked to engage in an online discussion concerning the following case study:

Tina Brown (age 32), a single mother of Christie (age 5) and Kyle (age 2), is subject to a 6-month supervision order with the condition that she attend Healthiest Babies Possible Program in Prince George. In addition, she must attend weekly alcohol and drug counseling sessions with the PG Alcohol and Drug Services Society. Tina has missed two sessions and has not yet made contact with Healthiest Babies Possible. What course of action, if any, would you take? Please explain.

Independent Study

Independent study refers to a set of guided instructional strategies provided to foster the development of individual learner initiative, self-reliance, and self-improvement. Each module contains a number of short answer assignment questions. Learners were required to research the content area by navigating a series of preassigned hypertext links. They could choose to work independently, in pairs or groups, by seeking input into their ideas through the discussion board, or connecting with others in small group chat sessions. The course included over 150 predefined hypertext links.

Interactive Instruction

Interactive instruction allows for a range of groupings and interactive methods. One such strategy is the use of debates. Debates require students to engage in research, encourage the development of listening and oratory skills, create an environment where students must think critically, and provide a method for teachers to assess the quality of student learning. Debates also provide an opportunity for peer involvement in course evaluation. Throughout the course learners are encouraged to debate issues from various theoretical orientations. A second interactive strategy linked students through an audio file transfer to the story of John Dunn: "Life in Foster Care is Like a Subway Ride." Students are also invited to participate in a follow-up online chat time with Mr. Dunn.

Grading

In keeping with the instructional design model, a mixed grading approach was employed. Grading criteria were equally distributed to assess participation, critical thinking skills, and the ability to provide correct answers. All criteria were applied to each module using multiple discussion board activities, short answer assignment tool activities, group chat, and a quiz.

Significance of the Formative Evaluation

The purpose of the formative evaluation was to assess the effectiveness of the computer-mediated learning process in facilitating learners, meeting the requisite course objectives, and acquiring related child welfare competencies. The specific objectives of the study were to (a) evaluate the perceived experience of the students, and (b) identify which online activities and instructional strategies within and across modules were considered effective in meeting the learning needs of the learners. Information was collected online through a semi-structured combined quantitative and qualitative survey. Two focus groups were also facilitated to probe into qualitative aspects of the students' online learning experience and solicit feedback regarding the usefulness of the course in meeting their learning needs.

The results of the evaluation were formulated on the responses of 16 out of 19 learners who completed the online child welfare practice course. In the area of curriculum content 73% of the learners rated the materials as either excellent or very good. In regards to online instructional strategies, 62% of the learners rated the design as excellent or very good with another 20% rating the instructional strategies as good. Learners were most varied in their responses in the area of online learning activities, with 57% rating the activities as excellent or very good, 24% as good, and 14% as fair. The assignment tool that directed learners to complete short answers to a series of predefined questions was rated by 45% of the learners as the most helpful strategy in their online learning process. This was closely followed by the discussion board at 42%. In addition, 75% of the learners reported they received helpful feedback regarding the online activities on a regular basis.

SUMMARY

The Web-based distributed learning environment offers a unique and enhanced set of options for constructivist instructional design, taking advantage of human and technology-enhanced content delivery formats and learning interactions. Learners "converse" with each other and the instructor as they do in conventional classrooms, but they also interact via the structure of the course design with a host of other resources independent of time and place constraints with immediate access to global net links, learning object repositories and digital libraries, and respected experts in the field.

From Pedagogy to Technology

Given recent advances in information and communication technologies (ICT), growing numbers of educators are employing constructivist principles in the design of computer-mediated learning (Blanchette & Kanuka, 1999; Gunawardena, Lowe, & Anderson, 1997;). Modern ICT options provide interactive environments supportive of instructional methods required to facilitate constructivist principles. For this reason, constructivism has become a popular epistemological position for many educators who are using technology-mediated learning. This course demonstrated that a Web-based constructivist design could enhance learners' interactive experience within a structurally focused social work educational program.

Doubters may still maintain that the networked computer is a non-pedagogical technology, that it is only a vehicle for the transmission of information. We submit that when learners take charge of their learning in ways that only online environments can make possible, the instructor's use of computer-enhanced interaction and content distribution creates an effective facilitative medium that expands constructivist learning options. The exponential growth in 2004–2005 of the open-source learning management system "Moodle" (www.moodle.org) based on a set of constructivist underpinnings, is one source of supportive evidence. Learner access to collaborative online tools in a Web-based learning environment enables activities that cannot be duplicated in the traditional classroom teaching approach.

This requires a transformational shift on the part of the "chalk on the sleeve" educator community, beginning with a fundamental reevaluation of the function of pedagogy in the instructor's constructed understanding of their relationship with learners. A helpful point of departure is an examination of the original Greek sources of the word pedagogy, which is derived from two word roots: "paed," meaning children, and "agogos," meaning to lead. In ancient Greece, a "pedagogue" was a slave who led children to school. Pedagogy was the act of leading children to school.

While we have great respect for the craft and profession of effective teaching that many educators equate with "pedagogy," we believe it is timely to reconsider its currency and utility in an evolving educational arena. Given the recent rapid expansion in the areas of midcareer adult and distributed learning, we suggest incorporation of both androgogical (Knowles, 1984; Tough, 1979) and effective online teaching/learning concepts (Harasim et al., 1996; Tiffin & Rajasingham, 1995) into a descriptor which more accurately captures emerging educational practice.

Returning to the word root approach, another useful concept is that of "technos," from the Greek meaning for skill. We propose the incorporation of the concept of "Technagogy," meaning "the skilled leadership of learners," as a more accurate descriptor for the emerging role and functional activity of educators in a constructivist, online paradigm. For human services professionals, adoption of the term opens doors to productive discussion around learner—educator and learner–learner dynamics in the vastly expanded exploratory realm offered by digital technologies integrated with effective teaching practice.

REFERENCES

Bednar, A. K., Cunningham, D., Duffy, T. M., & Perry, J. D. (1998). Theory into practice: How do we link?. In T. M. Duffy & D. H. Jonassen (Eds.), *Constructivism and technology of instruction: A conversation* (pp. 17–35). Hillsdale, NJ: Erlbaum.

Blanchette, J., & Kanuka, H. (June, 1999). Applying constructivist learning principles in the virtual classroom. In *Proceedings of Ed-Media/Ed-Telecom 99 World Conference*. Seattle, WA.

Brown, A. R., & Voltz, B. D. (March, 2005). Elements of effective e-learning design. *The International Review of Research in Open and Distance Learning*. http://www. irrodl.org/content/v6.1/brown_voltz.html

Collins, A., Brown, J. S., & Newman, S. E. (1989). Cognitive apprenticeship: Teaching the craft of reading, writing and mathematics. In L. B. Resnick, (Ed.), *Knowing, learning, and instruction: Essays in honor of Robert Glaser*. Hillsdale, NJ: Erlbaum.

Dede, C. (1996). The evolution of distance education: Emerging technologies and distributed learning. *The American Journal of Distance Education, 10*(2), 4–36.

Dodge, B. (1996). *Distance Learning on the World Wide Web*. Retrieved January 12, 2003 from http://edweb.sdsu.edu/people/bdodge/ctptg/ctptg.html

Duffy, T. M., & Jonassen, D. H. (1992). *Constructivism and the technology of instruction: a conversation*. NJ, Hillsdale: Erlbaum.

Eastmond, D., & Ziegahn, L. (1996). Instructional design for the online classroom. In Z. L. Berge & M. P. Collins (Eds.), *Computer mediated communication and the online classroom* (pp. 59–80). Cresskill, NJ: Hampton.

Gove, Justice Thomas (1995). *Report of the Gove inquiry into child protection*. British Columbia: Ministry of Social Services.

Gunawardena, C. N., Lowe, C. A., & Anderson, T. (1997). Analysis of a global online debate and the development of an interaction analysis model for examining social construction of knowledge in computer conferencing. *Journal of Educational Computing Research, 17*(4), 397–431.

Hannafin, M., Land, S., & Oliver, K. (1999). Open learning environments: foundations, methods, and models. In C. M. Reigeluth (Ed.), *Instructional-design theories and models: A new paradigm of instructional theory* (pp. 105–140). Mahway, NJ: Erlbaum.

Harasim, L. (1996). Online education: the future. In T. M. Harrison & T. Stephen (Eds.), *Computer networking and scholarly communication in the twenty-first-century university* (pp. 203–214). Albany, NY: State University of New York Press.

Harasim, L., Hiltz, S. R., Teles, L., & Turoff, M. (1996). *Learning networks: A field guide to teaching and learning online.* Cambridge, MA: The MIT Press.

Jonassen, D., & Mayes, J. T. (1993). *A manifesto for a constructivist approach to technology in higher education.* Retrieved May 20, 2003, from http://apu.gcal.ac.uk/clti/papers/TMPaper11.html and http://apu.gcal.ac.uk/clti/papers/TMPaper11.html.

Jonassen, D. H., Peck, K. L, & Wilson, B. G. (1999). *Learning with technology: A constructivist perspective.* Upper Saddle River, NJ: Prentice-Hall.

Knowles, M. S., & Associates (1984). *Androgogy in action: Applying modern principles of adult education.* San Francisco: Jossey-Bass.

Lebow, D. (1993). Constructivist values for instructional systems design: Five principles toward a new mindset. *Educational Technology Research and Development, 41*(3), 4–16.

Martin, J. (1983). *Mastering instruction.* Toronto: Allyn & Bacon.

Presseley, M., Hogan, K., Wharton-McDonald, R., & Mistretta, J. (1996). The challenges of instructional scaffolding: the challenges of instruction that supports student thinking. *Learning Disabilities Research and Practice, 11*(3), 138–146.

Reigeluth, C. M. (1991). Reflections on the implications of constructivism for educational technology. *Educational technology,* 34–37.

Ricks F., (2002). *The paradox: Research as practice and practice as research.* Paper presented at the 2nd National Research Symposium on the Voluntary Sector Coalition of National Voluntary Organizations, Toronto, ON.

The Cognition and Technology Group at Vanderbilt. (1993). Designing learning environments that support thinking: The Jasper series as a case study. In T. M. Duffy, J. Lowyck, D. H. Jonassen, & T. Welsh, (Eds.), *Designing environments for constructive learning* (pp. 9–36), Berlin: Springer-Verlag.

The E-Learning Guild (2004). *The E-learning insider: How do people learn? Some new ideas for e-learning designers.* Retrieved April 6, 2005, from http://www.e-learning guild.com/pbuild/linkbuilder.cfm?selection=doc.703#Three.

Tiffin, J., & Rajasingham, L. (1995). *In search of the virtual class: Education in an information society.* London: Routledge Press.

Tough, A. (1979). *The adult's learning projects. Research in Education Series no. 1.* Toronto: Ontario Institute for Studies in Education.

Twigg, C. (2001). *Innovations in online learning: Moving beyond no significant difference.* Center for Academic Transformation, Rensselaer Polytechnic Institute. Retrieved May 12, 2003 from http://www.center.rpi.edu/PewSym/mono4.html and http://www.center.rpi.edu/PewSym/mono4.html

Von Glasersfeld, E. (1995). A constructivist approach to teaching. In L. Steffe & J. Gale (Eds.), *Constructivism in education* (pp. 3–16). Hillsdale, NJ: Erlbaum.

Wild, M., & Quinn, C. (1997). Implications of educational theory for the design of instructional multimedia. *British Journal of Educational Technology, 29*(1), 73–82.

Willis, J. (1995). A recursive, re?ective instructional design model based on constructivist-interpretivist theory. *Educational Technology, 35*(6), 5–23.

Winfield, W., Mealy, M., & Scheibel, P. (1998, August). Design considerations for enhancing confidence and participation in Web based courses. In *Proceedings of the Annual Conference on Distance Teaching & Learning*. Madison, Wisconsin.

Winograd, T., & Flores, F. (1986). *Understanding computers and cognition: A new foundation for design*. Norwood, NJ: Ablex.

Woolfolk, A. E. (1993). *Educational psychology*. Boston: Allyn & Bacon.

ABOUT THE AUTHORS

Gerard Bellefeuille is an associate professor of social work at the University of Northern British Columbia in Canada. bellefeg@unbc.ca

Luca Botturi is an instructional designer with eLab, Università della Svizzera italiana, Lugano, Switzerland. luca.botturi@lu.unisi.ch

Alain Breuleux is an associate professor in the Department of Educational and Counselling Psychology, McGill University, Montreal, Quebec, Canada. ED13@MUSICA.MCGILL.CA

Martin Paul Buck is a professor at Camosun College in Victoria, British Columbia, Canada. buck@camosun.bc.ca

Katy Campbell is interim dean of the Faculty of Extension, University of Alberta, Canada.

Tara Campbell-Ray is in the Department of Kinesiology and Health Education at the University of Texas at Austin.

Lorenzo Cantoni is with eLab, Università della Svizzera italiana, Lugano, Switzerland. lorenzo.cantoni@lu.unisi.ch.

David Cavallo is a research scientist in the Epistemology and Learning Group at the Massachusetts Institute of Technology Media Laboratory. cavallo@media.mit.edu

Katherine Cennamo is an associate professor of instructional technology at Virginia Tech University. cennamo@vt.edu

Frank R. Dinter is a professor in the Fakultät Erziehungswissenschaften of the Institut für Schulpädagogik und Grundschulpädagogikat the Dresden University of Technology, Germany.

Elizabeth Edmundson-Drane is in the Department of Kinesiology and Health Education at the University of Texas at Austin. edmundson@mail.utexas.edu

Alexandra E. Evans is an assistant professor in the School of Public Health at the University of South Carolina. sevans@sph.sc.edu

Michael Hannafin is a professor in the Department of Educational Psychology and Instructional Technology at the University of Georgia. hannafin@uga.edu

Karol Kaye Harris is in the Department of Kinesiology and Health Education at the University of Texas at Austin. kk.harris@mail.utexas.edu

Richard Kenny is an associate professor in the Centre for Distance Education at Athabaskca University in Alberta, Canada. rikk@athabaskau.ca

Elizabeth Ann Kinsella is an assistant professor in the School of Occupational Therapy at the University of Western Ontario, Canada. akinsell@uwo.ca

Robert Martin is at FTR and Associates/eLearnerWorks in Victoria, British Columbia, Canada. rrmartin@shaw.ca

Bernedetto Lepori is with eLab, Università della Svizzera italiana, Lugano, Switzerland. Bernedetto.lepori@lu.unisi.ch

Karen E. Norum is an assistant professor of leadership studies at Gonzaga University in Spokane, Washington. norum@gonzaga.edu

Virginia Richardson is professor emeritus, College of Education, University of Michigan.

Roderick Sims is an adjunct professor at Capella University in the United States as well as principal consultant at KnowledgeCraft in Australia. rodsims@knowledgecraft.com.au

Richard Schwier is a professor of educational communications and technology in the Department of Curriculum Studies at the University of Saskatchewan. richard.schwier@usask.ca

Marcos Silva is at the Instituto de Computação—Universidade Federal de Alagoas (UFAL) Campus A. C. Simões, Brazil

Marueen Tam is director of the Teaching and Learning Centre at Lingnan University in Hong Kong. mtam@In.edu.hk

Stefano Tardini is with eLab, Università della Svizzera italiana, Lugano, Switzerland. Stefano.tardini@lu.unisi.ch

Feng Wang is director of distance education at Mount St. Mary College in Newburg, New York. wang@msmc.edu

Karen Wiburg is an associate dean in the College of Education, New Mexico State University. kwiburg@nmsu.edu

Jerry Willis is professor and director of the doctoral program at Manhattanville College in Purchase, New York. jwillis@aol.com